Kessinger Publishing's
Rare Mystical Reprints

THOUSANDS OF SCARCE BOOKS ON THESE AND OTHER SUBJECTS:

Freemasonry * Akashic * Alchemy * Alternative Health * Ancient Civilizations * Anthroposophy * Astrology * Astronomy * Aura * Bible Study * Cabalah * Cartomancy * Chakras * Clairvoyance * Comparative Religions * Divination * Druids * Eastern Thought * Egyptology * Esoterism * Essenes * Etheric * ESP * Gnosticism * Great White Brotherhood * Hermetics * Kabalah * Karma * Knights Templar * Kundalini * Magic * Meditation * Mediumship * Mesmerism * Metaphysics * Mithraism * Mystery Schools * Mysticism * Mythology * Numerology * Occultism * Palmistry * Pantheism * Parapsychology * Philosophy * Prosperity * Psychokinesis * Psychology * Pyramids * Qabalah * Reincarnation * Rosicrucian * Sacred Geometry * Secret Rituals * Secret Societies * Spiritism * Symbolism * Tarot * Telepathy * Theosophy * Transcendentalism * Upanishads * Vedanta * Wisdom * Yoga * *Plus Much More!*

DOWNLOAD A FREE CATALOG
AND
SEARCH OUR TITLES AT:

www.kessinger.net

THE SEA OF GALILEE

Frontispiece.

THE PILLAR
IN THE WILDERNESS

BY

BENJAMIN JOHN

LONDON
WILLIAMS AND NORGATE, LTD.
36 GREAT RUSSELL STREET

First published in January 1936

Printed in Great Britain
All rights reserved

CONTENTS

CHAPTER I

The Logos; the Universe; philosophic hypotheses; the earth and its evolution; man; intuition and reason; the emergence of the Adamic race; the locality of Eden; freewill; His Divine Purpose 1

CHAPTER II

The Plan of His-Story: the Adamic "earth"; the Flood; the House of Cain—founding of first world civilisation, as the Sumerians; and of polytheism; Sargon-Marduk; Ur and Elam 29

    Appendix A: Identity of Cain; the children of the Sun . 53
    Appendix B: Later mythology 66

CHAPTER III

The House of Seth; Druidism, its teaching—monotheism; Stonehenge; the Great Pyramid: its Divine Message; its geometrical measurements and symbolism; "its displacement factor" 80

CHAPTER IV

The Noachic race: Japheth, Shem, Ham; early history; the Divine Covenant, the Mosaic Covenant; the symbolism of the Exodus; Israel in Canaan; the Assyrian and Babylonian captivities; the era of Babylon 105

CHAPTER V

Israel in prophecy; Israel in history—the Sacæ and the Beth-Khumri; the Massagetæ on the Caspian; Cyrus; the Getæ; Ezekiel; "these bones" 140

CHAPTER VI

The Massagetæ migration—the Asar; the Getæ on the Danube; 4 B.C.; the Getæ against Greece and Rome; the Ostrogoths; the migration to Britain as Angles, Saxons, Danes, Normans; the pre-Captivity migrations of Ephraim, Dan, Zarah; Jeremiah to Ireland; the Royal House of Zarah-Pharez; the ancient Britons, Cæsar's campaign . . . 170

CHAPTER VII

The Anglo-Israelite theory—the answer; the return of the Jews and Benjamin from Babylon; Apocalyptic eschatology—2 Esdras 204

CHAPTER VIII

The Christ 231

Appendix: Modern scientific theory and religious thought; the atomic make-up of freewill; action by the Logos . . 260

CHAPTER IX

Greek philosophy; the founding of Christianity in Britain; Caradoc's defence; his family in Rome; St Paul; the defence of Arviragus; Menai and Boadicea; Britain under Lucius Christian; Constantine; Augustine's attempt . . . 271

CHAPTER X

The Great Pyramid—its symbology of descent, the "rhomboid of displacement," the era of Inductive Science; the "Times" ending and the United Kingdom emerges A.D. 1800; the English Church and nineteenth-century science; the Pyramid's symbology of the low passages—the tribulation of the "elect" 299

CHAPTER XI

Post-War Britain; eschatology; Chaos—the fall of the economic system, and gold standard; the purpose of Britain; the founding of the Totalitarian State; sin; the Divine system; the Sabbath, a sign; Whither Britain? the finger . . 326

CONTENTS

CHAPTER XII

The cry at midnight; the Great Pyramid—symbology of the King's Chamber; the *"mitring"* of the Apex Stone; the Anglo-Saxon and the Continental outlooks; Italian action; Germany and Russia; the Logos intervenes—*"that great day of God Almighty"*; and afterwards; Eden revived and Jerusalem; the Theocentric Kingdom; the Logos sums up A.D. history 372

INDEX 431

ILLUSTRATIONS AND CHARTS

THE SEA OF GALILEE	*Frontispiece*
	PAGE
KEITH'S "ANTIQUITY OF MAN" CHART	12
THE GREAT PYRAMID (ITS CASING STONES LOOTED)	*facing* 84
Its apex stone displacement	90
And passage system	93
THE GENEALOGY OF ABRAHAM	110
THE SYMBOLISM OF THE EXODUS	125
"THESE BONES" OF BRITAIN	164
THE LINEAGE OF OUR ROYAL HOUSE	194
A FIELD OF MIXED GROUND	*facing* 236
SHEEP AND GOATS	,, 240
*"WITHOUT A CITY WALL"	,, 250
*GOLGOTHA	,, 252
*THE GARDEN TOMB	,, 254
*THE GARDEN TOMB	,, 256
*"IN THE HEART OF THE EARTH"	,, 258
"ATOMIC" MAKE-UP OF FREEWILL	267
PYRAMID PASSAGE SYSTEM	300
THE DATINGS OF "SEVEN TIMES"	312–316
"THE CONSUMMATION OF THE AGE"	321
"LIGHT RETURNED TO ITS SOURCE"	353
THE "MITRING" OF THE APEX STONE	381

* *Reproduced from postcards (see footnote p. 257).*

INTRODUCTION

To meet a pillar when wandering in a wilderness suggests the work of some mind, superior to the environment, that has passed this way. To interpret the design and the meaning contained therein the traveller requires a similar type of mind or way of thinking.

Science to-day has taught a new conception of the physical world. To think of Matter in terms of non-material entities requires a mind literally born again.

With the highest concepts of the human mind alone can man seek to see the design of the Pillar—Reality as Perfect Oneness.

The present times require the restatement of the meaning of Christ in modern scientific terms. Such conception has been present for years in the Logos idea but not developed: knowledge has now advanced for such physical explanation to be suggested.

The Space-Time continuum is now the limit of the human intellect as the One Reality, a finite conception of a Spirit realm of perfect Oneness of Divine Harmony, where God is All in All.

It is the source of creative activity, from it "world-stuff" comes into being, in it things happen.

The Ether is the medium between Spirit and Matter.

Radiant or etheric energy is essential for the manifestation of life in Matter. In other words, Divine Spirit "breathed" into man, the highest form of evolution, is essential to manifest freewill or personality.

In the creation of the universe, Spiritual agents deviated in action from the Divine Oneness of Harmony—the tendency of freewill in operation being to deviate from Divine principles that have to be taken on faith—such deviation in a perfect realm leads to its own discord, chaos and inevitable destruction.

Thus when the make-up of the atom was put together, and when combined with "world-stuff" in the carbon atom which strung itself to protoplasm, life emerged to die.

In Divine Purpose, action by the Logos is

(i) to show to freewill, from the lowest spirit realms to the highest, how to attune to Divine Harmony, the highest form of which is love, by the highest form of action, which is sacrifice;

(ii) as "*by Whom all things continue to consist*," to delay the inevitable annihilation until His work is accomplished to restitute the displacement in the created earth;

(iii) in physical terms of reality, at present beyond the human mind, to regroup the various forms of cosmic energy in the make-up of the atom so that the "atomic" make-up of freewill is restituted to life everlasting.

Such action could be achieved only by His Incarnation on earth. And "*when for three days and nights in the heart of the earth*" He acted on the indeterminancy of the atom, the universe, and man's freewill, He risked annihilation not only for Himself but for all mankind if He failed.

His Resurrection proves the efficacy of His action.

Meanwhile for nearly 6000 years is seen action by man's freewill in the history of his civilisations.

Divine Purpose requires attuned freewill to absorb and emanate Divine Love, and He "chose" such as Noah, "*a just man and perfect in his generations*," and Abraham, "*because thou hast obeyed My voice*," that from their seed should emerge a nation "*to be a blessing to all the families of the earth*," to stand against the divergence founded by Cain-Bel in man's first historic civilisation, the Sumerian.

The training of this nation, Israel, is the theme of the Old Testament; and because the will of her people deviated from His commands she has been punished for a definite "seven times" period (2520 years), moulded unknown to history but never destroyed, while the rule of the world in a succession of kingdoms was permitted to the Gentiles.

INTRODUCTION

The "consummation" of this Time and the "*restitution of all things*" are now at hand.

The restlessness of the present age denotes a premonitory "shaking."

How is the Chaos of man's proud civilisation, deviating to its own destruction, in these days to be restituted?

"*Watch.*" The Logos for the second time intervenes, "*for in Him all things consist.*"

THE PILLAR IN THE WILDERNESS

In the Wilderness of all things, in Brython and in Babylon, is
   heard a Voice crying:
   "*The Kingdom of God is at hand, make straight His paths.*"
But how, and how?
"*Seek ye first the Kingdom of God, and His righteous laws.*"[1]

CHAPTER I

¶ 1. The Universe—the immensity of space with nebulæ and stars—is His: He works outside Time and Space, which also are part of His creation.

"*In the beginning was the Word,*" John i, 1–3 (Logos,[2] signifying in John's time the supreme creative power of deity, polytheistic in Greek and Roman mythology as in the earlier Babylonian and Egyptian religions). "*And the supreme creative power was with God*" (the Son of Man, the Messiah-King, *vide* Prov. viii, 23–29, John xvii, 5), "*and the supreme creative power was God*" (John maintains, monotheistic).

"*All things were made by Him, and without Him was not anything made that was made.*" Christ—"*the image of the invisible God; all things were created by Him and for Him; and He is before all things, and by Him all things consist*" (Col. i, 15, 16).

---

[1] Matt. vi, 33.
[2] Logos:
in modern times, Science, or the study of;
theologically, "the personification of the eternal creative power of God," in John's times;
"the active force inherent in physical nature, as a whole and in its different parts" (Stoics);
"the Good regarded as creatively active, instead of being merely immanent in the Cosmos, it has an independent existence" (Plato);
"an aspect of the Divine activity, a mediator between the infinite Eternal God and the finite, transitory world" (Philo) (Sullivan and Grierson, *Outline of Modern Belief*, part vi, p. 367, Geo. Newnes, Ltd.).

Paul perceives a mystery:

"*The mystery, which from the beginning of the world, hath been hid in God, who created all things by Jesus Christ*" (Eph. iii, 9).

"*The mystery of God, and of the Father, and of Christ, in whom are hid all the treasures of wisdom and knowledge*" (Col. ii, 2, 3).

This mystery "*hid in God*," with all its "*treasures of wisdom and knowledge*," Man seeks.

Man "wishes to explore the Universe, both in space and time, because he himself forms part of it, and it forms part of him." [1]

Say—the Son of man forms part of the Universe, and it forms part of Him.

Thus within the Space-Time continuum, in the creation (as some distant past in Time), in the continuance (as the "consisting" of present Time), and in the ending (as future Time) of the Matter and Radiation that make up this Universe—there is action by the Logos, the Son of Man.

And how?

Amid the wilderness of voices innumerable of scientists, philosophers and theologians, the heart of man still seeks for His righteous Laws and for His Kingdom, more restlessly urgent in these times as His Kingdom draws near.

And in this wilderness, where is the Path?

Seek—present tense to every age: reward not an end to be sought: every finding mere incentive to re-search:—justifies the scientific method. Whereas creed backed by dogmatic power limits the domain of the King and the finite human intelligence of each succeeding generation concerning it.

¶ 2. *The Creation of Matter.* All this makes it clear that the present matter of the universe cannot have existed for ever; indeed, we can probably assign an upper limit to its age; say . . . 200 million million years. . . . Our next step back in time leads us to contemplate a definite event, or series of events, or continuous process, of creation of matter at some time

---

[1] Sir J. H. Jeans, *The Universe Around Us* (Cambridge Press).

THE CREATION OF THE UNIVERSE

not infinitely remote. In some way matter which had not previously existed came or was brought into being.

If we want a naturalistic interpretation of this Creation of Matter, we may imagine radiant energy of any wave-length of less than 1.3×10^{-13} cm. being poured into empty space; this is energy of higher "availability" than any known in the present universe, and the running down of such energy might well create a universe similar to our own. . . .

If we want a concrete picture of such a creation, we may think of the finger of God agitating the ether.

We may avoid this sort of crude imagery by insisting on space, time and matter being treated together and inseparably as a single system, so that it becomes meaningless to speak of space and time as existing at all before matter existed. Such a view is consonant . . . with the modern theory of relativity.

The universe now becomes a finite picture whose dimensions are a certain amount of space and a certain amount of time; the protons and electrons are the streaks of paint which define the picture against its space-time background (Sir J. H. Jeans, *The Universe Around Us*, pp. 354, 355).

Life may exist in some imperceptible form unknown to us, but without radiant or etheric energy, it could not enter into relation with matter, it could not grow and develop and become conspicuous. Energy may not be necessary to abstract Life, but it is essential to the display or manifestation of life in matter (Sir Oliver Lodge, *Ether and Reality*, p. 114).

Gen. i, 1, 2 (translated for origin):

"*By headships, or series of processes, God created that which produced the visible heavens, and that which produced the earth*" (Spirit becomes the origin of the space-time continuum, and the source from which appears Radiation and Matter).

"*And the ether was without form, and void; and darkness was upon all Space. And the Spirit of God moved upon Space*" (the suggestion of Space-Time, the making of the first ripple; the birth of radiant energy; an idea of waves, a stirring in the protons and electrons, separated by immense distances, that causes them to move into the vicinity of each other, and then, themselves combined with radiation, to acquire motion round each other—the foundation of the quantum theory).

Thus the Ether is the medium between Spirit and Matter; in it is created a reservoir of energy, for some Purpose to be revealed, and in it a part of Perfect and Divine Essence, "*by Whom all things consist,*" the Christ, whom we seek.

"*I am the Alpha and the Omega, the First and the Last*": I am the First Cause [1] and the Ultimate Reality.

Let Eddington [2] continue:

> The beginning of the world—a primeval chaos. . . . Its vastness appals the mind, space boundless though not infinite, according to the strange doctrine of science. The world was without form and almost void.
>
> At the earliest stage we can contemplate the void is sparsely broken by tiny electric particles, the germs of the things that are to be; positive and negative they wander aimlessly in solitude, rarely coming near enough to seek or shun one another . . . all space is filled, and yet so empty that in comparison the most highly exhausted vacuum on earth is a jostling throng.
>
> Darkness was upon the face of the deep, for as yet there was no light.
>
> The years rolled by, million after million. Slight aggregations occurring *casually* in one place and another drew to themselves more and more particles . . . until the matter was collected round centres of condensation leaving vast empty spaces from which it had ebbed away. Thus gravitation parted the primeval chaos.
>
> These first divisions were . . . "island universes," each ultimately to be a system of some thousands of millions of stars . . . they acquired rotation (*we do not yet understand how*) which bulged them into flattened form and made them wreathe themselves in spirals (the spiral nebulæ).

Let Jeans interpose and amplify:

> *Each failure to explain* the spiral arms makes it more and more difficult to resist a suspicion that the spiral nebulæ are

---

[1] "In man's experience . . . every event is determined by a cause. But that cause is itself an event determined by a cause. There must therefore be an uncaused Cause, which is the ultimate cause of the whole series of events which proceeds from it—the First Cause."

[2] Sir A. Eddington, *Science and the Unseen World* (George Allen & Unwin).

THE CREATION OF THE UNIVERSE 5

the seat of types of *forces entirely unknown* to us, forces which may possibly express novel and unsuspected metric properties of space . . . that the centres of the nebulæ are in the nature of "singular points" at which matter is poured into our universe from some other, an entirely extraneous, spatial dimension, so that to a denizen of our universe they appear as points at which matter is being continually created. . . .

Everything points with overwhelming force to a definite event, or series of events, of creation at some time or times, not infinitely remote.

The universe cannot have originated *by chance* out of its present ingredients, and neither can it have been always the same as now. . . .

The main mass and the main energy of the universe do not exist in the form of atoms but of intangible radiation. We may say the universe is mainly a universe of radiation, combined, in a far lesser degree, with the atoms out of which radiation is continually being formed.

Can we regard this new universe as a fortuitous concourse of atoms and radiation?

Eddington continues:

As it had divided the original chaos, so gravitation divided the island universes. First the star clusters, then the stars themselves were separated. And with the stars came light, born of the fiercer turmoil which ensued when the electric particles were drawn from their solitude into dense throngs. . . .

In the vast expanse of the heavens the traffic is so thin that a star may reasonably count on travelling for the whole of its long life without serious risk of collision . . . a star journeying through space *casually* overtook the Sun (our Star), not indeed colliding with it but approaching so close as to raise a great tidal wave. By this disturbance jets of matter spurted out of the Sun; being carried by their angular momentum they did not fall back again but condensed into small globes—the planets . . . a lump of matter small enough and dense enough to be cool.

The design of the first stage of evolution seems to have been that matter should ordinarily be endowed with intense heat . . . a temperature of 10 million degrees or more prevails through the greater part of the interior of a star . . . cool matter appears as *an afterthought*. The provision of certain cool planetary globes was the second impulse of evolution, and it has exhausted itself in the formation of inorganic rocks and ores and other materials.

And Jeans:

The planets are the only places we know where life can exist. The stars are too hot, even their atoms are broken up by intense heat . . . nebulæ are unsuitable, . . . so drenched with highly penetrating radiation as to render life impossible. . . .

It does not look at present as though Nature had designed the universe primarily for life. . . . Life is the end of a chain of by-products: it seems to be *the accident*, and torrential deluges of life-destroying radiation the essential. . . .

These protons and electrons (from the Sun) are pure bottled energy; the continuous breakage of these bottles in the sun sets free radiation which warms and lights our earth, and enough unbroken bottles remain to provide light and heat for 15 million million years to come. . . .

An electron may be annihilated not only inside a star, but in free space, or in one of the almost transparent nebulæ. . . . In a sense this radiation is the most fundamental physical phenomenon of the whole universe, most regions of space containing more of it than of visible light or heat; our bodies are traversed by it night and day. . . . It may be essential to life or it may be killing us.

Now Eddington:

Out of the electric charges dispersed in primitive chaos, 92 different kinds of matter, chemical elements, have been built . . . whilst other atoms organise themselves in twos and threes or it may be in tens, Carbon atoms (element No. 6) organise themselves in hundreds and thousands . . . *love to string themselves* in long chains. . . . From this potentiality of carbon atoms to form more and more elaborate structure a third impulse of evolution arises . . . the beginning of life.

. . . The history of the development of living forms extending over a thousand million years is recorded, though with many breaks, in fossil remains . . . seemingly Nature made nearly every possible mistake . . . to provide a form fitted to endure and dominate . . . at last tried a being of no great size, almost defenceless, defective in at least one of the more important sense-organs; one gift she bestowed to save him from threatened extinction—*a certain stirring, a restlessness*, in the organ called the brain, and so we come to Man.

¶ 3. The continuity of Evolution being admitted, Philosophy submits hypotheses to explain the inexact and non-scientific language of the phrases above in italics, and postulates:

the "world-stuff" of Eddington;
the "élan vital" of Henri Bergson;
the "emergent quality of the natural" of C. Lloyd Morgan;
the "world-stuff original of Mind-Body" of Wells and Huxley;
the "common ancestor of Mind-Matter" of Bertrand Russell;
the "unknown third" of Bernhard Bavink;
the "universal mind" of Jeans;

each explanatory of each one's mode of thought, a philosophic Bridge between Science and Spirit.

Darwin himself in the *Descent of Man* says: "This grand sequence of events the mind refuses to accept as the result of blind chance. The understanding revolts from such a conclusion."

And even Haeckel, the arch-materialist, says: "Our cosmology knows only one sole God, and this Almighty God rules the whole of Nature without exception. We contemplate his operation."[1]

The brilliant observation and lightning (and enlightening) progress in astro-physical knowledge in recent years are so amazing that these uncertain factors will in the near future be clarified by further research, so the Bridge is a contracting one.

> The ultimate realities of the universe are at present quite beyond the reach of science, and may be—and probably are—for ever beyond the comprehension of the human mind. It is a priori probable that only the artist can understand the full significance of the picture he has painted . . . and the apparent contradictions.

---

[1] Quoted by Alfred Noyes, *The Unknown God*, pp. 116–121 (Sheed & Ward).

The higher unity of ultimate reality must no doubt reconcile them all, although it remains to be seen whether this higher unity is within human comprehension or not.

. . . The message of astronomy is of obvious concern to philosophy, to religion, and to humanity in general, but it is not the business of the astronomer to decode it . . . it is for others to try to understand and explain the ultimate decoded meaning (Jeans, *The Universe Around Us*, pp. 356–357).

Science . . . accepts no creed that conflicts with the truth . . . but quietly and unexpectedly it has met Religion at the cross-roads. . . . Each is standing with a new humility in the other's presence, before an unfathomable and eternal mystery (Alfred Noyes, *The Unknown God*, p. 13).

Modern physical science has restated the terms of Matter in mathematical symbols which when put into word-pictures only roughly indicate the reality. This new conception of Matter involves a change in our mode of thought about the world that can almost be called "*born again.*"

Will the Bridge eventually disappear?

The gulf thus fixed has been the incentive for "seeking" in every age of man, throwing his mind ever toward the Logos.

And within the past twenty years man

has looked inside the atom, a body one-millionth the diameter of a pin-head, and found an infinitely small nucleus one ten-thousandth the diameter of the atom, and arranged about it as many as ninety-two electrons (in uranium) each playing its appropriate rôle in a symmetrical co-ordinated atomic system.

He has then looked inside that nucleus and counted in uranium 238 positives and 146 negatives, and he has found that the atom changes to something else if any one of these positives or negatives drops out.

He has watched the interplay of radiation upon these electrons, both within the nucleus and out of it, and found everywhere amazing orderliness and system.

He has learned the rules of nature's game in producing the extraordinarily complicated spectrum like iron.

Man has turned his microscope on the living cell and found it even more complex than the atom, with many parts each performing its function necessary to the life of the whole; and again, he has turned his great telescopes upon the spiral nebulæ a million light-years away and there also found system and order.

After all that, is there anyone who still talks about the materialism of science?

Rather does the scientist join with the Psalmist of thousands of years ago in reverently proclaiming *"the Heavens declare the glory of God and the Firmament showeth His handiwork."*

The God of Science is the spirit of rational order and of orderly development, the integrating factor in the world of atoms and of ether and of ideas and of duties and of intelligence (Professor Millikan).

Through Einstein's theory of relativity . . . the four-dimensional world is conceived as a sort of blend of space and time, and that matter is not an independent enduring substance inhabiting space and time but is intimately connected with the four-dimensional compendium . . . (popularly called the four-dimensional world) which postulates a continuous thing or continuous chain of "events" . . . the space-time continuum is not something one can see or handle or picture in the mind even.

The theory of the electrical constitution of all matter has abolished matter . . . the old picture of a mechanistic world has gone; there is nothing now but energy. . . .

Man must try to forget his three-dimensional world; think of possibilities right outside actual human experience, think of the non-material shadowy four-dimensional continuum as a never-never land, a never-get-at-able place where the Great Operator works with entities a human being cannot see nor handle, nor as yet dimly understand.

To do so perhaps we would need to be more than human; need to have other senses and more perfect eyes, a better brain and a different body. . . . We have to acquire a new point of view, a new habit of thought, a new kind of consciousness . . . almost as if a new faculty of the mind had to be born and developed.[1]

To Nicodemus, a Jewish intellectual and a Master of Israel, Christ speaks of the same conception (John iii, 1–12), though to His audience on earth, limited by human environment which lacked the scientific knowledge attained in present times, He went straight to the essentials, which

[1] Sullivan and Grierson, *Outline of Modern Belief*, ch. iv, pp. 148–150.

man seeks (though he calls it to-day "mind-stuff," "mental reality"), the abode of Spirit, as the source of energy—

"*Except a man be born again, he cannot see the Kingdom of God. That which is born of the Flesh is Flesh; and that which is born of Spirit is Spirit*" (John iii, 3-6).

Translated to these times: except a man "acquire a new point of view, a new habit of thought, a new kind of consciousness . . . a different body . . . a new faculty of the mind to be born and developed"—for that which is born of Matter is of atoms, molecules, "world-stuff," combined in long carbon chains into animal life: but the Spirit in Man is born of the source of all energy which is but "a reservoir of Matter and Radiation"—is born of the Spirit of God.

A scientific mind, deep in thought for days, on the verge of solution of a problem which just eludes—"*the wind bloweth where it listeth and thou hearest the sound thereof*" —so in a flash [1] comes a new conception, a new way of thinking about it, which is discovery—"*so is every one that is born of the Spirit.*"

In their (poet, musician, artist, novelist, man of science, the sophisticated philosopher) best moments something veritably new does just come, and is accepted as a gift to be greeted with glad surprise. Cognitive data are selectively collected with a definite end in view; reflective thought plays round them and sets them in order: and then, often quite suddenly, often in a flash, what comes just comes. . . .

This, though not the end of the matter, lies close to the heart of the matter. . . .

There is a factor that in the midst of much conscious guidance is none the less beyond and deeper than conscious guidance, though its coming is accompanied by a peculiarly rich thrill of joy, conscious in the wider sense of the word.[2]

---

[1] Concepts usually come as a flash of intuition after sleep following a "hectic" day, when one action has followed another, event on event without intermission, with scarce time for food, where spirit has given out thought at high speed till the brain reels—

"*If you can fill the unforgiving minute
With sixty seconds' worth of distance run*" (Kipling)

—and not after a comfortable day paddling idly down a backwater.

[2] C. Lloyd Morgan, *Life, Mind and Spirit*, pp. 134-136.

Light . . . the direct intuitions which lift untaught genius to the heights of philosophy and beyond them. . . . God is Light (Alfred Noyes).

"For the invisible things of Him from the creation of the world are clearly seen, being understood by the things that are made, even His eternal power and Godhead."

¶ 4. Two thousand million years ago our earth took shape, shot from a spiral wreath of the sun.

At first for the most part gaseous, it begins to lose heat, and its gases liquefy; then slowly cooling it begins to solidify. . . .

A monotonous smoking desert, cindery underfoot . . . here and there out of a crack comes a crawl of molten rock, like a very coarse-grained tar, hardening and blistering on its surface as it cools. . . . No sun by day, no moon by night, nor any stars, but a thick curtain of cloud over everything and hiding the heavens. And beneath the cloud a dense and dusty unbreathable air, with carbonic acid gas and water-vapour. . . .

There was no sign of life at all, nor any sound save crackling and hissing, and now and then a big explosion.[1]

How did the first moisture come? As rain? Science is uncertain, but supposes that from somewhere below the surface, accumulated perhaps in some ocean, the water first came up to the earth's surface: and probably not as rain.

"The Lord God had not caused it to rain upon the earth. . . . But there went up a mist from the earth and watered the whole face of the ground" (Gen. ii, 5, 6).

Ages pass: mountains, valleys, "clouds form and storms begin to darken the heavens. There are winds and rushing rivers and waterfalls, lakes and shallow seas, soil formed by the rushing torrents carrying great loads of sediment and thick mud. Then slowly there come primeval vegetation and floating plants. . . .

"During all this time . . . the earth has been continually shrinking, owing to the contraction of its internal mass and the compression of the material by the pressure of the overlying rocks; periodic shrinkages bringing

[1] Sir J. A. Thompson, *Modern Science*.

about the buckling up of successive series of mountains and new physical features. Atmospheric influences have been at work disintegrating rocks and changing the land surface. . . ." [1]

A thousand million years have passed, and we come to the Palæozoic era (ancient life)—"a richer vegetation, a jungle growth or shrubby bush, primitive forests of giant horse-tails, tree-ferns and club-mosses, and the peopling of the seas" [1]—fishes, and then wallowing amphibia venturing on to the dry land (the first creatures to use a voice); then land animals.

More millions of years to the Mesozoic era (mediæval life)—"higher flowering plants, birds, the first mammals, flying dragons, and great dinosaurs—reptiles of gigantic size." [1]

More millions of years to the Cenozoic era (modern life), three to five million years ago—"the rise of the higher mammals, and the emergence of the dawning ancestors of man . . . the lands begin to bloom with cereals, and fruit-bearing, flowering plants and grand hardwood forests, the atmosphere is scented with sweet odours; a vast crowd of new kinds of insects appear, and the place of the once dominant reptiles of the lands and seas are taken by the mammals.

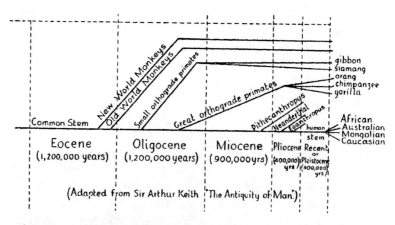

"Out of these struggles there rises a greater intelligence,

[1] *Outline of Modern Belief*, ch. v, pp. 154-157.

THE EMERGENCE OF LIFE

seen in nearly all of the mammal stocks, but particularly in one, the monkey-ape man."[1]

The particular gift which has been given to man is that of making a deliberate choice; he can turn a matter over in his mind and after looking at it from all sides determine his course of action.

What is it that has given man this power of choice? . . . it is the enormous development of the cortical areas of his brain.

This "power of choice" appeared late in the history of living things. The intelligence of bees, ants, and of all lower forms of vertebrate animals is of a "press-the-button" kind: under a given set of circumstances the animal can act in only one way. Every predicament calls forth its one particular reaction.

With the evolution of mammals cortical areas began to be added to the surface of the brain. In man this "grey matter" of the brain has reached colossal proportions . . . 14,000,000,000 nerve cells—not one of these cells is isolated, all are joined and connected by fine nerve fibres (Sir Arthur Keith).

Something of the same general nature as consciousness accompanies the activities of all living matter, it may be of all matter . . . we cannot think of our living brains and bodies apart from our minds. If the world-stuff is organised in a particular way, in the form which develops into a human being, it will be both a body and mind. Man on this hypothesis is not Mind plus Body, he is a Mind-Body.[2]

In all manners of ways we are pressed to the view that biosis (protoplasmic life) and psychosis (mental life) are everywhere associated throughout living nature. If they were not there in co-ordinated interaction from the first germ of life onward, how and when can they ever get together and work together?

Mentality cannot be juggled out of mechanism.[3]

It is probable that neither is the mind a mere function of the brain, nor the brain a mere tool of the mind, but that both are functions of an unknown "third," and are therefore connected with one another in some way—how, we do not know.[4]

---

[1] *Ibid.*, ch. v, pp. 154–156.
[2] Wells and Huxley, *The Science of Life*. On this subject see pp. 265–268.
[3] Thomson and Geddes, *Life: Outlines of General Biology*.
[4] Bernhard Bavink, *The Anatomy of Modern Science*. The "way" is suggested on p. 267.

The trend of modern thought is towards the belief that mind is essentially free or creative: a *vera causa*: that consciousness is not a mere function of the brain, rather that consciousness uses the brain as an instrument.

What we have to take into account is the living organism as a complete whole, and the key to the whole has not been found; vital reactions are not to be explained by physical mechanistic activities alone.

Those who think that a disbelief in immortality is justified by science and philosophy are dupes of their own cleverness or erudition.[1]

From reflex action and tropisms evolution continues to instinct and intelligence.

The two are divergent processes of evolution; Henri Bergson regards instinct and intelligence as opposite and complementary kinds of knowledge. They represent two different diverse expressions of divergent currents of evolution.

Instinct is something immanent in the creature's constitution, almost infallible, but limited in its scope: intelligence, on the other hand, we regard as something exterior.

The two powers immanent in life it may be surmised were originally intermingled as they still intermingle; although they have parted company in the evolutionary process they are not entirely divorced. Both have their advantage in life.[2]

Instinct as "congenital behaviour (the predetermined paths in the nervous system laid down in the course of the creature's development and come to form a part of the animal's hereditary constitution)"[3] is seen in "the spirit of the hive" (Maeterlinck) in the bee, and in the co-operative system of their colonies in the ant.

And as finally developed—"Intuition is instinct raised to its highest power" (Bergson).

On the other line, intelligence developed by experiment and experience evolved finally in man to Reason.

We have here the contrast between two different kinds of knowledge—two kinds which may indeed co-exist in the same

---

[1] *Outline of Modern Belief*, Sullivan and Grierson, who discuss this whole subject clearly on pp. 202-218, 261-268, 325-344.
[2] *Ibid.*, p. 268.
[3] *Ibid.*, p. 265.

REASON AND INTUITION

living creatures but which are essentially antithetical, or at least complementary in their nature—the knowledge that is innate and intuitive, and the knowledge that is begotten of experience. And these two different kinds of knowledge are the expression of, or are due to, two diverse principles of faculties: the faculty of instinct and the faculty of reason.[1]

In the evolution of animal life there has been described the "dismal cockpit" in the "struggle for existence"—the "savage warfare of beak and claw," "the crafty running fight or the set ordeal of battle."

Kropotkin gives the other side:

> There is as much or even more of mutual support, mutual aid, and mutual defence among animals belonging to the same species.
>
> Sociability is as much a law of Nature as mutual struggle.
>
> Who are the fittest? Those who continually war with each other, or those who support one another?
>
> Self-sacrifice is no less primordial than self-preservation. From the dawn of life, altruism has been no less essential than egoism (Herbert Spencer).
>
> Animals have become organically interested in working for their species, and even though they know it not, their individuality completed itself in the larger life of their race.
>
> What it seems to mean, according to current evolutionary theory, is that variations (probably altogether germinal to begin with) in directions that made for the welfare of offspring, family, society or species, have been established in the course of selection no less securely than those which made for self-preservation.
>
> Metaphorically speaking we may say that this has been Nature's way of setting her seal of approval on altruistic behaviour (Sir J. A. Thompson, *The System of Animate Nature*).

Some mammals and birds play for the love of playing; the flying sports and the song of birds express their *joie de vivre*, their sense of

> well-being, joy of existence, happiness at feeling in his place in a chosen corner of nature. . . .

---

[1] *Ibid.*, p. 268, quoting Prof. Lloyd Morgan.

Then see the Beauty in Nature—the bluebells, the violets, the honeysuckle, the butterflies flitting from flower to flower, the beauty of insects' wings, the colouring and plumage of birds, the glory of a yellow poplar in spring, when it rises like a golden fountain in the air.

No less the beauty in wild animal nature, the beauty of form, of strength, of symmetry, of curves and shapes and rhythms, and of movement.[1]

Beauty is merely the spiritual making itself known sensuously.[2]

In the evolution of the Universe by the Will of God, the order and arrangement of its unfolding should bear the intelligible characteristics and reflect the beauty of the Supreme Mind in which it all originated.

Look but on the wings of butterflies, the glorious mathematics underlying their symmetries of colour, their exquisitely elaborate designs, a work of art as artists understand by the phrase "intellectual schemes," the absolute perfection of the colour harmonies, the loveliness of their gradations.

Look at beauty in ferns, sunsets, flowers—where little or none of sex in Nature.

The miraculousness of the world that surrounds us—the apple bough in blossom, the up-breaking of Spring, the song of the bird, the vastness of the clouds, the strength of the winds, the grandeur and generative power of the sun, the heavens glittering with stars, the wonderful variations of the moon, its increase and waning—are all dulled to our accustomed eyes.[3]

"When I consider Thy Heavens, the work of Thy fingers, what is man, that thou art mindful of him? For Thou hast made him a little lower than the angels; Thou hast made him to have dominion over the work of Thy hands."

¶ 5. In the consciousness of man has been seen the evolving of two faculties of knowledge—emerging as Intuition and Reason. Where, in the animal world, Intuition became its guard in danger for the preservation of its animal life, and its altruistic behaviour; at a certain stage in man (Gen. ii, 7) Intuition was transformed into

---

[1] *Outline of Modern Belief*, p. 406.
[2] Hegel, *Philosophy of Religion*.
[3] Alfred Noyes, *The Unknown God*, pp. 155–180.

perception of the Divine, as immanent in Matter, and guard of its spiritual life—mere "animal" instinct in man then degenerating.

This knowledge of the Divine presence is intuitively spiritual, and receptive of Divine revelation or "voice."

Reason, its partner, may or may not acquiesce.

The two are component parts of Freewill, the essence of all Spirit, the power of choice in which way to operate.

Thus man is lifted up out of the animal creation, into a kingdom of his own, which gives him dominion over all created things.

In what way man was endowed with Spirit is given in Gen. ii, 7: "*The Lord God . . . breathed into his nostrils the breath of life and man became a living soul.*"

Nay more, it signifies Spirit's power over Matter, and man as partaker of that reservoir of energy which is the Ether, and so attuned to the use of that power, immortality.

"*All things were made by Him and without Him was not anything made that was made. In Him was life, and the Life was the light of all men*" (John i, 3-4).

In Him (the Son of man, with supreme creative power) originated life, that which astrophysics and biochemistry seek, nay more—Life, the light of all men who had become "living souls."

And thus—just as some 50,000 years or more ago there evolved on the mammal-humanoid stem a cave-man Palæolithic people using stone implements to carry Homo Sapiens safely through the four great Ice Ages; just as from these, some 25,000 years ago when the ice-fields had cleared, there emerged Cro-Magnon and Neolithic man; and as from these, 15,000 years ago, there arose the four races of man (given by Keith, see p. 12): so out of, or among, the Caucasian, in 4000 B.C., there was "created" or emerged the Adamic race, selected and elected for His Divine Purpose, and endowed with Freewill, the essence of Spirit, to operate to the Divine Will of Harmony, and with special faculties to originate Personality.

With Spiritual Intuition, that has supplanted mere animal instinct, and with Intelligence-Reason, "the living soul" is able to associate ideas; to think to the past, and to the future, to choose which way to act, to appreciate "forms"—beauty and goodness and truth—to receive and emanate Divine Love, and (summing up) "*to see God.*" [1]

Thus endowed, in some spot named Eden where it had emerged, the Adamic race was placed in supreme domination on the Earth, to adjust itself to the passive working of Natural Law and Order now seen in visible effect.

And the light of all His righteous laws, in which Reason and Intuition go hand in hand perfectly attuned, is summed up in one, His first law of LOVE—not the sentimental, cover-for-lust sort of thing which a type of cinema and jazz has made it; more than the charity of Paul, which makes of it something human, something theological: but in an infinite way expressive of the Logos in Oneness in all Vital Evolution, which includes the highest human emotion as the immensity of the Universe includes the smallness of earth, expressive of harmony in all Creation —God in all, as God is Love . . . something so awe-full and holy, that discord and deviation from it means but one thing—destruction and death, which is annihilation: a concept at the limit of human intelligence that "*except a man be born of water and of Spirit, he cannot enter into the Kingdom of God*" (John iii, 5).

"*And the light shineth in darkness; and the darkness comprehended it not*" (John i, 5; *vide* also John viii, 12).

The Law of God's Universe of Spirit, Energy and Matter, all three but varying expressions of the One Divine Energy, Love, has shined in this world since "the beginning," but man comprehended it not; instead he "has in the course of his existence upon this planet tried every form of government and mode of life to ensure peace and contentment, but to no purpose . . . from the most rigid autocracy to sheer utopianism" (Basil Stewart, *The Witness of the Pyramid*). And he has trusted every god of his own making, the polytheistic trinity of Romans and Greeks (Zeus, Apollo, Athena, Jupiter, Mars, and the rest) with their mystical cults and mystery religions and lesser

[1] See p. 266.

deities; Assyro-Babylonian perversions of Divine history and Egyptian imitations of the Messianic idea; even gods of wood and stone, the animism of the heathen.

And His righteous Laws:

(i) immanent in Nature, the Universe, our Earth; Newton's of gravitation, Einstein's of relativity; the electron theory of matter, and the quantum theory of the electron; the mathematical of numbers, and the geometrical of dimensions; the biological of heredity and natural selection; and the host of others searched and yet to be sought—wherein the scientist trusts to make sense and not chaos, knowing that theories which do not fit must be laid aside:

"*For the invisible things of Him from the creation of the world are clearly seen, being understood by the things that are made, even His eternal Power and Godhead*";

and (ii) transcendent, revealing His purpose for man, "*by the mouth of all His holy prophets since the world began,*" His promise of a Messiah King to replace Creation's displacement, His covenant with Abraham and our forefathers, and foreknowing that the words of His Book can be disputed by clever brains, even enshrining them in Stone in mathematical symbols (which are also used for His Laws of Matter) in the Great Pyramid, standing for 4560 years, and still standing, the only one of the Seven Wonders of the World, six built by merely earthly hands having disappeared.

¶ 6. And 1939 years ago, "*The Logos* [the Supreme Power, as Creation's promised Messiah] *was made flesh, and dwelt among us (and we beheld His glory, the glory of the only begotten of the Father), full of grace and truth*" (John i, 4).

And He chose therefore to be born precisely there where Adam was created, on the primeval and yet untrodden site where Bethlehem was long afterwards built, . . . in order to fully realise the life of man in the aggregate whom He came to save.

And exactly there in the choice garden where had been the tree of forbidden fruit, on that spot, Jerusalem, He decided that the atonement should be made by Himself . . . in order that it might be complete and perfect in its primeval historical reference as well as each chorographical particular.

And nothing which occurred at the beginning however remote will be without confirmation and explanation at the coming end.[1]

For where was Eden, in which freewill in man first deviated, there—and not at some vague spot in remote Armenia with no Divine association—was the Holy Land in historic times.

Jerusalem was first founded by Melchisedec, "*King of Salem, priest of the Most High God, without descent, made like unto the Son of God*" (Heb. vii, 1, 7; Gen. xiv, 18); who, Manetho states (as Shem, but more probably one of the House of Seth), after building the Great Pyramid went to Judea and founded Jerusalem.[2]

To such a Priest-King, knowing the significance and direction of the Pyramid's ascent angle, the site for his city must be a memorial to commemorate some past event, shown to him in a Divine message—"*a city which I have chosen Me to put My name there*" (Kings xi, 36).

In later years when "*They have defiled My land, they have filled Mine inheritance with the carcases of their detestable and abominable things*" (Jer. xvi, 18), the Logos says:

"*O Jerusalem, Jerusalem, thou that killest the prophets and stonest them which are sent unto thee, how often would I have gathered thy children together, even as a hen gathereth her chickens under her wings, and ye would not!*

"*Behold your house is desolate.*

"*For I say unto you, Ye shall not see Me henceforth, till ye shall say, Blessed is He that cometh in the name of the Lord*" (Matt. xxiii, 37–39).

And as Adam for deviation from Harmony was driven from Eden eastward beyond Assyria, so, 3230 years after, Israel for deviation from His commands was driven into captivity eastward to Assyria.

[1] Piazzi Smyth, *Our Inheritance in the Great Pyramid*, pp. 629–631, developing the suggestion first given in W. Henderson's pamphlet of 1853.
[2] See pp. 85, 111, 258.

Geology has yet to verify the site by search.

"*And a river went out of Eden to water the garden; and from thence it was parted and became into four heads*" (Gen. ii, 10) called

    (1) Pison.
    (2) Gihon, usually identified as the Nile.
    (3) Hiddekel, ,, ,, ,, Tigris.
    (4) Perath, ,, ,, ,, Euphrates.

Then in those days, before the earth's surface altered, Pison must have connected, and can but be the Jordan valley. For it is known that a geological fault runs from the Caspian Sea, down the valley to the Dead Sea, on to the Gulf of Akaba, and by the Nile to the African lakes.

The remains of this river system is seen in the vast underground lake under the Mount of Olives, discovered by Kitchener, and in the mighty subterranean river heard under Mount Moriah.[1]

Thus the river which went out of Eden is the same river revived at His Second Coming as "*the river from the sanctuary eastwards*" when the land is again restored [2] (Ezek. xlvii, 1–12), so that "*they shall say, This land that was desolate is become like the Garden of Eden*" (Ezek. xxxvi, 35).

"*For God so loved the world* [surely Nature as well as man] *that He gave His only begotten Son* [in Whom the continuance of all things consist] *that whosoever believeth in Him, shall* . . ." (John iii, 16).

For what? Nay, in seeking, leave out the "shalls"—man knows them, for Theology has emphasised them as rewards; and the idea of reward makes impure the motive.

Man well understands the law of downward movement—the stone that is rolling down a smooth plane shall ultimately come to rest at the bottom, but he does not interpret rest as "reward" for the stone because it rolled.

When He speaks of His Kingdom's Laws He understands the natural sequence; yet, in a human way, the miracles He performed and the rewards (little of material value)

---

[1] See p. 410.   [2] See p. 419.

He offered express His eagerness for man to understand that

"*For God so loved the world, that He gave His only begotten Son.*"

—for what? His Purpose, not only for man, but for His whole spirit realm, in the heavens above and in the earth beneath.

¶ 7. For this purpose, here consider Davidson's monumental work, *The Great Pyramid: its Divine Message*[1]—a work equal in conception and implication to any in recent astronomy and physics: a future text-book when Einstein's Law is a theory.

On p. 422, "Summary and Conclusions" (another 100 pages of detail follow):

> The first Book of Genesis (Gen. i–xi) begins with the narration of a vision of six consecutive days, compressing in epic form the events of six long successive periods of a finite stage of God's Creation . . . (like a series of six historical acts of a drama performed in one night).
>
> In the single initial verse of Genesis is included everything that preceded in the indefinite past the final phase of earth's material development.
>
> The marvel lies in the fact of Creation, not in the time in which it was effected. The concentrated mental idiom of the first eleven chapters is analogous to the expressive phraseology of modern science and technology, which compresses the subject matter of generations of experience and research into a single comprehensive expression.

Here is described the special creative selection or election of the first Adam and his seed—the Adamic race—as the final and objective process of six finite "active stages" of the Earth's creation. Thus a definite "passive" and objective stage of law and order was attained in relation to the Earth and Man, described as God resting upon the "seventh day" (in Gen. i, 26; ii, 7).

"*Thus the heavens and the earth were finished and all the*

---

[1] D. Davidson and H. Aldersmith, *The Great Pyramid: its Divine Message* (Williams & Norgate).

FREEWILL

host of them . . . and He rested on the seventh day, from all His work which He had made" (Gen. ii, 1–2).

"Man, provided with his special faculties, was placed in supreme dominion on the Earth, to adjust himself to the passive working of the Laws of God as translated into visible effect in man's dominion."[1]

These Science continually seeks, and knows as Truth.

"The Lord God formed man of the dust of the ground" . . . (Can science describe in four words and in more vivid terms?— "dust," suggestive of bacteria, minutest life; "ground," earth-matter from which came life) . . . *"breathed into his nostrils the breath of life, and man became a living soul"* (now more than animal, a Spirit with Freewill—the power of choice in which way to act, the essential attribute of Spirit, "in the image of God").[2]

"And God said, Let us make man in our image, after our likeness: and let them have dominion over . . ." (Gen. i, 27). (Is there a distinction intended between the singular "man," as a single race, the Adamic, and, in the following sentence, the plural "them," the evolved races of mankind? And what "image"?)

"So God created man in His own image" (Gen. i, 27).

Christ *"who is the image of the Invisible God, the firstborn of every creature: for by Him were all things created, that are in heaven, and that are in earth . . . all things were created by Him, and for Him: and He is before all things; and by Him all things consist"* (Col. i, 15–17).

"Who verily was fore-ordained before the foundation of the world" (Peter i, 20).

Let Davidson continue (p. 450):

The ideal type, Christ, was not adhered to in the developing, by God's spiritual agents, of certain details of creation relating to the Earth and Man. The operation of Freewill was known by Him in His supreme wisdom to form the essential element in the created state of spiritual being for the reception and emanation of Divine Laws, which is Love.

The tendency of Freewill in operation being to deviate from principles of operation that have to be taken on faith, the

---

[1] *Ibid.*, p. 424. [2] See p. 267.

spiritual agents of creation deviated from God's principle (Christ) in certain details of creation relating to Man.

Thus is indicated a displacement that occurred in the spiritual conditioning of the heavenly hosts—

"And the angels which kept not their first estate but left their own habitation, He hath reserved in everlasting chains under darkness into the judgment of the great day" (Jude 6) (that is, their power restrained until judgment given; deviation from Divine Law of harmony in creation bringing discord, means death, *i.e.* annihilation).

—prior to, or coincident with the final stage of Earth creation—man; and that as a result of this heavenly displacement, the displacement of man followed, more or less, as a natural consequence.

But these displacements had been eternally foreknown by God to be the necessary consequences of the operation of Freewill in His scheme of creation. To correct the Freewill tendencies of His agents both in heaven and on earth, He had already with eternal fore-knowledge—before the processes of active creation began—appointed that He Himself should suffer to effect the reconciliation and to illustrate the highest form of the outpouring of Divine Love.

God therefore cut off from Himself the One Fundamental Essence [1] of His Creation, the Highest Creative Expression of Himself, who is the Fountain Source of Love.

By projecting this Spiritual Essence of His Own Eternal Being as Christ—not into the high spiritual realms where the displacement in creation had been first effected, but into the lowest realm of Earth's creation—He set in action the fore-ordained cycle of events, that, reacting upwards from the lowest Spiritual realms to the highest, would restore the free-will balance of Love in all God's creation, for them to realise the magnitude of Divine Love as expressed in the highest form of sacrifice.

" . . . For this reason has He been chosen and hidden before God before the creation of the world" (Enoch xlviii, 6: of the Son of Man).

In John v–x He develops the source of Spirit and the purpose of God: and these chapters read with the above conception are strikingly vivid, even allowing for the

---

[1] The Logos.

FREEWILL

imperfection of human interpretation (especially chapter viii, which some Modernists[1] find difficult to accept as authentic and in keeping with the rest of His sayings), and clarify Paul's Christology:—

"The mystery of God, and of the Father, and of Christ, in whom are hid all the treasures of wisdom and knowledge" (Col. ii, 2, 3).

"The mystery, which from the beginning of the world hath been hid in God, who created all things by Jesus Christ. To the intent that now unto the principalities and powers in heavenly places might be known the manifold wisdom of God, according to the eternal purpose which He purposed in Christ Jesus our Lord" (Eph. iii, 9-11).

"God made not death: neither hath He pleasure in the destruction of the living. . . . For God created man to be immortal and made him to be an image of His own eternity. Nevertheless through envy of the devil came death into the world" (Book of the Wisdom of Solomon i, 12; ii, 23, 24).

"For man was created exactly like the angels to the intent that he should continue righteous and pure; and death, which destroyeth everything, could not have taken hold of him; but through this their knowledge they are perishing, and through this power [of knowledge] *it* [death] *is consuming me"* (Enoch lxix, 11).

Gadreel, the Satan, *"who has taught the children of men all the blows of death, and he led astray Eve."*

Also repentance, the coming back of the sinner to his God, existed in ideal, before the creation of the world; as it is written (Ps. xc), Before the mountains were brought forth, from that very hour Thou turnest man to contrition; saying, Return, ye children of men (R. Ababah Bar Rabbi Zengirah).

Six things preceded the creation of the world. Some of them were created; some existed as ideals, as part of the thought of the Creator, to emerge created in the future; so that their real being (*noumena*) was in existence, although ages should

---

[1] The Modernist on the Fourth Gospel: "The narratives of John are not, properly speaking, history, but a mystic contemplation of the Gospel. The speeches contained in his Gospel are theological meditations about the mystery of salvation void of any historical truth."

pass before their appearance as phenomena (Beresbith Rabbah, ch. i).

Long before man or the higher animals appeared on earth, lower organisms were forming and developing on lines and principles foreshadowing the higher organism, Man. Through all the species of organic life, the same underlying scheme is progressively apparent, and the same principles are seen to apply that culminate in the structure of man (Davidson).

"I am fearfully and wonderfully made; marvellous are Thy works, and that my soul knoweth right well. My substance was not hid from Thee. When I was made in secret, and curiously wrought in the lowest parts of the Earth, Thine eyes did see my substance, yet being imperfect; and in Thy book all my members were written, what days they should be fashioned, when as yet there was none of them" (Psalm cxxxix, 14–16).

A theory of evolution,[1] written 3000 years ago, which science to-day would uphold. Nay more, it enlightens the modern conception in which man, endowed with brain, thinking back along millions of years on the path of his history, seeks to unveil the cause of his birthday.

"When I consider Thy heavens, the work of Thy fingers, the moon and the stars which Thou has ordained, what is man, that Thou art mindful of him, or the Son of Man that Thou visitest him? For Thou hast made him a little lower than the angels, and hast crowned him with glory and honour. Thou madest him to have dominion over the work of Thy hands; Thou hast put all things under his feet."

---

[1] Nearest perhaps to C. Lloyd Morgan's "emergent evolution," which "is from first to last a revelation and a manifestation of . . . Divine Purpose. . . . What comes by nature is in my belief a Divine gift" (*Life, Mind and Spirit*, p. 145).

Also: "In so far as there is progress and advance in the cosmos, God is All in All, but in diverse modes and degrees of manifestation" (p. 302). . . . "The emergence of the religious attitude is a mental attitude toward the acknowledged reality of Divine Purpose . . . the rational order of the cosmos, no less than Divine Purpose, is dependent on a mind" (p. 303).

". . . (1) that the rational order has being independently of the reflective mind that is evolved within it, and (2) that Divine Purpose has being independently of the spiritual attitude through which it is revealed in this or that individualised person" (p. 304).

"In our passing life we touch the fringe of immortality when we acknowledge God as Ultimate Substance" (p. 313).

Also, on the same theme, see Alfred Noyes, *The Unknown God*, pp. 70–80.

THE DIVINE PURPOSE

¶ 8. Thus the Bible is made up of:[1]

(1) The Prologue—a vision of past history whereby the Spiritual Power given to man was forfeited.

(2) The His-Story of the fore-ordained cycle of events, of Abraham and his seed, Israel, of the promise of the Messiah, and ending after the death of the Testator; and of the means whereby the spiritual power was redeemed.

(3) The Epilogue (Revelation)—a vision of the future,[2] in which the spiritual power, the original Gift, will be restored to man; "*to be a blessing to all families of the earth.*"

While man thinks and talks, evolution stands not still.

Whereas in Creation, Freewill in the Spirit realm had in some way deviated from Divine Principle, then the Logos had delayed the consequent chaos and destruction (for "*in Him all things consist*") to show to all Spirit, in heaven above and earth beneath, His Father's Divine Purpose for Harmony in all action by Freewill—Sacrifice, as the highest form of action; Love, as the highest form of harmony: thus in and by His action to redeem man.

Our civilisation has created knowledge, but has used cosmic energy to Chaos,[3] in spite of the Light shown by the Logos—that the Logos again intervenes, for in Him the continuance of all things consist.

Thus, *circa* A.D. 2001,[4] or earlier, there is marked for appearance a mind of higher capacity for Truth; an intellect which will understand "good and evil," the emergence of life and the physical proof of Resurrection; a body "incorruptible" which will operate cosmic energy; a Freewill that having attuned by Faith will in action accord with His Divine Law of Harmony.

The Divine Will is that spiritual man, accepting the Law of Love as the first principle of creative energy, should control the realm of energy in relation to God's Kingdom on earth.

---

[1] See Davidson, p. 422.   [2] Isaiah xxxv, 1-10; lv, 12, 13; lxv, 17-23.
[3] "The International anarchy of a world administered by some sixty sovereign independent States, nursing unlimited national ambitions, was a disease carrying the seeds of death" (Prof. Gilbert Murray, in the Romanes Lecture, 14th June 1935).   [4] Refer pp. 250, 259, 372, 418.

"*The wages* [penalty] *of sin* [non-love] *is death* [withdrawal of spirit* [1]*], but the gift of God is eternal life* [retention of Divine Spirit, the source of all energy, and the dominion over Matter]."

Then, with action by the Logos accomplished—

"*Then cometh the end, when He shall have delivered up the Kingdom to God, even to the Father, when He shall have put down all rule and all authority and power.*[2]

"*For He must reign, till He hath put all enemies under His feet.*

"*The last enemy that shall be destroyed is death.*" [3]

"*And I saw a new heaven and a new earth, for the first heaven and the first earth were passed away, and there was no more sea. . . .*

"*Behold, the tabernacle of God is with men,*[4] *and He will dwell with them, and they shall be His people. . . .*

"*He that overcometh shall inherit all things; and I will be His God, and he shall be My Son. . . .*

"*. . . that God may be All in All*" [5] (1 Cor. xv, 24–28; Rev. xxi, 1–7).

This is then His Purpose, for His whole Spirit Realm, in the heavens above, and for man in the earth below, where "*whosoever believeth in Him should not perish, but have everlasting life.*"

---

[1] See pp. 249, 266: without spirit the "ego" dies (Gen. iii, 19); with spirit the "ego" acts.

[2] Freewill of man, free to act outside material conditions, henceforth in the spiritual realm where God is All in All.

[3] *I.e.* Freewill that has deviated and would still deviate from His Divine Will of Love, in spite of the action of the Logos.

[4] Men restored, redeemed, to the Spiritual realm.

[5] Freewill restored to perfect harmony in His Universe.

CHAPTER II

¶ 1. As Jude 6 indicates the deviation of freewill in His Spiritual realm at or before Earth's creation, so Genesis iii indicates the deviation of freewill in His selectively created Adamic race.

Thus in His Perfect Oneness of Harmony such deviation, as evil and sin, leads to discord and chaos, and by its own working-out to ultimate annihilation.

Now this discord born of deviating freewill and bred in the earth is seen passing down the centuries in the history of man: and alongside it His Purpose for man's redemption. His Purpose has been defined in Chapter I.

And all through time the Most High who "*so loved the world*" declares to the spiritual intuition that is in man, whose freewill is attunable, some "light" as to His Purpose, as in the mouth of His prophets throughout the Old Testament; and in case their words are questioned by Reason He has embodied this Purpose as universal law in the stone make-up of the Great Pyramid.[1]

The first civilisation of the world, then, is that founded by Cain—this is verified as the Sumerian in the present chapter.

The pre-diluvian migration of Seth is also noted (see Chapter III).

From the Adamic race, still deviating and so destroyed by the Flood, the family of Noah, "*a just man and perfect in his generations*," reaches the land of Cain, and overspreads the earth as the white races.

Out of the discordant mass of now mixed races, He "*elects*" from one freewill, Abraham, tested by Faith and who "*hast obeyed My voice*," a seed, chosen "*to be a blessing*

[1] So through His-Story, the Biblical quotations and Pyramid symbology are seen running together.

to all the families of the earth," a *"people"* with whom He establishes His Covenant (see Chapter IV).

His Israel, trained and guided, reject His Covenant; so to the land of Adam's eldest son, now Babylon, as *"this head of gold,"* is permitted the rule of the kingdoms of this world for *"the time of the Gentiles"*—a definite period of 2520 years, during which Israel, though dead to history, is undergoing its punishment and being trained to become the nucleus of His coming Kingdom (see Chapters V and VI).

Esdras questions His Purpose, and sees no *"light"* in all man's history, until finally shown in a vision (see Chapter VII).

Until in 4 B.C. there appears the *"Light"*—the Divinely promised Messiah, earth's Redeemer, freewill's Saviour—who taught how freewill is to be attuned to Harmony, which is Love, the first Law of the Universe. Nay more, since the Most High created the Space-Time Continuum and had "cut off" Divine Essence as the Logos to be the active principle in it, then the redemption of Freewill could be achieved only by action in and on the proton-electron + radiation make-up of the Universe — some physical use of cosmic energy as yet beyond man's intelligence, thus stressing Faith to Freewill in action (this is indicated in Chapter VIII and Appendix).

During the Christian era, still continues the history of man's Freewill,

(i) the deviation of its action from the *"Light"* shown, to the State idea and Chaos,

(ii) the emergence of Israel as an Empire, accepting the *"Light"* (see Chapters IX–XI).

The stage is set for the final *"restitution of all things," "the consummation of the age,"* when *"after those days"* His Kingdom will come on earth (Chapter XII).

Thus let His-Story unfold.

¶ 2. Genesis iii gives the parable-story of "Adam's failure to retain his place in God's created scheme: . . . the tendency of freewill in operation being to deviate from

THE EDEN STORY

principles of operation that have to be taken on faith"—so bringing discord in harmony, the result of such being death.

"*And the Lord God commanded the man, saying, Of every tree in the garden thou mayest freely eat :—but of the tree of knowledge of good and evil, thou shalt not eat of it, for in the day that thou eatest thereof thou shalt surely die*" (Gen. ii, 16, 17).

In iii, 8–10, is the indication of the replacement of animal instinct, or intuition in man, by Spirit Intuition, Conscience; of the presence in the world and in man of consciousness of the Divine. The effect of deviation is given—

iii, 15, to the serpent,[1] "*And I will put enmity between thee and the woman ; and between thy seed and her seed.*"

Later it will be seen that Cain's seed adopted by Bel, the Devil, founded the earliest civilisation of Babylon: while the woman's seed, by Seth, becomes Israel, from the Pharez-Judah line of which is born our Christ.

iii, 16, to the woman, "*I will greatly multiply thy sorrow and thy conception, in sorrow thou shalt bring forth children.*"

(In modern times it is asked why, in woman of the white, or Adamic, race the process is more of pain and risk, from puberty to menopause, "the curse of Eve," that effects her life and mental equilibrium, filling the hospital wards; and the usual answer is that she is more highly civilised.)

"*And thy desire shall be to thy husband*" (physiologically it is so) "*and he shall rule over thee*" (so it was in history until modern times, and still is in the ideal partnership, to the satisfaction of both).

iii, 19, to the man, "*In the sweat of thy face shalt thou eat bread*" (modern economics is still busy adjusting it), "*for out of dust wast thou taken ; for dust thou art, and unto dust*

---

[1] The word used for serpent is "Thuban," meaning subtle: the primary meaning of which is fine, or delicate, relics in the gracefulness of movement and beautiful markings in many snakes; the secondary meaning is cunning or deceitful. "*Yea* [a diplomatic affirmative] *hath God said ?*" (an artfully expressed doubt, negativing the yea).
Such may have appeared to Eve as an *angel of light* (2 Cor. xi, 14), and, as the effect of the curse, turned into a writhing reptile (Gen. iii, 14) (Basil Stewart, *The Witness of the Pyramid*).

shall thou return." (Spirit dominion over matter is forfeited.[1])

Driven from the realm of his first dominion, he was forced to migrate eastwards—since from the east he was prevented from returning.[2]

"So he drove out the man; and he placed at the east of the garden of Eden Cherubims and a flaming sword which turned every way, to keep the way of the tree of life" (Gen. iii, 24)—the position of Eden being given in Genesis ii, 8–14, as interpreted on pages 19–21.

In the new realm in which he was forced to settle the conditions were such that he could sustain himself only by the most arduous labour (Gen. iii, 17–19, 23; viii, 21, 22).

Here is pictured the existence and development of the Adamic race in an isolated world of its own, shut off by natural barriers from the external world, and therefore remote from the general races of mankind, and still under Divine tuition.

Such is indicated in Genesis iv, 14–16, where Cain says on his expulsion, *"Behold, Thou hast driven me out this day from the face of the earth; and from Thy face shall I be hid; and I shall be a fugitive and a vagabond in the earth; and Cain went from the presence of the Lord and dwelt in the land of Nod, on the East of Eden."*

The narrative of the First Book of Genesis clearly implies that the Adamic "world" was a sterile mountain-encircled basin, or system of basins, situated some considerable distance East of the River Tigris. The only locality satisfying these conditions—within the historical period—is that of the mountain-encircled and landlocked system of basins lying in Eastern Turkestan and Tibet.

---

[1] And the Spirit? to the God who gave it, And judged? according to whether its Freewill is in accord with the Divine Law of Harmony. This having "deviated," its redemption becomes a necessity of Divine Love, and the purpose of the Messiah, as "the displacement factor" (Chapter III). In this is the germ of the idea that when Spirit Freewill redeemed by Faith to operate in harmony, then to it restored, by the inexhaustible source of Divine Energy, immortality.

"Our wills are ours to make them thine."
"Whose service is perfect freedom."

[2] Davidson, p. 424.

THE ADAMIC WORLD

These territories comprise an earthquake area that is still very active, and are characterised by a continuous change in physical features and conditions, due principally to desiccation, on a large scale and at a rate practically unknown elsewhere on earth.[1]

Though it is in reality an elevated plateau . . . it is nevertheless a depression when compared with the girdle of mountains which surround it on every side except the East, and even on that side it is shut in by the crumbling remains of a once mighty mountain system, the Pe-shan. . . . The mountain ranges which shut off East Turkestan from the rest of the world rank among the loftiest and most difficult in Asia and indeed in the world. . . . During the later tertiary period (Davidson suggests the inundation may actually have occurred within the historical period) all the desert regions would appear to have been covered by an Asian Mediterranean, or, at all events, by vast fresh-water lakes, a conclusion which seems to be warranted by the existence of salt-stained depressions of a lacustrine character, by traces of former shore lines, more or less parallel and concentric, by discoveries of fresh-water mollusc shells, the existence of belts of dead poplars . . . the presence of ripple-marks . . . it is perfectly evident not only that this country is suffering from a progressive desiccation, but the sands have actually swallowed up cultivated areas within the historical period (*Ency. Brit.*, vol. xxvii, pp. 422, 423, "East Turkestan").

Professor Osborn (American Museum of Natural History) states: "Central Asia was the original home of mammal life which presents similar types in regions thousands of miles to the east and to the west." Dr R. C. Andrew (of same, leading expedition to Gobi Desert) confirms: "Central Asia would prove to be the original home of the human race, as well as of a great many of other higher vertebrates."

One remarkable feature, implied rather than directly stated, concerns the relations between the external world and the world of the Adamic race.

The barriers of isolation, although hindering intercourse with the external world, did not prevent murderers from being

---
[1] *Ibid.*, p. 425.

expelled (as Cain, westward to land of Nod, on the east of Eden, and there married with the external world, Gen. iv, 16, 17). Intermarriages with the external world, at first limited to outcasts and renegades, finally became the custom.[1]

"There were giants in the earth in those days.

"The sons of God saw the daughters of men that they were fair; and they took them wives of all which they chose.

"And they bare children to them, the same became mighty men which were of old, men of renown" (Gen. vi, 2–4).

Wives then were taken by the Adamic race from the races of the external world, resulting in a raising of the physical, and a lowering of the spiritual standard.

"And God saw that the wickedness of man [the Adamic race, as well as the external world] *was great in the earth, and that every imagination of the thoughts of his heart was only evil continually"* (Gen. vi, 5).

Christ amplifies that which is now appearing in man:

"For out of the heart proceed evil thoughts, murders, adulteries, fornications, thefts, false witness, blasphemies" (Matt. xv, 19).

Spiritual control gave place to physical violence:

"The earth was corrupt before God, and the earth was filled with violence."

Natural law of the survival of the fittest (for succumbing to which in the Adamic world Cain had been expelled) has spread throughout the Adamic world:

"Violence filled the earth . . . for all flesh had corrupted his way upon the earth" (Gen. vi, 11).

"And the Lord said, My Spirit shall not always strive with man, for that he also is flesh, yet his days shall be an hundred and twenty years" (Gen. vi, 3).

Here the emphasis (*also*) relates to the Adamic race having partaken of the physical nature of the races external to the Adamic world, or *Earth*.

By having partaken of this nature, the Adamic race had still further debased and subverted its special spiritual faculties. It had lost the power of controlling "flesh" by becoming "flesh," and by thus losing, had still further lost the power

---

[1] Davidson, p. 425.

THE FLOOD

of its spiritual dominion over the Earth, and all the physical kingdoms and entities of the Earth.

The narrative concerning our Lord—as the "Last Adam"—rebuking the spirit of the storm, and by so rebuking, calming the storm, describes one application of the special spiritual faculties operative in the power of spiritual dominion over the Earth, bestowed on the "First Adam" (Davidson, p. 427).

¶ 3. Davidson further shows that in the genealogies of Genesis

the dynasties of the Adamic world were judged as to their fitness at the termination of calendar cycles, and of outstanding intercalary periods, such as those occurring at intervals of 30 and 120 years, and 97, 103 and 329 years. The tradition was followed in ancient Egypt and Babylon,—The Sed Festival and its rites were institutions that the early Dynastic Egyptians derived from an earlier civilisation . . . when the king and his selected co-regent presented themselves before the shrine of the "presence form" of the Dynastic god, as to his fitness to reign, and there to renew his spiritual and physical strength from the god.[1]

"Noah was a just man, and perfect in his generations." (His dates, A.K. 1056–1656, 20 Sed periods.)

"And it repented the Lord that He had made man on the earth, and it grieved Him at His heart. And the Lord said, I will destroy man [the Adamic world] whom I have created from the face of the earth; both man and beast, and the creeping thing and the fowls of the air.

"But Noah found grace in the eyes of the Lord.

"And God said to Noah . . . But with thee will I establish My covenant" (Gen. vi, 6–9).

The personal family of Noah, alone of the isolated land-locked world of the Adamic race, was saved from the catastrophe that visited the whole race, destroying all its creatures and engulfing its civilisation, with all its possessions and works.[2]

Thus the Flood (2345–2344 B.C.) did not (and was not meant to) apply to the pre-Adamite world, the races of

---
[1] *Ibid.*, p. 425, with pp. 27, 36.   [2] *Ibid.*, p. 425.

man evolved nor the lands beyond; such were not at all affected, except parts of China in the east.

"*All the foundations of the great deep were broken up and the windows of heaven were opened*" (Gen. vii, 11), describes a general earthquake shock with coincident downpouring of rain;

> the collapsing of the crust of the earth that roofed over the great submerged lakes of Eastern Turkestan and Tibet. Such conditions cannot have failed to extend in their effect, to a minor extent, into China, . . . where an inundation of the land by a tidal wave, and the overflow from the mountains was bound to swell to an abnormal extent the flow of the Hwāng-Ho (Davidson, p. 428).

W. G. Old, *The Shu-King*, pp. 301, 302, Section V, states that:

> During the reign of the Chinese King Yaou, 2356–2254 B.C., Yu, the Chinese engineer and administrator, reports, "When the floods were lifted to the heavens, spreading far and wide, surrounding the hills and submerging the mounds, so that the common people were bewildered and dismayed . . . I drained off the nine channels, directing them into the Four Seas; I dug out ditches and canals and brought them into the rivers."

Yu also refers to years of work in cleaning and draining the valleys and ravines in the mountains of the interior.

Tsae-chin, the commentator of the Shu-King, thirteenth century A.D., gives the names of the eight rivers which, with the Hwang-Ho, formed the "Nine Rivers"—all to be found between latitudes 37° and 39° North, to the east of Peking.

> The Chinese Noah—the Fu-hi of Chinese mythology—is not only mentioned in relation to the Deluge, but in connection with his selection of the animals by *sevens* (Gen. vii, 2), in connection with his sacrifice, after the Deluge, of certain of the animals saved with him (Gen. viii, 20), and in connection with the rainbow covenant (Gen. ix, 8–17) . . . that he made his appearance on "Mountains of Chin" after the Deluge.[1]

Ferrar Fenton has shown: "the compound Hebrew word 'Ararat' means simply and literally 'the Peak of the High

[1] Davidson, p. 429.

THE ADAMIC RACE

Hills' . . . so the special topographical feature connected with the preservation of Noah would be carried in the subsequent migration of race and connected with the highest peak in Armenia."

After the Deluge "*they journeyed from the east, that they found a plain in the land of Shinar* [*Ency. Brit.*, "the word identical with Sumer"]; *and they dwelt there*" (Gen. xi, 2).

Professor Sayce states the tenth chapter of Genesis[1] is not "an ethnological table," but

> a descriptive chart of Hebrew geography, the various cities and countries of the known world being arranged in it genealogically in accordance with Semitic idiom. . . . We are not to look, then, to the tenth chapter of Genesis for a scientific division of man into their several races. We are not to demand from it that simple and primitive division according to colour. . . . As a matter of fact, *all the tribes and nations mentioned in the chapter belonged to the white race.* Even the negroes are not referred to, though they were well known to the Egyptians, and the black-skinned Nubians are carefully excluded from the descendants of Cush (*Higher Criticism and the Monuments*, pp. 119-123).

The whole context of the First Book of Genesis—studied in the light of the new evidence, and considered in relation to the origin and the physical and ethical characteristics of the white race—confirms Sayce's conclusions.

The restless energy and the "spiritual" stamina—as distinct from purely bestial stamina—characteristic of the white race, their dominating influence over other races, and their so-called "conquest" of Nature and of Nature's science and elements, all indicate their retention, to a certain extent, of the power—not yet entirely latent—and the faculties—not yet completely atrophied—attributed in the First Book of Genesis to the founder of the Adamic race, and in a lesser degree to the Adamic race.[2]

The Adamic race is the white race, and the cultural civilisation of the world from the earliest times has been due to the white race.

What we do not know is the extent to which the stock of the white race has influenced the stock of other races, or of all

---

[1] See also Ragozin, *Chaldea*, pp. 128-143. [2] Davidson, p. 431.

38 THE PILLAR IN THE WILDERNESS

races. The scriptural promise of redemption, by its being extended to "all families of the earth" in the days of Abraham, seems to imply that the white race has influenced, to some extent, the stock of all races, and that in consequence, by having become "all flesh," and thereby having lowered its spiritual "potential," it has also raised the spiritual "potential" of "all flesh." This would go far towards explaining the mystery of the foreknowledge of God as related to His permitting a sequence of events—that from the beginning presupposed (and, in fact, asserted) the pre-election of a saviour before there was either the active necessity to save, or the entities to save.[1]

¶ 4. Now to be considered:

(1) The House of Cain, after his expulsion.

(2) The descendants of Seth, who migrated from the Adamic world many years before the Flood, 2345 B.C.

These carried into all the countries of the earth—

(1) the first civilisation of the world, and the polytheistic idea of God;

(2) having been under Divine tuition in the isolated Adamic world, the monotheistic idea of God.

(1) The House of Cain:

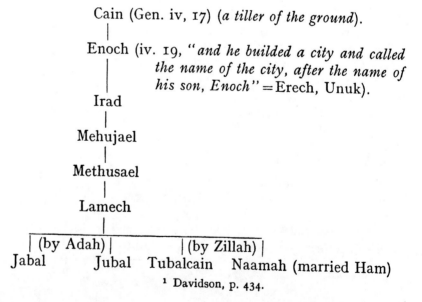

[1] Davidson, p. 434.

iv. 20. Jabal, "*he was the father of such as dwell in tents, and of such as have cattle*" (Agriculture).

iv. 21. Jubal, "*he was the father of all such as handle the harp and organ*" (Art and Music).

iv. 22. Tubalcain, "*an instructor of every artificer in brass and iron*". (Industry)—"*improver in every work of copper and iron*" (Ferrar Fenton).

This narrative clearly intends to account for the origin of the various arts as they existed in the narrator's time . . . an ancient poem (Gen. iv, 23, 24) is connected with this genealogy:

> *Adah and Zillah, hear my voice;*
> *Ye wives of Lamech, give ear unto my speech.*
> *I slay a man for a wound,*
> *A young man for a stroke;*
> *For Cain's vengeance is sevenfold,*
> *But Lamech's, seventy-fold and seven.*

—expressing Lamech's exultation at the advantage he expects to derive from Tubalcain's new invention; the worker in bronze will forge for him new and formidable weapons (*Ency Brit.*, vol. xvi, p. 122).

Gen. iv, 16, "*And Cain went out . . . and dwelt in the land of Nod, on the east of Eden*"—that is, he went westward from the Adamic world, towards Eden, perhaps even searching for it, journeying south of the Caspian, towards Palestine.

Bearing above in mind, read Leonard Woolley's unconscious testimony for identity as to facies and physique, industry and fighting qualities:

By the side of the Persian Gulf [1] hillmen [2] had settled in the first years of the world. They had big heads, and broad, thickset bodies. They were dogged and hard-working; expert husbandmen,[3] capable metal workers [4] and first-rate fighting men [5]; they had brought to that region the first seeds of the civilisation of the world: their ancient cities were Lagash, Erech, Eridu and Ur.

---

[1] C. Leonard Woolley, *The Sumerians*, pp. 2–6.
[2] *Ibid.*, pp. 6–16.
[3] *Ur of the Chaldees*, p. 20.
[4] *The Sumerians*, p. 42.
[5] *Ibid.*, pp. 49–60; *Ur of the Chaldees*, p. 87.

The Sumerian genius had invented, in particular, the cuneiform alphabet and the art of building the dome, developed a civilisation which lasted 1500 years, and had its influence on all the early empires.[1] From this part of the world . . . Sumer spread by the valleys of the Tigris and Euphrates as far as Syria and even beyond the Taurus, so that all the barbarous countries received a veneer of Sumerian civilisation.[2] This civilisation was so fine that it did not suffer by the extinction of the people. The Code of Hammurabi was based on the old laws of Sumer,[3] the Sumerian gods only changed their names to become Chaldean gods: the Sumerian language was forgotten, but thanks to its alphabet, it was translated into Semitic, into Assyrian; Assyrian sculpture of the eighth century is the same as that of Sumer.

The civilisation of Sumer was already brilliant in 3500 B.C. The Ist Dynasty of Ur (3100–2930 B.C.), contemporary of Menes (3100 B.C.), then founding at Thinis the Ist Dynasty of the Pharaohs [4] . . . the underlying foundations of Egyptian civilisation were of Sumerian origin.[5] Ur and Thinis were the two lamps of civilisation in a world still plunged in savagery [6] (Madame Tabouis, *Nebuchadnezzar*, p. 19, summing up C. Leonard Woolley, *The Sumerians*).

Into these valleys of the Tigris and Euphrates down to the Persian Gulf this "people came from a hill country"; "judging by their physical type they were of the Indo-European stock, in appearance not unlike the modern Arab . . . well-developed intellectually." [7] Tradition would make them come from the East, the study of their bones and skulls shows that they were a branch of the Indo-European stock of the human race, resembling what is called Caucasian man, a people who in stature and appearance might pass as modern Europeans rather than

[1] *The Sumerians*, p. 189.
[2] *Ibid.*, pp. 79–82; *Ur of the Chaldees*, pp. 107–112.
[3] *Ibid.*, pp. 90–94.
[4] *Ibid.*, p. 44; *Ur of the Chaldees*: "when civilisation of Egypt was a novelty, that of Sumer, ancient," p. 186; "the influence in Egypt . . . directly or indirectly came from Sumer," p. 188.   [5] *Ibid.*, 188–192.
[6] *Ur of the Chaldees*, p. 88: "to the Sumerians we can trace much that is at the root not only of Egyptian, but also of Babylonian, Assyrian, Hebrew and Phœnician art and thought, and so see the Greeks also were in debt to this ancient and for long forgotten people, the pioneers of progress of Western man."
[7] *The Sumerians*, pp. 6, 7.

CAIN IN SUMER

as Orientals. Geographically it is likely that their original homeland was in a hilly country (*Ur of the Chaldees*, p. 117).

The Sumerians believed that they came into the country with their civilisation already formed, bringing with them the knowledge of agriculture, of working in metal, of the art of writing, since then, they say, "no new inventions have been made" . . . later research may well discover some site (between the Euphrates and Indus valleys) where the ancestors of our Sumerians developed the first real civilisation of which we have any knowledge [1] . . . possibly towards Afghanistan and Baluchistan." [2]

Of the aborigine inhabitants of the land, "Mankind, when created, did not know of bread for eating, or garments for wearing. The people walked with limbs on the ground, they ate herbs with their mouths like sheep, they drank ditch-water," says a Sumerian hymn. [3]

In the midst of peoples physically more powerful and addicted to war as a pastime, the Sumerians built up an empire because they had a better army, and better weapons than their neighbours." [4]

The Army must have been more than a match for anything that could be brought against it at that time. The chariotry was to inspire an almost superstitious terror . . . it is not surprising that until they had taught their neighbours to profit by their example, they found no opponent to withstand their advance." [5]

The former inhabitants were not driven out, but, inferior as they were in all the arts of life, must have sunk to the condition of serfs. [6]

On the "standard" found in a royal grave at Ur [7] (date *circa* 3500 B.C.) the chariot is depicted as drawn by four asses (horses not known until 2000 B.C.), having four solid

---

[1] *Ur of the Chaldees*, p. 20.
[2] *The Sumerians*, pp. 8, 9. Thus confirming the civilisation of the selectedly created Adamic race in the isolated "earth" of East Turkestan —as origin.
[3] *Ibid.*, p. 13. This the opinion of Cain on meeting aborigine, evolved man.
[4] *Ibid.*, p. 49, confirming "*there were giants in the earth in those days*" (Gen. iv, 23, 24).
[5] *Ur of the Chaldees*, p. 87.
[6] *Ibid.*, p. 20.
[7] *The Sumerians*, pp. 50–60. Dungi, of IIIrd Dynasty of Ur (2260 B.C. ?), says he "enrolled the sons of Ur as archers."

wheels and carrying a driver and a fighting man, with quiver attached for spears. The infantry is shown advancing in close order, meant as a phalanx (meaning some military drill), wearing conical copper helmets with chin straps, and armed with short-handled spear, axe, scimitar and dagger. From arrow-heads and fragments of decorated bows, at this early period archers may have been known. Also were carried "large rectangular shields of wood ornamented with metal bosses."

A regular paid army was probably not kept till the time of Sargon, of Agade (? 2630 B.C.), who speaks of 5400 men, who "ate daily at his cost."

> The contents of the tombs illustrate a very highly developed state of society . . . the architect; the artist; the craftsman in metal possessed a knowledge of metallurgy and a technical skill which few ancient peoples ever rivalled; the merchant carried on a far-flung trade and recorded his transactions in writing: agriculture prospered, and great wealth gave scope to luxury. Tombs' date . . . 3500 B.C. . . . this civilisation was many centuries old.[1]

The contents of the tombs are described: "copper and silver spears, spears with heads of gold, a gold cup near the hand of the body of Queen Shub-ad, gold ornaments, a broad gold ribbon round the hair . . . of a wig padded out to almost grotesque size, with a gold 'Spanish' comb and ear-rings[2]; the remains of a wonderful harp . . . the wooden upright beam was capped with gold, gold-headed nails, the sounding box edged with a mosaic in red stone, lapis lazuli and white shell, projecting from it a splendid head of a bull wrought in gold . . . across the ruins of the harp lay the bones of the gold-crowned harpist.[3] In the furnishing of the royal graves, one constant feature is the harp, or lyre" (*Ur of the Chaldees*, p. 65). Also is mentioned the sistrum, a musical instrument which was introduced into Egypt.[4]

In trade exchange from Akkad and Sumer came grain

[1] *Ur of the Chaldees*, p. 87.
[2] *Cf.* Gen. iv, 22, *Tubalcain*, and as above, Lamech's weapons.
[3] *Ur of the Chaldees*, pp. 46–49 (*cf.* Gen. iv, 21, *Jubal*).
[4] *The Sumerians*, p. 186.

THE SUMERIAN CIVILISATION

and dates, probably garments and woven stuffs; herds of oxen, flocks of sheep, lambs and goats [1] (L. King, *Sumer and Akad*, p. 237).

As regards agriculture and trade: "Rich as were the irrigated plains of Lower Mesopotamia, their wealth is purely agricultural [2] . . . no metal, nor stone found. Bitumen was brought down the river from Hit, copper from Oman and probably from the Caucasus; silver from northern Cilicia and hills of South Elam; gold from Elam, Cappadocia, Khabur district, Antioch region of Syria; limestone from 100 miles south and from upper Euphrates valley, diorite by sea from Magan on Persian Gulf; alabaster or calcite and lapis lazuli from Persia." [3]

Regarding the mass death of men and animals found in the pit graves: "According to Sumerian beliefs a king was the vice-regent of God upon earth, 'tenant farmer' they call him." [4] "In much later times Sumerian kings were deified in their lifetime, and honoured as gods after death . . . looked upon as superhuman, earthly deities . . . as a god, he did not die as men die but was translated, and it might therefore be not a hardship, but a privilege for those of his Court to accompany their master and continue in his service." [5]

¶ 5. Running into the Persian Gulf, besides the two great rivers, in those days were two others—the Karun, fast-running and carrying as much silt as both these rivers, from the Persian hills; and the Wadi al-Batin, more sluggish but also rich in silt, from the heart of Arabia. Thus is formed a bar from shore to shore, enclosing a great lagoon, on the western edge of which is Ur, "waters gradually turning from salt to brackish, from brackish to fresh, a vast delta of clay and sand and mud, marshes and reed-beds." [6]

---

[1] *Cf.* Jabal, Gen. iv, 20.
[2] Cain, "*tiller of the earth*" (Gen. iv, 2).
[3] *The Sumerians*, p. 45.
[4] *Ur of the Chaldees*, p. 145.
[5] *Ibid.*, p. 65.
[6] *The Sumerians*, p. 2. "The walled town was reserved for the moral conquerors of the country, the Sumerians; while the barbarians put up their mud huts at the foot of the mound's slope, or on flat cultivated land" (*Ur of the Chaldees*, p. 21). "The Sumerians were pre-eminently town dwellers" (*The Sumerians*, p. 16).

Legends assign to the Babylonian god, Marduk, the work of reducing the primeval chaos to order by the separation of the land from the water, and the first foundings of the homes of men . . . we evidently have here a vivid recollection of the time when the whole of South Babylonia was a swamp, and in these islands, towns arose (as in the fens of England), dykes were heaped up and the shallows were gradually reclaimed, till the demon of the watery chaos, Tiamat, finally vanquished, retreated from the land. Marduk had created the land and the two great rivers [1] (Hall, *History of the Near East*, p. 175).

"The conquest of the monster Tiamat, symbolising the chaos prevailing in primeval days, was ascribed to Marduk . . . with this stroke Marduk became the creator of the world, including mankind" (*Ency. Brit.*, "Marduk").

Merodach, "the donor of fruitfulness, the founder of agriculture, the creator of grains and plants, who causes the green herb to spring forth" (Clay, *Origins of Biblical Tradition*).

Merodach, "the gladdener of the corn, and the creator of the wheat and barley, renewer of the herd" (*Hibbert Lectures*, p. 537).

Merodach, or Marduk, is always called the first-born of Ea (the god of the water, "Lord of the Deep," the Fish god), also the first-born of Ishtar (the female form of Ea, the ocean mother). "He possessed all his, Ea's, wisdom; Ea said to him, 'My child, what I know, thou knowest, also'" (Delaporte, *Mesopotamia*).

Berossus, writing in the third or fourth century B.C., describes a race of monsters, half man and half fish, which, led by one Oannes, came out of the Persian Gulf and, settling in the coast towns of Sumer, introduced the arts of writing, agriculture and working in metal, "in a word all the things that make for the amelioration of life were bequeathed to men by Oannes, and since that time no further inventions have been made" (*The Sumerians*, p. 189).

The religion of the Sumerians was polytheistic (*The*

---

[1] Thus Cain, the agriculturist, by irrigation of the land, becomes the benefactor and god, Marduk. See later proof of identity, and Ea, the perverted form for Eve.

THE SUMERIAN FLOOD

Sumerians, p. 128); "from astronomical knowledge, the sun and moon and the planets were identified with the gods" (p. 128). "Throughout the religion of the Sumerians is one not of love, but of fear, fear of Beings all-powerful, capricious, unmoral" (p. 125). "All sickness was brought about by the malignant spirits which thronged the universe and preyed on man" (p. 127). Thus arose in the priesthood, magicians and soothsayers, and, with such a false view of Deity, great opportunity for preying on man, of which these took the fullest advantage.

¶ 6. The King-Lists (*The Sumerians*, pp. 21–26) given show:

A. The Kings before the Flood

The eight antediluvian kings are credited with reigns which, added together, make the modest total of 241,200 years . . . behind these grotesque sums there may be some confusion between different systems of notation (p. 29).

"The Flood came. After the Flood came, Kingship was sent down from on high."

B. The Kings after the Flood

Follow 23 kings (24,510 years) of the Ist Dynasty of Kish, in Agade, and 12 kings (2310 years) of the Ist Dynasty of Erech (perhaps some of these more or less contemporary). These succeeded by the Ist Dynasty of Ur (3100–2930 B.C.), and a whole series of dynasties down to *circa* 2470 B.C., when the dynasty of Gutium destroyed the dynasty of Agade, begun by Sargon (given as 2630 B.C.[1]) (Woolley).

Now giving the eight antediluvian kings a reasonable reign of thirty to fifty years each, and taking Cain's entry with the commencement of Sumerian civilisation at some time after 3940 B.C., the Sumerian Flood is dated between 3700–3540 B.C.

[1] Some authorities give the date of Sargon as 3800 B.C. In which case the House of Cain may have entered Sumer after the subsidation of the Flood, and directed the reclamation of the land in South Mesopotamia, thus considered its benefactors and gods.

The Flood, if at all worth mentioning, filling up the lagoon and overflowing, must have preceded the pit graves of the royal cemetery at Ur, dated 3500 B.C., though the high-walled cities of lower Mesopotamia may have survived.

This gives a minimum of 440 years (3540–3100 B.C.) until the Ist Dynasty of Ur, for the 23 kings of Kish and 12 of Erech—some of these were probably reigning more or less as contemporaries.

Now the date of the Genesis Flood is fixed at 2345 B.C. (Davidson). It occurred in the mountain basin of Turkestan and overflowed eastward into China. Therefore its effects were not felt nor recorded in Sumer. Also the flood waters of Babylon would follow the valleys of the Euphrates and the Tigris from the mountains of Asia Minor.

Again, the Sumerian Flood was such as happens to river basins, as in the Yangtse valley in China (1931), in the Mississippi valley (1927), and there is no geological evidence that it was accompanied by earthquake, as in the Genesis Flood, "*all the fountains of the great deep were broken up*" (Gen. vii, 11).

Thus the Sumerian Flood[1] was an earlier, less extensive and destructive flood than the Genesis Flood, which destroyed every evidence of the Adamic civilisation but the Noachic family.

Thus the House of Cain persisted for about 840 years[2] (3940–3100 B.C.), when it splits up amongst his descendants, each adopting and ruling a separate town—Akkad, Erech, Ur, Lagash, Opis, Mari; each ever striving for suzerainty over the whole,[3] each adopting its own god—at Babylon, Marduk; at Larsa, Shamash; at Ur, Nannar; at Nippur, Enlil; at Erech, Innini (or Ishtar); at Eridu, Enki.[4]

[1] Davidson, in *Early Egypt, Babylonia and Central Asia* (Charts 15, 22, 30), shows that "the Chaldean flood, evidenced by the recent excavations at Kish and Ur, was an earlier flood than the Biblical deluge . . . and was merely an abnormal inundation due to the lack of proper irrigation channels, before the advent in Chaldea of colonies of the white race that introduced the later Chaldean system of flood relief channels."

[2] On analogy of the antediluvian dynasties of Genesis v. (Davidson, p. 36).

[3] *The Sumerians*, p. 19: "Civil War was the rule rather than the exception."

[4] *Ibid.*, p. 119.

FOUNDING OF THE EGYPTIAN DYNASTY 47

"The seat of the first post-diluvian dynasty is at Kish, in Agade . . . then the focus of power shifts to the south, Erech; and the next, south again, Ur" (*The Sumerians*, p. 48). Thus follows the degeneration of Sumerian civilisation.[1]

One of these sons migrated to Egypt, and founded the Ist Dynasty of the Pharaohs, at Thinis, Menes (3100 B.C.). "The Egyptians traced back the beginning of their history to Menes, before whom came darkness and demigods, and the discoveries of archæology have justified their belief; for the Sumerians, the Ist Dynasty of Ur came at the end of a period of civilisation whose duration was to be reckoned in thousands of years."[2]

"A scene engraved on an ivory knife shows the reception, far from friendly, given by the Egyptians to the Sumerians on the shore of the Red Sea" (Madame Tabouis, *Nebuchadnezzar*, p. 31).

Of the first three Egyptian dynasties, Professor Petrie says: "The rapid rise of Art is the most striking activity of this age . . . so soon as the dynastic race comes in, there begins the enormous step in Art, rapidly developing to perfection within its natural limits."

Intruding within these warring dynasties of Sumer, after 3100 B.C., appear two from outside—Awan, a people from the land which became Elam, and Khamasi, a city east of the Tigris. These may have been colonies founded by later renegade Adamites.

After the Genesis Flood (2344 B.C.) there appeared in the land of Sumer the family of Noah (Gen. xi, 2).

Nimrod, son of Ham by marriage with Naamah, the only woman descendant of Cain mentioned (iv, 22), evidently by this right succeeded to the throne of Agade (x, 10).

Asshur, son of Shem, founded Nineveh in Gutium (x, 11), and the destruction of the dynasty of Agade by the Guti [3]

---

[1] "As far as we know, the 4th millennium B.C. saw Sumerian art at its zenith. By the Ist Dynasty of Ur . . . a decadence set in" (*The Sumerians*, p. 44).

[2] *The Sumerians*, p. 186. Thousands of years are assumed, if the source of this civilisation is not the Divinely selectively created race of Adam.

[3] Woolley's date for this, *circa* 2470 B.C., is more correctly placed *circa* 2186 B.C.; see page 49.

is thus due to him, Nimrod [1] escaping to Egypt. Gutium is the ancient name for Assyria.

At this point, therefore, is the first appearance of Semitic influence in the Sumerian civilisation.

Woolley continues the history:

> The temple hymns bewail the violation wrought by the Guti in the shrines of Nippur, Adab, Erech and Kish . . . during the Guti period of 125 years, business documents and works of art are missing. . . . Before long the Guti kings were dedicating their offerings in the temples of the Sumerian gods which the first invaders had despoiled [2]. . . . Later only a loose control was exercised over Lagash by the Guti suzerain [3] . . . (established at Kish, of Akkad, p. 130).
>
> In the business tablets from Tello, Semitic names occur freely, and even in the religious texts Semitic phrases are introduced; it is clear that so far south as Lagash the population was becoming more and more mixed. . . . [4]

"Utu-Khegal, the patessi of Erech, revolted against 'the enemy of the gods.' . . . In the battle that followed, Tirigan, the Guti king, was abandoned by his troops and made prisoner. Utu-Khegal assumed the title of king . . . and under a Sumerian dynasty revived the imperial organisation of Sargon" (p. 89), ". . . and under him the Sumerian revival saw the evolution and development of civil law . . . and the reform of abuses and reinforcement of the law" [5] (which later led to their codification by a Semitic ruler, Hammurabi).

> Amongst the local governors installed by Utu-Khegal was Ur-Nammu, patessi of Ur. . . . Ur-Nammu revolted, conquered Erech (p. 130), and subdued Kish (the old Semitic

---

[1] Or a son? as Shargalisharri, of whom was written despairingly, "Who was King, who was not King" (*The Sumerians*, p. 83).

[2] Thus the House of Asshur, of Shem, accepts polytheism and by intermarriage with Cain-Ham stock causes deterioration of the Noachic race, and incidentally leads to the downfall of the Guti (see Chapter IV).

[3] *The Sumerians*, pp. 83-85.

[4] *Ibid.*, p. 88.

[5] *Ibid.*, p. 90 and pp. 91-129 give the description of the laws and customs.

ENTRY OF THE SEMITES

capital of the Guti and still under Semitic influence), and made himself master of the whole country, so that once again there was a "King of Sumer and Akkad," founding the IIIrd Dynasty of Ur. . . .[1]

At Ur he built the city wall, the great Ziggurat [2] Tower, the temples of Nannar and Ningal, and the royal palace . . . and was active in the digging of canals (p. 153).

In every city there was at least one such tower—using "brick and slime" (bitumen), crowned by a sanctuary forming a temple complex: of them all the biggest and most famous was the Ziggurat of Babylon, which in tradition became the Tower of Babel.[3]

Now the IIIrd Dynasty of Ur is dated 2152–2035 B.C. by Davidson,[4] by comparing the chronology of Babylon, Egypt and the Hebrews. And the dynasty of Gutium, 2186–2100 B.C. (86 years plus 40 days' reign of Tirigan) —i.e. for some years co-existent—was not so much a supremacy as a series of Guti raids, "establishing a supremacy of terror in Sumer and Akkad." This dating fits the entrance of Asshur, son of Shem, to introduce the first Semitic influence among the Sumerians.

The IIIrd Dynasty of Ur was continually at war, contending the supremacy with Babylon in the north, until Dungi (2134–2077 B.C.)[5] sacked Babylon, looted its treasury, and extended his domain.

In his time, palm groves and dates are mentioned as being cultivated by artificial fertilisation.

Now appears a fresh invasion from outside. "An Amorite, Isbi-Irra of Mari, rose in revolt, and marching

---

[1] Woolley gives its dates as 2278–2170 B.C.; whereas Davidson gives 2152–2035 B.C.

[2] Ziggurat—the mountain of God, or the hill of heaven (*Ibid.*, p. 133).

[3] *Ur of the Chaldees*, p. 119: "If Ziggurats were the custom of the people, the huge affairs at Babylon and at Ur were built thus after the Semitic entrance."

Thus is confirmed the building of the Tower of Babel by the Noachic family after the Genesis Flood (Gen. xi, 1–9), evidently as a centre for their civilisation, but the Divine Plan was for the three sons to spread over the earth, and not to concentrate in the land of Cain.

[4] Davidson, *Early Egypt, Babylonia and Central Asia* (Charts 24–27), and quoting Dr S. H. Langdon, *Weld-Blundell Collection*, vol. ii, W-B. 444, and *Camb. Anc. Hist.*, vol. i.

[5] These and subsequent dates are given as Davidson's.

down from the middle Euphrates invaded Akkad, occupied Isin, and then opened negotiations with Elam,"[1]

In 2068 B.C. Libit-Istar, of Isin, overthrew Bursin II of Ur, and in turn was defeated by Gungunu of Larsa. Gimilsin II reigns at Ur as his vassal.

> The country was plunged into a welter of civil war, and division again occurred into numerous city-states—Isin, Larsa, Erech, Sippar, Babylon, Kish: the power which was with Isin was taken by Larsa. . . . This was ill-tolerated by Elam, now in alliance with Isin.[2]

In 2035 B.C. Elam was ready to move, their forces crossed the Tigris and overran Sumer. Babylon and Larsa were taken, but as vassal states were allowed to start fresh kings; Ibisin, of Ur, was taken captive to Anshan.

> After two more short reigns, in 1977 B.C., we find an Elamite, Warad-Sin, on the throne of Larsa, and with Larsa went Ur, Eridu, Lagash, Nippur and Erech. The Delta was now divided into three states, Larsa, Isin, and Babylon.
>
> Rimisin (who succeeded his brother, Warad-Sin) finally captured and laid waste the city of Isin, and continued the policy of his fathers of placating Sumerian feeling, in order to win their support in the final struggle with Babylon, by strengthening Ur's defences and restoring temples and ziggurats.[3]

The Semites, under the new dynasty of Babylon started by Sumurabi in 2035 B.C., had been rising to power and were pushing the claims of Marduk.

Thus had some part of Shem in contact with the land of Cain adopted polytheism and forgotten their Divine tuition in landlocked Turkestan. Yet their influence may be seen in the following:—

> Three or four tablets have been found, smothered as it were among thousands of polytheistic inscriptions, which clearly show that the knowledge of God existed in Babylonia before

---

[1] This land (the mountainous region east of the Tigris) was now occupied by Elam, son of Shem (Gen. x, 22), who thus introduces further Semitic influence into Sumer.
[2] *The Sumerians*, p. 168.   [3] *Ibid.*, pp. 176, 177.

THE RISE OF THE CHALDEES

the year 2200 B.C., where there is said to have been a literary revival, during which older writings and traditions were reproduced . . . these inscriptions telling about One Supreme Being, the rebellious angels, and the fall of Adam.[1]

In the decadence of the Sumerian nation we find the period of their greatest literary and historical activity. It was under the Larsa Dynasty (*i.e. circa* 2180 B.C.) that the Scribes took in hand historical works, as the King Lists, the Book of Omens, the old legends of the creation, the flood, and the official pantheon.[2]

In Babylon appeared a leader, Hammurabi (1933 B.C.), who wrested Isin and Erech from Larsa, and later . . . captured Larsa and so became master of the whole country of Babylon and Sumer.

Thus began the empire of the Chaldees and the fall of Sumer.

Of the Sumerians nothing more is heard. Their civilisation, lighting up a world still plunged in primitive barbarism, was in the nature of a first cause.[3]

This brings the story to the time of Abraham, and is continued in Chapter IV.

¶ 7. Thus, to sum up: Cain entered the land of Nod (Sumer) from the east, there married and built cities, Erech ("*called it after the name of his son, Enoch*"[4]), Lagash, Eridu, Ur. His House, as hillmen from the mountains of East Turkestan, founded the first civilisation of the world, having knowledge of the arts, music, writing, metallurgy, building, agriculture, from the special faculties Divinely given to Adam. With the superiority of his organisation and weapons he obtained the mastery over

---

[1] Bristowe, *Sargon the Magnificent*, p. 115; and p. 117, quoting Sayce, *Assyria, its Princes, etc.*, p. 85: "The higher minds of the nation struggled now and again towards the conception of One Supreme God and of a purer form of faith, but the dead weight of polytheistic beliefs and practices prevented them from ever really reaching it."

[2] *The Sumerians*, p. 178.

[3] *Ibid.*, p. 181. It will be interesting to see whether the discovery of the ancient civilisation at Mohenjo Daro and Harappa, in the Indus valley, can be linked up to the Sumerian as of Cainite, or of renegade Adamic, origin.

[4] Gen. iv, 17. Prof. Sayce, *Hibbert Lectures*, p. 185, identifies Unuk (or Erech) with the Enoch of Genesis iv, 17. P. 72: "Erech is called the old city," "the place of settlement," "Unuk is found on the oldest bricks."

52   THE PILLAR IN THE WILDERNESS

the "giants," a pre-Adamite race of men (possibly a black race). Knowing the "spirit dominion over matter" possessed by his race, he introduced the idea of the king as an earthly deity, and in the tombs of his House are the graves of his Court.

He himself, or one of his House, may have actually reigned as Sargon.[1]

He founded a false religion of polytheism based on his own ancestry, with deliberate intent and in compact with the Evil One, wherewith to befuzzle the minds of men in the worship of the One God, the Creator—Anu (Adam), Ea (Eve), Bel (the Devil); and after his death was himself deified as Marduk (a saviour!), and with the religion, he originated a priesthood which deliberately hid true knowledge in allegory, to be a prey on man.

And by the marriage of Ham and elements of Shem with this Cain-aborigine mixture—against Divine command as to pollution of race stock—the Noachic race partook of this spiritual degeneration.

The effects of Cain-Bel civilisation were felt in all the ancient civilisations, Egypt, Assyria, Babylon, the Hittites, Persia, Greece and Rome; and even in these days, in the philosophic doubts of the Most High in the minds of men, is seen the vintage of his cup.[2]

"Babylon hath been a golden cup in the Lord's hand, that made all the earth drunken; the nations have drunken of her wine, therefore the nations are mad" (Jer. li, 7).

Further inquiry as to the connection of Cain, on the one hand with Bel, and on the other hand with the land of Sumer, is well brought out in *Sargon the Magnificent*, by Mrs S. Bristowe, who establishes his identity in (i) the reign of Sargon; (ii) the founding of polytheism.

Quotations from her researches are given in Appendix A, which follows.

---

[1] Bristowe, *Sargon the Magnificent*, p. 31, identifies the root *Sar* with Shah, Czar, Sahib, Sire, meaning ruler or king, and *gon* with Cain.
"Names of antediluvian patriarchs correspond with Babylonian roots and words, such as Cain with gina and Kinn" (George Smith, *Chaldean Genesis*, p. 295).

[2] Refer pp. 135, 207, 237, 272, 318, 327, 351.

It might be more accurate to read "the House of Cain" where she writes "Cain," as persisting from *circa* 3940–3100 B.C.—just as the lives of the antediluvian patriarchs of Genesis v are interpreted as dynasties or reigning houses (as explained by Davidson, pp. 35–37, 425, 426).

In Appendix B, references are made to Sir J. G. Frazer's work, *The Golden Bough*, to connect this Cain-Bel story to Græco-Roman mythology, and through such perversion to its influence on early Christianity.

APPENDIX A TO CHAPTER II

Col. Conder, *The First Bible*, pp. 218–220, giving three inscriptions:

(1) King Sargina, King of Erech, having overthrown the world . . . has erected a temple this day for the god Enlil, King of all lands, to worship Enlil.

(2) To the god Enlil, the King of all lands, King Sargina, King of Erech, the world-king, the prince of God, the mighty man, the obedient son of the god Ea, the great ruler or patesi (priest-king) of the divine King of all lands, listening obediently to the god Enlil . . . having become the sole King of Erech, invoking Nina, the far-famed lady of Erech; through the mighty aid of his god, in the day that the god Enlil made to King Sargina the grant of royalty on earth, allotted to him in the sight of the world, the hosts of the lands being obedient from east to west, he has added every land by making conquest. . . . He has made the high place of Erech a shrine of Ea [on a votive vase of white calcite stalagmite].

(3) The divine Sargani, the illustrious King, a son of Bel the just, the King of Agade and of the children of Bel.

Ragozin (*Chaldea*, pp. 205–207), quoting inscription, in which Sargon says:

For 45 (undecipherable) years, the kingdoms I have ruled, and the black-heads (or black) race I have governed. In multitudes of bronze chariots I rode over rugged lands. I governed the upper countries (Assyria, etc.). Three times to the sea I have advanced.

Hibbert Lectures, Prof. Sayce, 1887, p. 185:

M. Dieulafoy's excavations on the site of Susa have brought to light enamelled bricks of the Elamite period on which a black race of mankind is portrayed, it may mean that the primitive population of Chaldea was blackskinned." [1]

Page 21:

It was known that the great temple of the Sun god at Sippara . . . had originally been erected by Naram-Sin, the son of Sargon, . . . Nabonidas, the last King of Babylon . . . lighted upon the "foundation stone" of Naram-Sin himself. This foundation stone, he tells us, had been seen by none of his predecessors for 3200 years . . . [so] Sargon I lived 3200 years before his own time, or 3750 B.C.

This date also verified by the American excavator, H. V. Hilprecht, but later Assyriologists deny this early date, holding "mistake in archetype, they wrote 3200 instead of 2200."

Prof. Leonard King, *Sumer and Akkad*, p. 216:

If any one point in early Babylonian history was to be regarded as certainly established, it was the historical character of Sargon of Agade. . . . Sargon's reign forms the most important epoch in the early history of his country.

Cambridge History, vol. i, p. 406:

Sargon divided his vast empire from the lower sea to the upper sea, from the rising to the setting of the sun, into districts of five double hours' march each, over which he placed the "sons of his palace." By these delegates of his authority he ruled the hosts of the lands altogether.

Page 404:

His career began with the conquest of Erech.

Page 405:

It seems impossible to explain away the voyage of Sargon across some part of the Mediterranean, and naturally Cyprus was his first objective.

---

[1] *Archæology of the Inscriptions*, Sayce: "The earliest Babylonian monument gives two types of man, one with oblique eyes and negrito-like face, the other heavily bearded."

SUMERIAN RELIGION

Col. Garnier, *The Worship of the Dead*, p. 398:

He is also stated to have made successful expeditions to Syria and Elam, and that with the conquered peoples of those countries he peopled Akkad, and built there a magnificent palace and temple, and that on one occasion he was away about 3 years when he advanced to the Mediterranean, and . . . left there memories of his deeds, returning home with immense spoils.

"Times" History, vol. i, p. 362:

Babylonian art had already a high degree of excellence; two seal cylinders of the time of Sargon are among the most beautiful specimens of the gem-cutter's art ever discovered. The empire was bound together by road, along which there was a regular postal service, and clay seals which took the place of stamps are now in the Louvre, bearing the names of Sargon and his son. A cadastral survey seems to have also been instituted. It is probable that the first collection of astronomical observations and terrestrial omens was made for a library established by Sargon himself.

Prof. Sayce, *Monumental Facts, etc.*:

Centuries before Abraham was born (about 2000 B.C.) Babylonia was full of schools and libraries of teachers and pupils, and poets and prose writers, and of the literary works they had composed.

Prof. Maurice Jastrow, *Religion of Babylonia and Assyria*:

It was through the temple schools and for the temple schools that the literature which is wholly religious in character was written . . . the power thus lodged in the priests was enormous. They virtually held in their hands the life and death of the people.

Prof. Darcy W. Thompson, *Transactions of the Royal Society of Edinburgh*, vol. i, part i, No. 3:

This dominant priesthood, whose domain was knowledge, holding the keys of treasured learning, opened the lock with chary hands, and veiled plain speech in fantastic allegory. In such allegory Egyptian priests spoke to Greek travellers.

Prof. Pinches on Babylonian literature in *Ancient Egypt* (edited by Sir F. Petrie):

> There is hardly any doubt that a desire existed to make things as difficult as possible.

And what was the religion founded by the House of Cain in the Sumerian civilisation?

Nothing less than the founding of polytheism, with its mythology of male and female gods, to befuzzle the minds of men.

"It is nothing more or less than the result of Cain's determination to counteract the worship of the One God, so faithfully preserved by the other branch of Adam's race" (Bristowe, p. 62).

The first Babylonian gods were a trio: "the supreme gods Anu, Mul-lil and Ea" (Prof. Sayce), ". . . they were the deified representatives of those Bible characters (Adam, the Devil, Eve), and that it was Cain who deified their memories by transferring to them some of the attributes of God, . . . an unholy compact between Cain and the Devil—the exiled man and the disgraced Spirit . . . to establish idolatry . . ." (Bristowe, p. 99) represented in the drawing in George Smith's *Chaldean Genesis* as "Sargon taking the hand of Bel."

Professor Sayce, describing the inauguration ceremony of the Babylonian kings, writes:

> The claimant to the sovereignty took the hand of Bel, as it was called, and then became the adopted son of the god. Until this ceremony, however much he might be a king *de facto* he was not so *de jure* . . . the legal title could be given by Bel and by Bel only (*Babylonian and Assyrian Life*, p. 36).

Madame Tabouis, *Nebuchadnezzar*, describes how Nebuchadnezzar breaks off his campaign against Egypt when in an advantageous position, on hearing of his father's death, to hurry by chariot across the desert to Babylon to "take the hand of Bel" before a relative should usurp the throne.

SUMERIAN RELIGION

G. Smith, *Chaldean Genesis*, p. 58:

Mul-lil was the original Bel of the Babylonian mythology, and was lord of the surface of the earth and of the affairs of men.

Hibbert Lectures, p. 147:

The supreme Bel was Mul-lil, who was called god of the underworld.

Prof. Jastrow, *Religion of Babylonia and Assyria*:

Ea,[1] the god of water, "the third in a great triad, of which the other two members were Anu, the god of heaven, and Bel, the god of earth.

G. Smith, *Chaldean Genesis*, p. 54:

At the head of the Babylonian mythology stands a deity ... considered as the ruler and god of heaven. This deity is named Anu ... (p. 53), the king of angels and spirits, lord of the city of Erech.

(A parody on the First Adam, and leaving no room for a Supreme Being.)

Ea is sometimes male, at other times female, as Isthar; named "Anatu" when wife of Anu; "Enki," "Belit-ili," or "Innana" when the wife of Bel: sometimes Isthar as daughter of Anu, sometimes daughter of Sin, the Moon-god: also named "Nina": Nannar, Nin-gal; also named "Davkina," when wife of Ea (!) and mother of Tammuz, and in same sentence, Istar is wife of Tammuz.

Hibbert Lectures:

When the legend of Tammuz got to Greece, his mother was said to be his sister.

Cambridge History, vol. i:

Tammuz, son of Innini (p. 413) ... and his sister Ishtar (p. 442).

And Ishtar is called "the Great Mother," the mother of

[1] C. L. Woolley, *Ur of the Chaldees*, p. 199 (in side door of a temple at Ur, a limestone relief of the god Ea):

"According to the old Sumerian convention, the god Ea is shown holding a vase from which two streams of water are pouring to the ground ... as Lord of the Waters of the Abyss, Ea holds the source (*i.e.* the Garden of Eden—Ea as Eve), from which rise the twin rivers Tigris and Euphrates, givers of life to the land of Mesopotamia."

mankind, the lady of Eden, the beloved of Anu, the goddess of Birth, the goddess of the Tree of Life, the lady of the Rising, the lady of the Deep, Mistress of the Abode of Fish, Istar the ocean-mother, and female form of Ia, Dea Myrionymous (goddess of 10,000 names, representing all the Babylonian ones, and most of later Egyptian, Greek and Latin ones).

And the following perversions of the Biblical story:

G. Smith, *Chaldean Genesis*, p. 109:

Isthar, with Anu, the king, into a noble seat they raised and in the government of heaven they fixed [in the Babylonian story of the Creation].

Hibbert Lectures, p. 241 (a myth translated by Sayce):

The altars amid the waters, the treasures of Anu, Bel and Ea, the tablets of the gods, the delivering of the oracle of heaven and earth, and the cedar tree, the beloved of the great gods, which their hand has caused to grow.

L. King, *Babylonia Religion*, p. 134:

Isthar cried aloud like a woman in travail, the Lady of the gods lamented with a loud voice, saying,—The old race of man hath been turned back into clay, because I assented to an evil thing in the council of the gods, and agreed to a storm which hath destroyed my people, that which I brought forth.

Prof. Sayce, *Religions of Ancient Egypt and Babylonia*, pp. 342, 343:

Unspeakable abominations were practised in the name of Isthar, which were outdone in horror in other Babylonian cities. The black Isthar, as we may call her, was the parody of the goddess of love, and the rites with which she was adored and the ministers by whom she was served were equally parodies of the cult which was carried on at Erech. Her priestesses were witches who plied their unholy calling under the shadow of the night and mixed the poisonous philtres which drained away the strength of their hapless victims.

Prof. Sayce:

The city of Erech was the seat of the gods Anu and Isthar (or Ea), who were afterwards adopted by the Hebrews [meaning prototypes of Adam and Eve].

Prof. Sayce in *Hibbert Lectures*, p. 29, *re* "Observations (or Illuminations) of Bel," being 72 books, part of Sargon's library:

It was later translated into Greek by the historian Berossos; and though supplemented by numerous editions in its passage through the hands of generations of Babylonian astronomers . . . It was remembered that the god Bel himself was its traditional author (p. 400). In the "Observations of Bel," the stars are already invested with a divine character. The planets are gods like the sun and moon, and the stars have already been identified with certain deities of the official pantheon. . . . The whole heaven, as well as the periods of the moon, has been divided between the three supreme deities Anu, Bel, and Ea. . . . This astro-theology must go back to the very earliest times. The cuneiform characters alone are a proof of this.

Ency. Brit., "Canis Major":

The Greeks, borrowing most of their astronomical knowledge from the Babylonians, held similar myths and ideas. . . . The Romans adopted the Greek ideas.

Bristowe, p. 64:

. . . but added to it some even worse features, . . . not of ignorance, but of evil.

P. 65:

. . . like the clinging weed which devastates a cornfield, mythology has obscured the whole of ancient history.

Max Müller, *Ency. Brit.*, "Mythology" (of Greek mythologians):

They would relate of their gods what would make the most savage of Red Indians creep and shudder . . . stories of the cannibalism of Demeter, of the mutilation of Uranus, the cannibalism of Cronus, who swallowed his own children, and the like. Among the lowest tribes of Africa and America, we hardly find anything more hideous and revolting.

From such inane impurity, scholars and Modernists hope to discover the source of the sublime First Book of Genesis.

Prof. Sayce, *Babylonia and Assyria*, p. 67:

Slavery was part of the foundation upon which Babylonian society rested.

Hibbert Lectures, p. 83, Appendix E:

Human flesh was consumed in Babylonia in the earliest times in honour of the gods . . . human sacrifices were part of their religion.

Bristowe suggests "cahna"=priest, "bal"=of Bel: thus the word "cannibal" derived from "priests of Bel."

P. 127:

Next to the worship of false gods, what more diabolical insult could be offered to the Almighty than . . . the eating of human flesh?

The Devil is represented in the names Bel, Mul-lil, En-lil, Akki, moon-god Sin.

A fourth god, Tammuz, is added.

Hibbert Lectures, p. 23:

The primitive home of Tammuz had been in the Garden of Eden, which Babylonian tradition placed in the immediate vicinity of Eridu, hence his mother (and wife) is called "the lady of Edin."

P. 245:

Like Abel, Tammuz was a shepherd, and was killed when young.

P. 212:

In the shepherd Tabulu, however, we have the double of the shepherd Tammuz himself. Tammuz, the son of the River-god Ea.

P. 153:

Another title connects Adar with the Ares of Greek mythology, who in the form of the wild boar slew the sun-god Tammuz. Adar's title "lord of the date" . . . the chief fruit of Babylonia . . . reminds us of Cain, who was a "tiller of the ground."

P. 154:

Adar bears the same relation to Mul-lil that Merodach bears to Ea. Each is the son and messenger of the older god. But whereas the errands upon which Merodach is sent are errands of mercy and benevolence, the errands of Adar are those that befit an implacable warrior.

Cain, as Merodach, has a convenient double, Adar, who does the bad things; just Abel has two doubles, Tammuz and Tabulu.

Bristowe, p. 76:

The legend of Isthar's descent into hell to bring back Tammuz (Abel) is evidently, as Prof. Sayce points out, the origin of the later legends of Isis and Osiris, of Demeter and Persephone, of Eurydice and Orpheus; and according to the same authority, Isthar became the Ashtoreth of the Canaanites, the Astarte of the Phœnicians, as well as Diana or Artemis, Venus or Aphrodite.

P. 91:

As Isthar descends into the underworld to rescue Tammuz, so Demeter descends into Hades to rescue Persephone: as Isthar's departure causes all fertility to cease, so it ceases upon Demeter's withdrawal into a hiding-place; as Ea, the male form of Isthar, bestows upon Mankind through Merodach the arts of agriculture, irrigation and law, so in Greece these arts are attributed to Demeter.

One of Demeter's emblems is a serpent (connecting her with Eve); another, a little pig (connecting her with Abel, Tammuz).

Sir J. Frazer, *Golden Bough*, p. 103, says that Adonis and Attes were later forms of Tammuz, and were sometimes regarded as boars and pigs—because slain by one.

"Attys was Tammuz" (*Hibbert Lectures*, p. 235); "Tammuz became Adonis" (G. Smith, *Chaldean Genesis*, p. 238).

Thus from these five gods evolved 6500 gods of Babylon —variations of each other, male and female forms mixed up.

On the death of Sargon I (C. L. Woolley, *Ur of the Chaldees*, p. 65, in which the custom described), his priests deified Cain as Merodach, or Marduk. (Cain, believing in the "Spirit dominion over Matter" of his race, but not of his aborigine subjects.) Marduk is always called the son or first-born of Ea (Ea being the male form of Eve), also the first-born of Isthar.

A. T. Clay, *The Story of the Creation—The Origin of Bible Tradition*, p. 203:

> Marduk is king . . . they bestowed upon him sceptre, throne and pala . . . by his side he slung the net, the gift of his father Anu.

Garnier, *The Worship of the Dead*, p. 399:

> Sargon, the mighty man, son of the god Ea, prince of the Moon-god, begotten of Tammuz and Isthar.

(Bristowe suggests Tammuz here introduced by the priests as a blind.)

From the *Hibbert Lectures*:

> The dignity of high priest in Babylonia was derived from Merodach (p. 551). Sargon on inscriptions, "the first high priest" (p. 26); "the deviser of constituted law . . . the very wise"(p. 28); "Merodach, the prince of the gods, the interpreter of the spirits of heaven and earth" (p. 128). Merodach, "the director of the laws of Anu, Bel (Mul-lil), and Ea" (p. 188).

Of the Code of Laws of Hammurabi (*circa* 1900 B.C., "Ampraphael," contemporary of Abraham):

> This is the oldest (known) code in the world . . . the laws themselves must have been in operation long before their codification and promulgation by Hammurabi (*One-Volume Bible Commentary*, p. 35).

> Fragments of law exist which antedate Hammurabi's age, which reveal an organised life not inferior in its cultural developments to that attested by his code. . . . We must not forget that previous to Hammurabi there existed a high culture and social developments underestimated (*Religions of Egypt and Babylonia*, Hugo Winckler).

This admission that Hammurabi's laws were a bad imitation of older laws agrees with my theory that the Divine rules given to Adam were to a certain extent the pattern upon which Cain founded his Babylonian laws. The fact that the greatest commandments—those against idolatry and murder—are omitted in the Babylonian code, while sorcery and witchcraft are encouraged, certainly seems to betray his editorship. Hammarubi's Code begins with what might be called a dedication to the gods Anu, Ea, Bel and Marduk (Bristowe, p. 124).

Cain as *tiller of the ground* is Marduk, the "organiser of irrigation," "the Sun God." [1]

Hibbert Lectures, p. 291: Merodach "the ploughman of the celestial fields, the Sun God, who trod his steady path through the heavenly signs like the patient ox dragging the plough through the fields below."

Professor Sayce, *The Religions of Ancient Egypt and Babylonia*, p. 22:

> The Pharaonic Egyptians . . . who embanked the Nile, who transformed the marsh and the desert into cultivated fields, who built the temples and the tombs, and left behind them the monuments we associate with Egyptian culture, seem to have come from Asia; it is probable that their first home was Babylonia.

Professor K. Douglas, *China*:

> The canals and the artificial waterways of China suggest a striking likeness to the canals with which the whole of Babylonia must have been intersected, and which must have been as characteristic a feature of that country as similar works in China at the present day. . . . There is nothing improbable in the supposed movement of the Chinese tribes from Mesopotamia to the banks of the Yellow River.

W. J. Perry, *The Children of the Sun*, p. 429:

> Those who inaugurated the irrigation system of Mesopotamia must have proceeded with deliberate intent.

Here consider the migrations of the House of Cain as "the children of Bel, or of the Sun." [2]

W. J. Perry, *The Children of the Sun* (1923), p. 141:

> Wherever it is possible to examine the ruling classes of the archaic civilisation, it is found that they were what are termed

---

[1] H. G. Wells, *World of William Clissold*, p. 217: "Wherever agriculture went there went with it traditions of a blood sacrifice, a human sacrifice. . . . The Maya, the Aztec religions were insanely bloody."

[2] And the origin of the Nazi swastika (p. 336).

gods, that they had the attribute of gods, and that they usually called themselves "the children of the Sun" . . . signs of them visible in India, in the Malay, in China, Japan, the Pacific Islands, in Central America (Maya), in Peru (Incas).

CHINA

F. Hawkes Pott, *Sketch of Chinese History*, p. 23:

The origin of the Chinese is shrouded in obscurity. Some suppose that the ancestors of the Chinese first lived in the territory south of the Caspian Sea and migrated eastward somewhere about the twenty-third century B.C. Others assert that their original home was in Babylon on the great Euphrates plain and that they derived many of the elements of their civilisation from the ancient Chaldeans . . . from the western part of Asia and made a settlement first of all in . . . modern province of Shensi in the valley of the Yellow River . . . took up agricultural pursuits [1] and ceased to be a pastoral people . . . they found aboriginal tribes already in possession and obtained the territory by conquest. These native tribes were pressed more and more south and west, but were never exterminated.

Ency. Brit., "China," 11th ed., vol. vi, p. 174:

The earliest traces of religious thought and practice in China point to a simple monotheism.[2] There was a Divine Ruler of the universe, abiding on high, beyond the ken of man. . . . Gradually . . . was added a worship of the sun, moon and constellations, of the five planets, and of such noticeable stars as Canopus . . . the home of the god of Longevity. . . . Side by side with such sacrificial rites was the worship of ancestors . . . not a memorial service in simple honour of the dead, but . . . worship in the strict sense of the term.[3]

---

[1] Sir John Frazer, *Golden Bough* (2nd ed., vol. ii, p. 12): "One of the earliest Chinese rulers known as 'the Divine agriculturist.' 'The emperor attended by the highest dignitaries of the State guides with his own hand the ox-drawn plough down several furrows, and scatters seed in a sacred field, a field of God.'"

[2] A monotheistic idea implanted by an earlier migration of the House of Seth (Druidism).

[3] As Cain was deified by his priests in Babylon, his descendants, migrating to China, institute the worship of their ancestor, and inculcate such of themselves by later generations.

P. 184:

Cannibalism . . . existed among the ancient Chinese.

Professor Elliot Smith, *The Evolution of the Dragon*, p. 102:

At a very early date both India and China were diversely influenced by Babylonia, the great breeding-place of dragons.

There can be no doubt that the Chinese dragon is the descendant of the early Babylonian monster, and that the inspiration to create it reached Shensi during the 3rd Millennium B.C.

The late Professor Terrein de Lacouperie claimed a great similarity between the Babylonian and Chinese beliefs and institutions, in their astronomy and medicine, and origin of cuneiform system of writing.

A Monthly Journal of Knowledge, Dec., 1926, re some pottery found in China:

The manner of its manufacture, its general appearance . . . all recall the pottery of neolithic and early Bronze Age in Eastern Europe and Western Asia . . . at Susa, at Ur—was there a common origin for the neolithic inhabitants of both extremities of the Asiatic continent?

Bristowe, p. 147:

Just as Cain's arrival in Babylonia seems the simplest explanation of the sudden advent in that country of civilisation and culture, so his influence would account . . . for the Chinese art, philosophy and science which are known to have existed at the beginning of their history. That a people who have never changed or advanced, and are in some ways so barbarous and so diabolically cruel should have possessed the knowledge of good and evil from the first is . . . explained by this theory about Cain.

She then suggests names, as China, Chang, Chuen, Kan, Kieng, Kha-Khan resemble Cain; and yellow being Cain's traditional colour [1]—the Ruler of the Yellow; Order of the Yellow Jacket; yellow, the Imperial colour; Yellow River; yellow tiles of Imperial palaces and temples.

Bel is represented as a dragon, and Sargon's subjects called "the children of Bel."

[1] Shakespeare wrote: "A little beard, a Cain-coloured beard."

W. J. Perry, *The Children of the Sun*, pp. 167-311:

Other evidence . . . suggests that the children of the Sun originally ruled over Tahiti. It is said that formerly some of the chiefs claimed descent from the great god Kane, evidently a sun-god. The Iku-pau were direct descendants of Kane, the god, or Kumuhenna, the first man . . . Kane being a sun-god, the Iku-pau would therefore be of the Sun, and this ancient Hawaiian society falls into line with that of the archaic civilisation in general.

W. J. Perry, *The Children of the Sun*:

The pioneers of iron-working in Borneo are called "the Kayan," their ancestors said to have been "a gang of criminals" who taught the aborigines of Borneo the art of working in metal (p. 110).

In any account of the arts and crafts of the Kayans, the working of iron claims the first place . . . of the skill and knowledge displayed by them in the operations by which they produce their fine swords. The origin of their knowledge of iron and of the processes of smelting and forging remains hidden in mystery, but there can be little doubt the Kayan were familiar with these processes before they entered Borneo (p. 122).

Their origin is answered—Tubalcain, of the House of Cain.

APPENDIX B TO CHAPTER II

From Sir J. G. Frazer's *The Golden Bough*,[1] quotations are given to show how the myth tale, invented by Cain from his own ancestry wherewith to found polytheism in the Sumerian civilisation, is carried through to Græco-Roman mythology and is infecting the early Christian Church of the second century, the Roman Church, and even influencing present-day Christianity.

Frazer set himself to answer, "First, why had Diana's priest at Nemi, the King of the Wood, to slay his predecessor? Second, why before doing so had he to pluck the branch of a certain tree?"[2]

The answer goes back to its origin in the story of Cain—

[1] Sir J. G. Frazer, *The Golden Bough*, abridged edition (Macmillan & Co.), 1922.
[2] *Ibid.*, p. 9.

ANCIENT MYTHOLOGY

(1) because Cain slew Abel, (2) because Eve plucked the apple—some historic event common to and known by the white race, and diversified by subsequent generations, when it became admixed with aborigine thought and absorbed into a universal mythology.

In Frazer's work there can be traced the influence of Cain's polytheism in the migrations of the Children of the Sun (or, of Bel) on the savage or heathen races of man, on their fetishism and animism, their belief in taboos and practice of magic.

To the Spanish conquerors of Mexico and Peru many of the heathen rites appeared to be diabolical counterfeits of the Christian sacraments. With more probability the modern student of comparative religion traces such resemblances to the similar and independent workings of the mind of man in his sincere, if crude, attempts to fathom the secrets of the universe, and to adjust his little life to its awful mysteries.[1]

More probable still is it that Cain-Bel influence was carried by the migrations of Naphtuhim (see p. 109) and the Children of the Sun (see p. 63), for where the white race (whether the House of Cain or of Seth) had not reached, as the Australian aborigine even recently discovered in central New Guinea, there the mind of man towards the secrets of the universe and life's mysteries remains just as it was 6000 years ago.

From the beliefs and practices of rude peoples all over the world . . . the savage fails to recognise those limitations to his power over nature which seem so obvious to us.

The conception of gods as superhuman beings endowed with powers to which man possesses nothing comparable in degree and hardly even in kind, has been slowly evolved in the course of history. By primitive peoples the supernatural agents are not regarded as greatly, if at all, superior to man; for they may be frightened and coerced by him into doing his will. . . .

The notion of a man-god, or of a human being being endowed with divine or supernatural powers,[2] belongs essentially to that earlier period of religious history in which gods and men are

---
[1] *Ibid.*, p. 358.
[2] This was exactly the origination of Cain with his higher intellect and knowledge on his arrival in Sumer.

still viewed as beings of much the same order, and before they are divided by the impassable gulf which, to later thought, opens out between them. Strange therefore as may seem to us the idea of a god incarnate in human form it has nothing very startling for early man.[1]

Thus Cain with his higher intellectual powers had an easy task to become accepted as a king and a god in Sumer, especially as he appeared after the Babylon Flood and with his agricultural knowledge reclaimed the earth and sowed it with corn to the benefit of aborigines there located.[2] Then, as King Sargon, he became on his death the god Marduk, the Divine agriculturist.

> Kings were revered, in many cases not merely as priests, that is, as intercessors between man and god, but as themselves gods, able to bestow on their subjects and worshippers those blessings which are commonly supposed to be beyond the reach of mortals, and are sought, if at all, only by prayer and sacrifice offered to superhuman and invisible beings.
>
> Thus Kings are often expected to give rain and sunshine in due season, to make the crops to grow, and so on.
>
> The early Babylonian Kings, from the time of Sargon I till the Fourth Dynasty of Ur or later, claimed to be gods in their lifetime.
>
> The union of sacred functions with a royal title occurs frequently outside the limits of classical antiquity, and is a common feature of societies at all stages from barbarism to civilisation.[3]

The marriage of women to gods (*The Golden Bough*, pp. 142–146) is an interesting comment on Cain's intermarriage with Caucasian-Turanian stock.

And one hundred pages (pp. 393–491) are given to corn myths—showing the importance to early mankind of agriculture—the special study of Cain.

Coming to the pantheon [4] that Cain founded out of his family:

---

[1] *Ibid.*, pp. 91, 92.
[2] From this also is derived the story of Osiris; for from Sumer, Thinis was founded.   [3] *Ibid.*, pp. 10, 104, 106.
[4] Bearing in mind:
(1) Eve—as Ea, sometimes male; when female the great mother-goddess, Ishtar, of Babylon; Aphrodite of the Greeks; Artemis of Ephesus;

ANCIENT MYTHOLOGY

The worship of Adonis was practised by the Semitic peoples of Babylonia and Syria, and the Greeks borrowed it from them as early as the seventh century B.C.

The true name of the deity was Tammuz; the appellation of Adonis is merely the Semitic "Adon," "lord," a title of honour by which his worshippers addressed him.

In the religious literature of Babylonia Tammuz appears as the youthful spouse or lover of Ishtar, the great mother-goddess, the embodiment of the reproductive energies of Nature. . . . Every year Tammuz was believed to die, passing away from the cheerful earth to the gloomy subterranean world, and every year his divine mistress journeyed in search of him. During her absence the passion of love ceased to operate, men and beasts alike forgot to reproduce their kinds, all life was threatened with extinction.

So intimately bound up with the goddess were the sexual functions of the whole animal kingdom that without her presence they could not be discharged.

A messenger of the great god Ea was accordingly despatched to rescue the goddess on whom so much depended. The stern queen of the infernal regions, Allatu, reluctantly allowed Ishtar to be sprinkled with the Water of Life and to depart, in company probably with her lover Tammuz, that the two might return together to the upper world, and that with their return all nature might revive. . . .

Mirrored in the glass of Greek mythology, the oriental deity appears as Adonis, a comely youth beloved by Aphrodite. Persephone refused to give him back to Aphrodite though the goddess of love went down herself to hell to ransom her dear one from the grave.

The dispute was settled by Zeus, who decreed that Adonis should abide with Persephone in the underworld for one part of the year, and with Aphrodite in the upper part for another part. At last the fair youth was killed in hunting by a wild boar, or by the jealous Ares, who turned himself into the likeness of a boar in order to compass the death of his rival.[1]

Cybele of Phrygia; Astarte of the Phœnicians. Abel—as Tammuz, Adonis, Attis. Cain—as Merodach, doing the good things; as Ares, when doing the bad things. And an ancient legend that Eve grieved for Abel. See also Ragozin, *Chaldea*, p. 368: "Ishtar's Lament."

(2) The possible addition by later migrations of renegade Adamic stock to the story common to both.

[1] *Ibid.*, pp. 325–327.

The great seat of the worship of Aphrodite and Adonis in Cyprus was Paphos. . . . Pygmalion was the father-in-law of Cinyras, the son of Cinyras was Adonis, and all three in successive generations are said to have been concerned in a love intrigue with Aphrodite, we can hardly help concluding that the early Phœnician kings of Paphos, or their sons, regularly claimed to be not merely the priests of the goddess but also her lovers.[1]

In Western Asia, Attis was to Phrygia what Adonis was to Syria. Attis was said to have been a fair young shepherd or herdsman beloved by Cybele, the mother of the gods, a great Asiatic goddess of fertility. Some held that Attis was her son.

Two different accounts of his death were current. According to one he was killed by a boar, like Adonis.[2]

Rome adopted the worship of Cybele in 204 B.C., when the prophecy of the sibylline books that Hannibal would be driven from Italy if the great mother of the gods were brought to Rome was fulfilled.

Attis was supposed to have died under a pine-tree, and each year on the 22nd of March his effigy was buried. Followed days of mourning, when the worshippers fasted from bread, for to partake of bread at such a season might have been deemed a wanton profanation of the bruised and broken body of the god. But when night had fallen, suddenly a light shone in the darkness: the tomb was opened: the god had risen from the dead: and as the priest touched the lips of the weeping mourners with balm he softly whispered in their ears the glad tidings of salvation. The resurrection of the god was hailed by his disciples as a promise that they too would issue triumphant from the corruption of the grave.

On the morrow, the twenty-fifth day of March, which was reckoned the spring equinox, the divine resurrection was celebrated with a wild outburst of glee. At Rome and elsewhere universal licence prevailed.[3]

The worship of the great mother of the gods and her lover or son was very popular under the Roman Empire (at Rome, in Spain, Africa, France, Germany, and Bulgaria). Their worship survived the establishment of Christianity by Constantine. . . . In Greece, on the other hand, the bloody

---

[1] *Ibid.*, pp. 329–332. [2] *Ibid.*, p. 347. [3] *Ibid.*, p. 350.

GRÆCO-ROMAN MYTHOLOGY

orgies of the Asiatic goddess and her consort appear to have found little favour. The barbarous and cruel character of the worship with its frantic excesses was doubtless repugnant to the good taste and humanity of the Greeks, who seem to have preferred the kindred but gentler rites of Adonis.[1]

In ancient Egypt the god whose death and resurrection was annually celebrated with alternate sorrow and joy was Osiris, the most popular of all Egyptian deities; and there are good grounds for classing him in one of his aspects with Adonis and Attis as a personification of the great yearly vicissitudes of nature, especially of the corn.

Osiris was the offspring of an intrigue between the earth-god Seb (or Keb) and the sky-goddess Nut. The Greeks identified his parents with their own deities Cronus and Rhea.

When the sun-god Ra perceived that his wife Nut had been unfaithful to him, he declared with a curse that she should be delivered of the child in no month and no year.

But the goddess had another lover, the god Toth (or Greek Hermes), and he playing at draughts with the moon won from her . . . five whole days which he added to the Egyptian year of 360 days.

On these five days, regarded as outside the year of twelve months, the curse of the sun-god did not rest, and accordingly Osiris was born on the first of them. On the second she gave birth to the elder Horus, on the third to the god Set (or Greek, Typhon), on the fourth to the goddess Isis (or Greek, Demeter), and on the fifth to the goddess Nephthys.

Afterwards Set married his sister Nephthys, and Osiris married his sister Isis.

Reigning as a king on earth, Osiris reclaimed the Egyptians from savagery, gave them laws, and taught them to worship the gods. Before his time the Egyptians had been cannibals. But Isis, the sister and wife of Osiris, discovered wheat and barley growing wild, and Osiris introduced the cultivation of these grains amongst his people, who forthwith abandoned cannibalism and took kindly to a corn diet.[2]

His brother Set by intrigue plotted his death, cut his body into fourteen pieces and scattered them abroad.

Isis sailed up and down the marshes and found the bits,

---

[1] *Ibid.*, p. 356.   [2] *Ibid.*, pp. 362–367.

and having moulded them entrusted them to the priests for burial, and for worship.

At the great festival of sowing, the priests used to bury effigies of Osiris made of earth and corn ("at Thebes with faces of green wax and interior full of grain"). When these effigies were taken up again at the end of the year, or of a shorter interval, the corn would be found to have sprouted from the body of Osiris, and this sprouting of the grain would be hailed as an omen, or rather as the cause, of the growth of the crops.

The corn-god had produced the corn from himself, he gave his own body to feed the people; he died that they might live.

And from the death and resurrection of their great god, the Egyptians drew not only their support and sustenance in this life, but also their hope of a life eternal beyond the grave.[1]

A god who thus fed his people with his own broken body in this life, and who held out to them a promise of a blissful eternity in a better world hereafter, naturally reigned supreme in their affections.[2]

In that welter of religions which accompanied the decline of natural life in antiquity the worship of Isis was one of the most popular at Rome and throughout the empire.

Some of the Roman emperors themselves were openly addicted to it.

Her rites appear on the whole to have been honourably distinguished by a dignity and composure, a solemnity and decorum well fitted to soothe the troubled mind . . . the serene figure of Isis with her spiritual calm, her gracious promise of immortality, roused in their breasts a rapture of devotion not unlike that which was paid in the Middle Ages to the Virgin Mary.

Indeed, her stately ritual, with its shaven and tonsured priests, its matins and vespers, its tinkling music, its baptism and aspersions of holy water, its solemn processions, its jewelled images of the mother of god, presented many points of similarity to the pomps and ceremonies of Catholicism. Ancient Egypt may have contributed its share to the gorgeous symbolism of the Catholic Church as well as to the pale abstractions of her theology. Certainly in art the figure of Isis suckling the infant

---

[1] *Ibid.*, p. 376.
[2] *Ibid.*, p. 382. The Egyptian priesthood here pervert the Seth tradition of the Messiah embodied in the Great Pyramid.

Horus is so like that of the Madonna and Child that it has sometimes received the adoration of ignorant Christians.[1]

Among the gods of eastern origin who in the decline of the ancient world competed against each other for the allegiance of the West was the old Persian deity, Mithra.[2] In respect both of doctrines and of rites the cult of Mithra appears to have presented many points of resemblance not only to the religion of the mother of the gods but also to Christianity. There can be no doubt that the Mithraic religion proved a formidable rival to Christianity, combining as it did a solemn ritual with aspirations after moral purity and a hope of immortality.

Indeed the issue of the two faiths appears for a time to have hung in the balance.[3]

It was adopted by the emperors Commodus, Marcus Aurelius, Licinius and Galerius; and was the main religion of the Roman army, from which it was finally displaced by the Emperor Constantine.

Of Mithraism, Sir Gilbert Murray, in *Christianity in the Light of Modern Knowledge*, says:

Mithras was a hero, a redeemer, a mediator between man and God, a champion ever armed and vigilant in the eternal war of Ormuzd against Ahriman, light against evil and darkness. Mithraism arose in the East, among the poor, among captives and slaves. It put its hopes in a Redeemer, a Mediator, who performed some mystical sacrifice. It held a Communion Service of bread and water.

It rested on the personal Faith of the convert to his Redeemer. It had so much acceptance that it was able to impose on the Christian world its own Sun-day in place of the Sabbath; its Sun's birthday, 25th December, as the birthday of Jesus; its Magi and its shepherds hailing the divine star, and various of its Eastern celebrations.

On the other hand, its Redeemer, Mithras, makes hardly any pretence to have had an earthly history. It is all myth and allegory; elaborate ritual, sacraments, and mystic names, with all its varied paraphrasing that is necessary for bringing

---

[1] *Ibid.*, p. 383.
[2] Zoroastrianism, see p. 169, for possible origin from Israel in captivity, and not Cainite.
[3] *Ibid.*, p. 358.

primitive superstitions up to the level which civilised men will tolerate.

In the Dacian [1] Revolt of 275, Mithras proved too weak to withstand the barbarian. He was no longer "The Unconquered." His cave-chapels, or Mithræ, were destroyed all along the frontier, where they had been at their strongest. The sect never recovered.

We can now come to Frazer's most interesting chapter, "Oriental Religions in the West," pp. 356–361:

Greek and Roman society was built on the conception of the subordination of the individual to the community; it set the safety of the commonwealth, as the supreme aim of conduct, above the safety of the individual whether in this world or in a world to come.[2]

Trained from infancy in this unselfish ideal, the citizens devoted their lives to the public services and were ready to lay them down for the common good; or if they shrank from the supreme sacrifice, it never occurred to them that they acted otherwise than basely in preferring their personal existence to the interests of their country.

All this was changed by the spread of Oriental religions which inculcated the communion of the soul with God and its eternal salvation as the only objects worth living for, objects in comparison with which the prosperity and even the existence of the state sank into insignificance.[3] The inevitable result of this selfish and immoral doctrine was to withdraw the devotee more and more from the public service, to concentrate his thoughts on his own spiritual emotions,[4] and to breed in him a contempt for the present life which he regarded merely as a probation for a better and an eternal.

The saint and recluse, disdainful of earth and rapt in ecstatic contemplation of heaven, became in public opinion the highest ideal of humanity, displacing the old ideal of the patriot and

---

[1] The Ostrogoths—of the tribe of Joseph, see p. 178.

[2] Such ethics was derived from Druidism, which covered the Continent and was the religion of the Kimmerian-Celtic migration that settled into Greece and Italy from the north before the seventh century B.C., see pp. 83, 107, 108.

[3] From such is derived the present attitude of the Conchie and maudlin pacifist (pp. 268, 389).

[4] Such is not the teaching of Christ; His Life stressed action to originate personality attuned to Divine Harmony (pp. 245, 269, 373). Nor is it that of Druidism, which taught that "byd bychan" was necessary before man could enter "gwynfyd" (p. 82).

ITS EFFECT ON EARLY CHRISTIANITY

hero, who forgetful of self lives and is ready to die for the good of his country.[1]

The earthly city seemed poor and contemptible to men whose eyes beheld the city of God coming in the clouds of heaven. Thus the centre of gravity, so to say, was shifted from the present to the future life, and however much the other world may have gained, there can be little doubt that this one lost heavily by the change.

A general disintegration of the body politic set in. The ties of the state and family were loosened: the structure of society tended to resolve itself into its individual elements, and thereby relapse into barbarism; for civilisation is only possible through the active co-operation [2] of the citizens and their willingness to subordinate their private interests to the common good.[3]

Men refused to defend their country and even to continue their kind. In their anxiety to save their own souls and the souls of others, they were content to leave the material world, which they identified with the principle of evil, to perish around them. This obsession lasted for a thousand years.

The revival of Roman law, of the Aristotelian philosophy, of ancient art and literature at the close of the Middle Ages, marked the return of Europe to native ideals of life and conduct, to saner, manlier views of the world. The long halt in the march of civilisation was over.[4]

Frazer's deduction is interesting. For with the fall of the Roman Empire [5] (the fourth and last of the Babylonian succession of kingdoms which accepted this Cain-Bel

---

[1] The two views are not incompatible, for at this time Druidism was established in Cambria and Siluria, and encouraged a stubborn and devoted resistance to this Roman Empire in the first century B.C. and A.D.; and also later, when Christian, to the successive invasions of the Angles, Saxons, Danes and Normans.

[2] On co-operation as mutual service in the State, see p. 368.

[3] What is the "common good"?—variable, according to the particular dictator.

For on this premise—"the subordination of the individual to the community"—the totalitarian State is to-day founded (p. 334). The idea of the "common good" requires a common and stabilised centre, attuned to Reality—i.e. the Law of Divine Harmony (pp. 23, 24, 265–269); not based on Reason, man-made and deviating.

[4] Ibid., p. 357.

[5] Whose last beneficent ruler was Theodoric, of the Ostrogoths, A.D. 488–526; A.D. 553, the last of the Ostrogoths expelled; these as part of the "stone smiting the feet of the image," the tribes of Joseph and Benjamin (see pp. 179, 180).

system), at the end of the sixth century A.D., the interval of a "thousand years," as "a halt in the march of civilisation," coincides with the power of the Roman Papacy. Until, at the end of the sixteenth century, there emerged the Stone Kingdom,[1] with action by the Logos during the rhomboid of displacement era (1558–1844) to stem the descent passage of the Cain-Bel system to the subterranean chamber, the Hall of Chaos (see Chapter X).

In the Julian Calendar the 25th December was reckoned the winter solstice, and it was regarded as the Nativity of the Sun, because the day begins to lengthen and the power of the Sun to increase from that turning-point of the year.

The ritual of the Nativity, as it appears to have been celebrated in Syria and Egypt, was remarkable. The celebrants retired into certain inner shrines, from which at midnight they issued with a loud cry, "The Virgin has brought forth! The light is waxing!"

The Egyptians even represented the new-born Sun by the image of an infant which on his birthday, the winter solstice, they brought forth and exhibited to his worshippers.

No doubt the Virgin who thus conceived and bore a son on the 25th December was the great Oriental goddess whom the Semites called the Heavenly Virgin or simply the Heavenly Goddess, in Semitic lands she was a form of Astarte.

Now Mithra was regularly identified by his worshippers with the Sun, the Unconquered Sun, as they called him; hence his nativity also fell on the 25th December.

The Gospels say nothing as to the day of Christ's birth and accordingly the early Church did not celebrate it.[2]

What considerations led the ecclesiastical authorities to institute the festival of Christmas? "The reason," says a Syrian writer, himself a Christian, "was this. It was a custom of the heathen to celebrate on the 25th December the birthday of the Sun, at which they kindled lights in token of festivity." In these solemnities and festivities the Christians also took part.

Accordingly when the Doctors of the Church perceived that the Christians had a leaning to this festival, they took counsel and resolved that the true Nativity should be solemnised on that day, and the festival of Epiphany on the 6th January. . . .

---
[1] Israel-Britain, in Elizabeth's reign. [2] *Ibid.*, p. 358.

Thus it appears that the Christian Church chose to celebrate the birth of its Founder on the 25th December in order to transfer the devotion of the heathen from the sun to him who was called the Sun of Righteousness.

The Easter rites still observed in Greece, Sicily and Southern Italy bear in some respects a striking resemblance to the rites of Adonis.

Now the death and resurrection of Attis were officially celebrated at Rome on the 24th and 25th March, the latter being regarded as the spring equinox, and therefore as the most appropriate day for the revival of a god of vegetation who had been dead or sleeping throughout the winter. But according to an ancient and widespread tradition Christ suffered on the 25th March and accordingly some Christians regularly celebrate the Crucifixion on that day without any regard to the state of the moon (in Phrygia, Cappadocia, Gaul and Rome).

The ecclesiastical historian, Mgr Duchesne, points out that the death of the Saviour was thus made to fall upon the very day on which, according to a widespread belief, the world had been created.

When we remember that the festival of St George in April has replaced the ancient pagan festival of the Parilia; that the festival of St John the Baptist has succeeded to a heathen Midsummer festival of water; that the festival of the Assumption of the Virgin in August has ousted the festival of Diana; that the feast of All Souls in November is a continuation of an old heathen feast of the dead; that the Nativity of Christ was assigned to the winter solstice in December, the day deemed the nativity of the Sun; then the solemnisation of Easter may have been from like motives of edification adapted to a similar celebration of the Phrygian god Attis at the spring equinox. . . .

From the testimony of an anonymous Christian, who wrote in the fourth century, Christians and pagans alike were struck by the remarkable coincidence between the death and resurrection of their respective deities, a theme of bitter controversy, the pagans contending that the resurrection of Christ was a spurious imitation of the resurrection of Attis, and the Christians asserting with equal warmth that the resurrection of Attis was a diabolical counterfeit of the resurrection of Christ.

Taken together, the coincidences of the Christian with the heathen festivals are too close and too numerous to be accidental.[1] The inflexible Protestantism of the primitive missionaries, with their fiery denunciations of heathendom, had been exchanged for the supple policy, the easy tolerance, the comprehensive charity of shrewd ecclesiastics, who clearly perceived that if Christianity was to conquer the world it could do so only by relaxing the too rigid principles of the Founder, by widening a little the narrow gate which leads to salvation. . . .

Such spiritual decadences are inevitable.[2] The world cannot live at the level of its great men.[3]

And the *Encyclopædia Britannica* (13th ed.) ends its article on "Greek Religion":

The indebtedness of Christianity to Hellenism is one of the most interesting problems of comparative religion; and for an adequate estimate a minute knowledge of the ritual and the mystic cults of Hellas is one of the essential conditions.

In seeking the Golden Bough of mythology, we find the golden cup of Babylon of which all nations have drunken, therefore the nations are mad; for thus the classical mind finds that Christianity has benefited from Cain-Bel polytheism.

We turn to the monotheism of the House of Seth—(1) in Druidism; (2) in the Great Pyramid. For in Druidism, British Christianity could have found a more sure basis, as that which Siluria accepted from Joseph of Arimathea, Aristobulus and Paul, and made Britain declare itself Christian before any other nation in A.D. 155, when the Continent was halting between Isis, Mithraism and Christianity.

As put by Morgan:

The ancient Briton has never changed the name of the God he and his forefathers worshipped, nor has ever worshipped but one God.

---

[1] The true dates of his Birth and Resurrection are given by the Great Pyramid's mathematics, respectively October 6th (a displacement of 286 days) and April 9th, Julian (pp. 97, 125).

[2] After comparing the histories of Christianity and Buddhism.

[3] *Ibid.*, pp. 359-361.

THE DIVINE MESSAGE

And by Taliesin, prince-bard and Druid:

Christ, the Logos from the beginning, was from the beginning our teacher, and we never lost His teaching. Christianity was a new thing in Asia, but there never was a time when the Druids of Britain held not its doctrines.[1]

And in the Great Pyramid in Egypt, there is embodied in stone and written in scientific laws a direct message from the House of Seth to the Anglo-Saxon race after 1844 to interpret.

[1] See p. 81.

CHAPTER III

The House of Seth

¶ 1. Quoting from R. W. Morgan, *St Paul in Britain, or the Origin of British as opposed to Papal Christianity*:

"Druidism was founded by Gwyddon Gahébon, supposed to be the Seth of the Mosaic genealogy, in Asia in the year . . . 3903 B.C. . . . 50 years after the birth of Seth.

"There can be no question that this was the primitive religion of mankind, covering at one period in various forms the whole surface of the ancient world."

(Migration of the House of Seth to China, before the advent of Cain, inculcating the idea of monotheism.)

"The astral bull of milk-white hue, its horns crowned with golden stars, became the symbol, or visible sacrament, of Druidism . . . preserved free, as far as we can judge, from idolatry by the Cymry of Britain.

"From Asia, Druidism was brought into Britain by Hu Gadarn, the Mighty, its first coloniser, a contemporary of the Patriarch Abraham; and under his successors . . . it assumed its complete organisation, becoming the ecclesiastical and civil constitution of the island. About the fifth century B.C. its civil laws were codified by Dunwal Moelmud, the British Numa, and have since that period remained the common, unwritten, or native laws of the land, as distinguished from the Roman, the canon and other codes of foreign introduction. These British or Druidic laws have been always justly regarded as the foundation and bulwark of British liberties."[1]

The civil code and the sciences were taught by the Druids—orally or in writing indifferently—to every citizen,

---

[1] See Sir J. Fortesque, *De Laudibus Legum Angliæ*: Coke, Preface to vol. iii of *Pleadings*: "Origin of the Common Law of England."

but the Druidic system of divinity was never committed to writing, nor imparted except to the initiated.

¶ 2. Druidism taught as follows:—

"In the infinite Deity, there exist in some incomprehensible mode indivisible from himself, infinite germs, seeds or atoms '*manredi*,' each in itself full and perfect deity, possessing the power of infinite creativeness. . . . Matter was created and systematised simultaneously by the Creator's pronouncing His own name. It cannot exist without God. . . . The laws of nature are, in the strictest sense, the laws of God; and that which is a violation of the laws of nature is necessarily a violation of the laws of God. The universe is in substance eternal and imperishable . . . every particle of matter is capable of all forms of matter,[1] and each form has its own laws of existence and action.

"The Universe is infinite, being the body of the Being, who out of himself evolved or created it, and now pervades and rules it, as the mind of man does his body. The essence of this Being is pure, mental light, and therefore he is called Du-w (the One without any darkness). His real name is an ineffable mystery, and so also is his nature. To the human mind, though not in himself, he necessarily represents a triple aspect in relation to the past, present and future—the Creator as to the past, the saviour or conserver as to the present, the renovator or re-creator as to the future. . . . This was the Druidic trinity, the three aspects of which were known as Beli, Taran, Yesu.

"When Christianity preached Jesus as God, it preached the most familiar name of its own deity to Druidism, and in the ancient British tongue Jesus has never assumed its Greek, Latin or Hebrew form, but remains the pure Druidic 'Yesu.' It is singular thus that the ancient Briton has never changed the name of the God he and his forefathers worshipped, nor has ever worshipped but one God."[2]

---

[1] Substitute "energy" for "matter," and to-day scientifically correct: the whole idea consistent with Genesis i, and modern science.

[2] Procopius, *De Gotticis*, lib. iii: "Hesus, Taranis, Belenus, unus tantummodo Deus Unum Deum Dominum universi Druides Solum

"There were originally but two states of sentient existence—God in '*Ceugant*' (the infinite space) and the '*Gwynfydolion*' (the beings of the happy or 'white state') in '*Gwynfyd*.' Certain of the Gwynfydolion, whose numbers are known only to God, attempted to do that which God can only do, to enter and sustain '*ceugant*,' and thus originated in themselves the state of '*Annwn*' (the lowest possible point of conscious existence, in which the evil is wholly unmitigated by any particle of good). The result was the inevitable consequence of their act itself,[1] not an external penalty imposed by God.[2] To restore them to the state of '*Gwynfyd*' God in His goodness created the third state of '*Abred*' which includes '*byd bychan*' (man's present state) in which he is a free agent, master of his own spiritual destinies.[3] If his soul willingly prefers good and abides by its choice, then, at the dissolution of the body, it re-enters '*gwynfyd*'[4] from which it fell."

In "*byd bychan*," there began liberty of choice and responsibility. Hence the essence of the soul was the will, and the essence of religion was willinghood. Without freedom of will, there was no "humanity" in its distinguishing sense from animal life.[5] Freedom of conscience was both the birth and breath of manhood, without which it was not manhood at all.

The knowledge and suffering of evil was held the *sine qua non* to the understanding and appreciation of good, being the only means whereby their difference could be realised to ourselves. Suffering was regarded as the pre-essential of enjoyment.

Man had the power by accepting every evil as his purification for "*gwynfyd*" to turn it to good. Hence willing suffering for our own good or that of others was the test-virtue of humanity.

agnoscunt." The saying of Taliesin, the prince-Bard and Druid, conveys a great historic truth, though over-strongly expressed: "Christ, the Word from the beginning, was from the beginning our teacher, and we never lost His teaching. Christianity was a new thing in Asia, but there was never a time when the Druids of Britain held not its doctrines" (Morgan, p. 73).

[1] Freewill of all Spirit being.
[2] Similarly in the Garden of Eden story.
[3] *Vide* Gen. i, 26; ii, 7, 16, 17; and iii; and pp. 17, 23.
[4] Eternal life.
[5] Verifying Gen. ii, 7.

DRUIDISM

The faculty of the soul which constituted more especially its eternity or imperishable self-identity was *"Cov"* (memory, the consciousness of self-identity).

A soul that had passed *"byd bychan"* might resume the morphosis of humanity for the good of mankind. The reincarnation of such was always a blessing.

Every soul guilty of crime, by voluntarily confessing it and embracing the penalty prescribed, expiated its guilt, and if in other respects good, re-entered *"gwynfyd."*

Cæsar's [1] words on this point are remarkable:

"The Druids teach that by no other way than the ransoming of man's life by the life of man is reconciliation with the divine justice of the immortal gods possible."

"Also the Druids make the immortality of the soul the basis of all their teaching, holding it to be the principal incentive and reason for a virtuous life."

"The Druids discuss many things concerning the stars and their revolutions, the magnitude of the globe and its various divisions, the nature of the universe, the energy and power of the immortal gods." [2]

Lucan states: "It is certain that the Druidic nations have no fear of death. Their religion rather impels them to seek it. Their souls are its masters, and they think it contemptible to spare a life the return of which is so sure." [3]

This teaching from the monotheism of Seth Britain has lost, and doubts an existence after death, through contact with Græco-Roman mythology derived from the poly-

---

[1] J. Cæsar, *Comment.*, lib. v.

[2] Cæsar has the polytheistic idea inculcated by Cain, when interpreting the One God, of the Druids, and human sacrifice attributed to them.

[3] "Safety first" was not meant for the men of Britain—"*He that loveth his life shall lose it*" (John xii, 25), "*For whosoever will save his life shall lose it*" (Matt. xvi, 25), "Blessed are the débonnaire for they shall inherit the earth" (as translated in Broadcast Sermon, December 1934). We, the descendants of Caswallon and Caradoc, Arthur and Richard Cœur de Lion, of Drake and Raleigh and Nelson, of Clive and Wolfe and Gordon —yes, in future generations of our glorious dead in France and Gallipoli, on the sea and in the air—verify this truth.

The safety of the road and the courtesy due from the driver of the combustion engine is based on a Druidic law nearly 3000 years old:

"Three things which belong to a country and its borders—the Places of Worship, the Roads and the Rivers—whoever draws a weapon in or on them is guilty of an offence against God and His peace, and shall be deemed a capital criminal."

theism of Cain-Bel "who has taught the children of men all the blows of death" (Enoch lxix, 11) and "through envy of the Devil came death into the world" (Wisdom of Solomon ii, 24).

Morgan continues: "The Druidic was essentially a priesthood of peace, neither wearing arms nor permitting arms to be unsheathed in its presence [1]; and though patriotism or defence of one's country in a just war was a high virtue in its system, we have no instance of Druidism persecuting or using physical force against any other religion or set of opinions."

¶ 3. "The temples of the Druids were hypæthral, circular and obelistic—*i.e.* open above and on every side, representing the dome of heaven—and composed of monoliths, or immense single stones, on which metal was not allowed to come." [2]

Stonehenge, the Gilgal of Britain, is the wreck of 4000 years' exposure to the elements.

"A circle of this diameter (1163 B″) falls precisely internal to the outer ring of stones forming the circle of Stonehenge . . . the circumference of this circle measured 3652·42 P″ (10 P″ = to a day of the solar year)." [3]

"Structures more or less similar to Stonehenge are found along a line from the East on both sides of the Mediterranean . . . in other words, they lie entirely along a natural sea-route. The Druidic culture (of Ancient Britain) had not passed through Gaul, and had therefore been water-borne to Britain, Stonehenge dating 1680 B.C. ±200 years." [4]

Thus another clan or tribe of Seth descent, who built the Great Pyramid, migrated along the Mediterranean

---

[1] Higgins' *Celtic Researches*, p. 196: "In the ancient world the Druids were the only priesthood of peace. Clad in his white canonicals, the Druidic herald presented himself between two armies, and every sword was instantly sheathed."

[2] And: "All the prehistoric temples of Palestine, Persia, Italy, Greece, commonly called Cyclopean or Pelasgic, were Druidic" (Morgan, p. 64).

[3] Davidson, *The Great Pyramid*, p. 66.

[4] *Ibid.*, p. 5, quoting Lockyer, *Dawn of Astronomy*, p. 90, and Prof. T. Eric Peet, *Rough Stone Monuments*, pp. 147-148, and Lockyer, *Stonehenge and other British Stone Monuments*, p. 323.

THE GREAT PYRAMID (ITS CASING STONES LOOTED)

THE GREAT PYRAMID

to Britain, there continuing their astronomical knowledge of the day-year circle in the stone monument of Stonehenge.

¶ 4. To a migration (after the Druidic) of the House of Seth [1] from the Adamic "earth" is attributed the building of the Great Pyramid; who made Cheops, his co-regent Surid, and the Egyptians work for him.[2]

Josephus (*Antiquities*, Bk. I, chap. iii) states that the descendants of Seth after perfecting their study of astronomy set out for Egypt, there embodied their discoveries in the building of two "pillars" (matsebhah), one of stone, the other of brick, in order that this knowledge "might not be lost, before these discoveries were sufficiently known, upon Adam's prediction that the world was to be destroyed by a flood . . . and in order to exhibit them to mankind. . . . Now this pillar of stone remains in the land of Siriad (Dogstar, land of Egypt) to this day."

The Coptic tradition (to whom the ancient Egyptian calendar and a year of 360 days is due) states that the Pyramid was built in antediluvian times, as result of a vision to King Surid (or Shaaru), contemporary and co-regent with Khufa, in which the flood was predicted (vision date 300 years before the flood) to enshrine "their wisdom and acquirements in the different arts and sciences . . . of arithmetic and geometry, that they might remain as records for the benefit of those who could afterwards comprehend them . . . also deposited . . . the positions of the stars and their cycles, together with the history and chronicle and times past, of that which is to come, and every future event which would take place in Egypt" (Davidson, p. 89).

---

[1] Or possibly by Shem, A.K. 1836-1856 (2164-2144 B.C.), latter date being the one defined by the scored line on floor of Entrance passage, defining the Alcyone alignment at that date; Herodotus states twenty years to build. This date, however, is too late for Cheops.

[2] Professor Piazzi Smyth, *Our Inheritance in the Great Pyramid*, pp. 527-530, quoting Cory's *Fragments* and Rawlinson's *Herodotus*, suggests that a shepherd-migration to Egypt (forerunner of the Hyksos) exerted "Mental control over King Shofo and his Egyptian people, not by the vulgar method of military conquest, but by some supernatural influence in connection with the service of the one and only true God"—and afterwards went to Judea, and there founded Jerusalem (refer p. 111).

Herodotus (who visited Egypt 450 B.C.) states that Khufa (or Cheops)

> closed all the temples and forbade the Egyptians to perform sacrifices, after which he made them work for him. Some were employed in the quarries of the Arabian hills to cut stones and drag them to the river and put them in boats; others being stationed on the opposite side to receive them . . . and the 100,000 men thus occupied were relieved by an equal number every three months. Ten years were occupied in constructing a causeway for the transport of the stones . . . besides the time taken to level the hill on which the Pyramid stands, twenty years more to construct the Pyramid.

Davidson gives:

(1) 3101 B.C. as first year of the sole reign of Menes.[1]

(2) 2645–2622 B.C. as dates of Khufu.

(3) Date of Flood, 2345 B.C.—thus, Surid's vision being 300 years before, the Pyramid was built 2645 B.C.

(4) 2645–2622 B.C.—also is date defined by position of Pyramid's entrance at 19th course.

Herodotus' twenty years gives Pyramid building dates as from 2645–2625 B.C.

The Pyramid is the only one of the Seven Wonders of the world still standing.

> An Omniscient mind, which foresaw in the beginning the whole history of man, ordained that the message, arguments and proofs of the Great Pyramid should not be expressed in letters of any written language, but in terms of scientific facts, or features amenable to nothing but hard science . . . the Pyramid's message was intended for all men, even as Christ's Kingly reign, at His Second Coming, is to be universal and not for the benefit of Israel only. . . .
>
> The idea of a Divine interposition in the planning and construction of the Great Pyramid . . . is perfectly rational and credible in the estimation of a rightly-instructed mind.
>
> Is it not most strange and significant that the ancient cubit of Egypt 20·7 B", or the Continental metre, if applied to G.P.'s base side or base diagonals, vertical height or arris-lines, or any other length of the building, brings out no notable

---

[1] Professor S. Langdon, from clay tablets found at Kish.

THE GREAT PYRAMID

physical fact, no mathematical truth? While the other length of 25·025 B" (of which the Egyptians, the Greeks and the Romans when in Egypt knew nothing) brings out so many of the most important coincidences with the laws of Heaven and the ordinances of the earth we inhabit, as to make the ancient monument speak, intelligibly, intellectually, religiously, to the scientific understanding of all Christian men . . . and preferentially of Great Britain [1] (Prof. Piazzi Smyth).

Re the Pyramid architecture:

The pavement, lower casing and Entrance Passage are exquisitely wrought . . . the means employed for placing and cementing the blocks of soft limestone, weighing from 12–20 tons each, with such hair-like joints, are almost inconceivable at present, and the accuracy of the levelling is marvellous (Sir F. Petrie).[2]

Nothing can be more wonderful than the extraordinary amount of knowledge displayed in the construction of the chambers over the roof of the principal apartment (the King's Chamber), in the alignment of the sloping galleries, in the provision of the ventilating shafts . . . nothing more perfect, mechanically, has ever been erected since that time (James Ferguson, *History of Architecture*, vol. i).[2]

The glory of the workmen who built the Great Pyramid is the masonry of the Grand Gallery. The faces of the blocks of limestone . . . have been dressed with a care not surpassed even by the most perfect examples of Hellenistic architecture (Perrot and Chipiez, architects).[2]

¶ 5. To attempt some condensation of the geometry of D. Davidson's *The Great Pyramid : its Divine Message*:

The unit is the Pyramidal inch, one-five-hundred-millionth part of the Earth's diameter from pole to pole (the Anglo-Saxon race holds the same unit measurement, 1 P" = 1·0011 B"), and a cubit of 25".

The Pyramid's base side = 9131·05" (about 760 feet) = 365·242 cubits.

---

[1] Professor Piazzi Smyth, Astronomer Royal for Scotland, *Our Inheritance in the Great Pyramid*, 1880.
[2] Quoted by Basil Stewart, *The Witness of the Great Pyramid*, pp. 129-131.

The perimeter of the four sides is thus 36524·2″ (100″ to the day for a year circle).

The height is then obtained:—

$$\frac{\text{Perimeter of base}}{\text{twice height}} = \frac{\text{circumference of circle}}{\text{twice radius (or the diameter)}} = \pi = 3\cdot 14159.$$

Therefore the height = 5813·01″ (about 484 feet, about 80 feet higher than the spire of Salisbury Cathedral).

Professor Piazzi Smyth estimated the weight of the masonry at 5,273,834 tons and the content at 161,000,000,000 cubic inches.[1]

The 201 courses of masonry were covered by blocks of limestone weighing tons, the edges of each so levelled to fit, that the whole Pyramid showed four smooth sides glistening white to the sun's light.[2]

> Some of the masonry finish is so fine that blocks weighing tons are set together with seams of considerable length, showing a joint of one ten-thousandth of an inch, and involving edges and surfaces equal to optician's work of the present day, but on a scale of acres instead of feet or yards of material (Sir F. Petrie).

The square (4) is symbolic of the world, the plane of created matter; the triangle (3) of divine elevation; so that the Pyramid symbolises completion of the perfect Man, the Spiritual Soul.

The Pyramid is perfectly oriented—its four sides north, south, east, west.

> We cannot safely come to any other conclusion than that the designer of the geometrical system of the Great Pyramid was also the Designer of the Universe (Davidson).

The summit of the Pyramid is truncated, a rough platform above the 201st course, so that the apex stone has never been placed.

From above, the Pyramid would look like four clear white cascades of water flowing to earth: looked at direct

---

[1] Later calculations give weight as 6,060,000 tons, and content as 85 million cubic feet.

[2] About A.D. 820 the limestone casing-stones were removed for a mosque in Cairo, the greatest piece of vandalism ever committed; only a few at the base of the north face now standing.

THE GREAT PYRAMID

in the sunlight, a plane flat surface would look bulged outwards owing to heat radiation. To correct this refraction each side of the Pyramid is recessed inwards at centre about 3 feet in base side of 760 feet.[1]

¶ 6. Now Davidson in the allegory of the Great Pyramid states: "The Design of the Great Pyramid was drawn up to counteract the effects of refraction."

Physical light signifies Spiritual Light: the Pyramid, the earth, created matter: physical reflexions symbolise witnessing by reflexion to the source of Spiritual Light: physical refractions (as some inherent factor in created matter) symbolise the operation of evil spiritual influences in distorting the image of Spiritual Light.

All the structural evidences existing show that the stepped core masonry, the lines for such along Pyramid base, the centre casing stones which cover the base side, the whole of the internal system of passages and chambers were laid out and built to the lines of the Design. The four points for the corner-stones were marked off. But there occurred some mistake (symbolising same mistake in Earth's creation) on the part of the builders in the erection of the casing stones at the corners—as regards the factor introduced to counteract the effects of refraction (*i.e.* of evil influence).

Thus as the summit was being reached the builders were not able to place the Apex Stone to conform to the Design.

"*The stone which the builders rejected*" (as the earth-created matter rejected its Messiah), "*the same is become the head of the corner*" (the Apex Stone which connects the whole and corners of a pyramid), "*the chief corner-stone*" (Matt. xxi, 42; Ps. cxviii, 22; Eph. ii, 20; Isa. xxviii, 16; Acts iv, 11).

The builder's error and the allegory of the Displacement Factor is best seen in a diagram of Davidson's [2]:

---

[1] This cannot be made out by the naked eye, but with optical instruments. It has also been verified recently by an aeroplane photograph, taken at 4000 feet, by light of the setting sun, which shows the half of one side in shade, the other half in light (Davidson, p. xxvi).

[2] Davidson, *The Great Pyramid's Prophecy concerning the British Empire and America*, pp. 25, 26.

90 THE PILLAR IN THE WILDERNESS

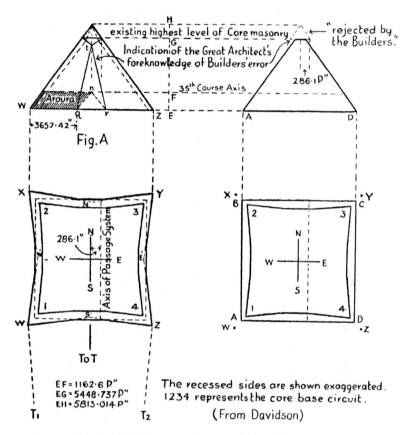

Square circuit ABCD = 36524·2465 − 286·1022 P″.

Square circuit WXYZ = 36524·2465 P″ = 100 × solar year (equinox to equinox) in days.

Hollowed-in circuit of base WXYZ = 36525·64715 P″ = 100 × sidereal year (position of earth relative to a fixed star at any time to its return to same position) in days.

Geometrical circuit WW′XN′YE′ZS′ = 36525·99 P″ = 100 × anomalistic or orbital year (perihelion to perihelion—*i.e.* earth's position nearest to the Sun, on January 2/3, to its return to the same position) in days.

T_1 and T_2 meet at T, such that 25 million × distance OT = major axis of the earth's elliptical orbit = twice the mean sun distance from the earth = 2 × 92,996,169 miles.

After numerous expeditions, science now gives this distance as 92,900,000 miles.

The sum of the base diagonals, WY + XZ, = 25826·52 P″, giving the value of the Precession of Equinoxes, of which the year 1844 is the mid year.

This year is marked by the Grand Step, in the central Pyramidal plane, but displaced 286·1 P″ to the east (or negatively), and commences the era to which the Pyramid's message is addressed, so called "the Epoch of Reference."

The Pyramid was left unfinished at a height of 5448·7377 P″ (=5454·73 British inches, of Petrie), when at this stage the builders realised their error.

The circuit of the platform thus left = 7 × 286·1022 P″; whereas as designed the base circuit at this point = 8 × 286·1022 P″.

"Fig. A shows the form, *qtr*, that the central hollowing-in effect took up the centre of each face slope, and how the Apex construction had to be mitred above the point *t*, to bring all surfaces to a common point of convergence at the Apex. This mitring detail is of great symbolic significance in the allegorical prophecy" (see Chapter XII).

For it symbolises the Great Architect's intention for the "mitring" of the rejected chief corner-stone to the perfect structure of the Stone Kingdom (Israel-Britain), the "wedding" of the bridegroom to His Theocentric State, to found the Kingdom of God on earth—the theme of Christ's parables.

Davidson,[1] in summing up the data given of the dimensions and motions of the earth and its orbit, says:

> The facts have proved to us that a certain stage of world civilisation ... in the past had evolved a geometrical system of Natural Law, in relation to the motions of the Earth and its orbit, equal to, superior to, or more comprehensive than the modern system of expressing this Natural Law ... we have not yet derived a single tangible indication as to how the savants of that period discovered their facts of science—

---

[1] *The Great Pyramid*, p. 138.

whether by methods of modern times, by methods unknown to modern times, or by the development of faculties now atrophied by long disuse. . . . In order to discover the scientific facts embodied in the Great Pyramid, it is essential that the investigator should have previously knowledge of these very facts.

(It is only since 1844 that modern science has had this knowledge, also it must use the inch as standard of measurement, not the metre.)

P. 139:

The Pyramid's external features are designed to direct attention to a further message of greater importance . . . to tell the future race of mankind what it could not possibly know . . . relating to a break in the continuity of something essential . . . that could not be restored otherwise than by being passed on from the former civilisation to the then remotely future civilisation. The inferred break in continuity can only be conceived as relating to some factor affecting the history of the previous civilisation and related to the history of the present stage of civilisation.

(Thus the message of the Pyramid is a direct one from the House of Seth of the ancient Adamic race to present-day Britain, who holds the inch measurement.)

The skill of this former civilisation in the science of gravitational astronomy—and therefore in the mathematical basis of the mechanical arts and sciences—means that it has taken man thousands of years to discover by experiment what he had originally and more precisely by another sure and simpler method . . . that the whole empirical basis of modern civilisation is a makeshift collection of hypotheses compared with the Natural Law basis of the civilisation of the past.

P. 138:

The rational development of Einstein's theory of Relativity now gives us reason to hope that these and the laws of other branches of science may be shown to be but varying phases of one Universal Law of Nature.

¶ 7. The system of internal passages and chambers has entrance on the north face of the Pyramid, but displaced

THE GREAT PYRAMID

286·1" to the east (or to left hand—that is, negatively) from the centre line.

Complicated calculations on the measurements and angles of this system are given by Davidson (pp. 196–208).

A concise picture is given by Professor Piazzi Smyth (Plate XXII and p. 438) as below (with an addition)—

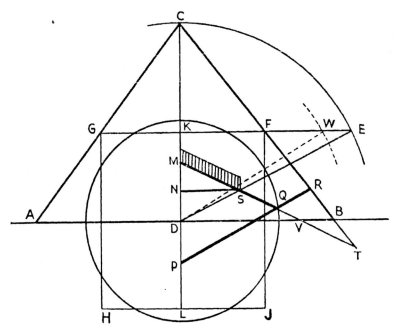

ACB is taken as the central vertical plane of the Pyramid, with height CD = 5813·01", and base AB = 9131·05" (as derived on p. 88).
The angle ABC = 51° 51' 14·3" (often called the π angle).
FGHJ = square and circle of area equal to area of triangle ACB.
Arc CE of circle of radius DC : KD trisected at M and N : DL bisected at P.
PQR drawn parallel to DE : join QM, and SN.
Then RQ represents the entrance passage, QP the descending passage, QS the ascending passage, SN the horizontal passage, and SM the Grand Gallery.
The angle of PQ and QM with horizontal is 26° 18' 9·63" (= angle BDE).
The angle BDW = 30° (the approximate latitude of the Pyramid), and DW = side of square FGHJ.
Area of circle with radius DW (or FG) = area of square on base AB.

On the floor of the Entrance Passage is a scored line, defining the year 2144 B.C. (autumnal equinox). In this year, the axis of passage pointed directly at the alpha Draconis star, then the chief star, but now the fourth in brilliance, of the Dragon constellation, which coils itself round the North Pole Star.

The angle of repose of hard limestone on hard limestone is 21°, tilted up to 26° 18' 9·63" limestone rolls down the plane. Thus the Dragon[1] looks at the descent of man down the descending passage to the subterranean pit (RQP.).

In that year, at right angles to this axis, so immediately above, was the star Alcyone (known as eta Tauri, of the constellation of the Pleiades)—thus control of man's descent by the "sweet influences." [2]

By the Precession of Equinoxes (the slow rotatory motion of the earth which causes the poles pointing to the heavens to alter position or to wobble like a top) these two constellations cannot come into the same position for 25,827 years (defined by the sum of the base diagonals).

Davidson (p. 216) states:

> These indications . . . were obviously directed to the future civilisations, together with the indication of the Epoch of reference for the Pyramid's message . . . as 5842·3397932 A.P. = 25/26th Jany, 1844 A.D.[3]

This date is also marked at the Grand Step at the end of the Grand Gallery, where the ascending plane changes to a horizontal plane.

The Great Pyramid at Gizeh is situated at the apex of the Nile Delta, at the "border" of it, but in the midst of the land of Egypt; if a line from it is drawn at an angle of 26° 18' 9·63", it passes through Bethlehem, and continued strikes the ford on the River Jordan, where Christ was baptized. It also represents the line which Israel might

---

[1] Isa. xiv, 13–14: *"I will ascend unto heaven, I will exalt my throne above the stars of God, I will sit also in the sides of the North, I will ascend above the heights of the clouds."*

[2] Job ix, 9; xxxviii, 31: *"Canst thou bind the sweet influences of the Pleiades?"*
Since the earliest times the Pleiades have been specially noted by mankind as bound up with human history and emotion.

[3] P. 216: "The Precessional rate for the Great Step Epoch is the known Precessional rate for A.D. 1844, and for no other year within a range of over ±30,000 years from A.D. 1844."

have taken at the Exodus, along the main military caravan route, but diverted at the "Sea of Reeds" through the Wilderness[1] : forty years later under Joshua, Israel entering Canaan crossed the Jordan on this same line.

Thus the ascending passage, diverting man from his course of descent, and meaning spiritual effort to ascend, points to Bethlehem, the place of the Nativity.

Isa. xix, 19, 20 (with 14-25): *"In that day shall there be an altar to the Lord in the midst of the land of Egypt, and a pillar*[2] *at the border thereof to the Lord. And it shall be for a sign and for a witness unto the Lord of Hosts in the land of Egypt; for they shall cry unto the Lord because of the oppressors, and He shall send them a Saviour, and a great One, and He shall deliver them."*

Also (Jer. xxxii, 20): *"Thou hast set signs and wonders in the land of Egypt . . . and has made thee a name,*[3] *which stands at this day."*

(Job xxxviii, 5, 6.) Jehovah, as Creator, is describing to Job His own Divine Action when the earth was formed, by comparing its foundation with some structure on the earth which was well known—*i.e.* the Pyramid; and referring to the ceremony which was known in ancient Egypt at the setting out of any building as "the stretching out of the cord or line"; the Great Pyramid is the only one where the corner sockets are let into the natural rock. *"Whereupon are the sockets thereof made to sink, or who laid the corner-stone thereof? Who hath laid the measures thereof, or who hath stretched the line upon it?"*

If the line MSQ is continued to strike base-line AB at V (in diagram, p. 93) and Pyramid side CB continued below rock, to meet each other at T;—

then the line TVQSM, on the scale of 1″ to the year,

---

[1] P. 372: "In the case of Israel's Exodus, the appointed access to the Land of Promise was hidden. The obvious route seemed easy, yet annihilation lay hidden at its end" (Davidson)—"*by way of the land of the Philistines,*" which "*was near*" (Ex. xiii, 17).

The remainder of this verse explains that the result of this course would be a return to bondage, the way of descent. Descent continued meant an increase in everything negative to spiritual progress.

[2] Hebrew word "matsebhah," meaning a monument, to record something remarkable, same word used by Josephus for pillar.

[3] Hebrew word means a monument, or something of renown.

represents the Biblical history of Israel (and the world) from the basal date at T as 4000 B.C.

In pp. 284-353, Davidson shows that the basal date of chronology—autumnal equinox of 4000 B.C.—was a universal one, and found in Hebrew, Babylonian, Egyptian and Chinese systems of chronology, having been derived originally from the Adamic race, as the parent of all these civilisations.

Then V (floor-level) = 1913 B.C.—the confirmation to Abraham of His everlasting Covenant. Q = 1486 B.C.—the date of the Exodus. S = A.D. 30—His Crucifixion and Resurrection, when the ascending passage height of 47·31 P" is raised by 286·1 inches at the commencement of the Grand Gallery, the Gospel Era; and M = A.D. 1844 (Jan. 25/26)—the Great Step Epoch, where the ascending plane changes to the horizontal.

¶ 8. The factor, 286·1", comes in constantly in the Design of the Pyramid.

So far mentioned:

(1) The square on the recessed circuit, made to counteract the effects of refraction (*i.e.* of evil influences inherent in matter), is 286·1" less than true square circuit; the same mistake of the builders in construction caused them to "reject" the Apex Stone.

(2) The system of passages and chambers are displaced 286·1" to east (or negatively) from the central vertical plane. This, as will be seen later, is replaced only on entering the last chamber, the King's Chamber.

(3) At the junction of Ascending Passage and Grand Gallery the height is raised 286·1" (at the date of His Crucifixion, A.D. 30).[1] In the Egyptian *The Book of the Dead* this junction is called "The Crossing of the Pure Waters of Life" and the two passages respectively "The Hall of Truth in Darkness" and "The Hall of Truth in Light."

Also consider the factor in the two following:—

(4) In the symbolism of the Grand Gallery, Davidson

---

[1] "*I, if I be lifted up from the earth, will draw all men unto Me; this He said, signifying what death He should die*" (John xii, 32, 33).

describes "the rhomboid of displacement," the end year being 1844 (the Grand Step), minus 286·1″ = Dec. 1557, "defining the creation of a 'field of influence' in the history of mankind; this being the 'influence' of the Messiah who 'purchased'[1] the right to exercise this influence by this 'displacement' or 'cutting off.'[2] . . . The symbolic and astronomical indications depict the hastening[3] of "man's material development to effect God's purpose, through Christ, in relation to spiritual development" (Davidson, p. 375).

(5) The Nativity of Christ was on the day of the "Feast of Tabernacles" (15th Tisri) in 4 B.C., date 6th October (Julian = 4th October, Greg.).

This date, 6th October 4 B.C.,[4] is marked at the floor-line of the Queen's Chamber produced to floor of Ascending Passage as A.P. 3996·027655. The date recognised by all Christendom was the winter solstice of 5 B.C.—Dec. 25. The displacement is 286 days.

Similarly the Annunciation of the Virgin, 25th March, should be advanced 286 days, to 3rd Jan. 4 B.C.

Birth of John the Baptist, summer solstice, 24th June, should be advanced 286 days, to 10th April 4 B.C.

Conception of Elizabeth, to 9th July 5 B.C. (Luke i, 24.)

Now let Davidson sum up the meaning of the displacement factor in these and other instances as

> the mathematical constant of the Creative Law of Relativity . . . all variations in astronomical time rates, and angular

---

[1] "*Thou hast led forth the people which Thou hast redeemed . . . the inhabitants of Palestina . . . shall be as still as a stone, till thy people pass over, which Thou hast purchased*" (Exod. xv, 16).

[2] "*. . . shall Messiah be cut off, but not for Himself*" (Dan. ix, 26).

[3] At "*the time of the end, many shall run to and fro, and knowledge shall be increased*" (Dan. xii, 4). "*I the Lord will hasten it in His time*" (Isa. lx, 2).

[4] Davidson, *The Great Pyramid*: "A prophecy, 5000 years old, proclaims the birth of the Messiah would be accomplished on Saturday, 6th October 4 B.C., a date corresponding to the 15th day of the Hebrew month Tisri, the day of the Feast of the Tabernacles, when '*the Word became flesh, and tabernacled among us*' (John i, 14); and that the Passion of the Messiah would be accomplished on Friday 7th April A.D. 30 (A.P. 4028·531789), a date corresponding to the 15th day of the Hebrew month Nisan, on which our Lord was crucified."

The Julian Calendar dates differ by two days from the equivalent Gregorian dates, instituted at the end of the sixteenth century A.D.

and linear distances, and their respective rates of change (earth's polar diameter, mean sun distance, the year), are expressed in terms of the single mathematical constant defined by the Pyramid's displacement factor (286·1 P").

It has therefore proved the Key value to the precise representation of the following:—

(1) The hollowing-in feature of the reflecting surfaces to prevent diffusion of the reflected light and to stabilise the reflected rays under the variable surface refractions due to heat radiations; for the purpose of defining the principal points of the solar year as astronomically defined, and of the solar year as defined by the phenomena of the seasons: and for the additional purpose of defining in relation of the latter to the prophetical or sacred calendar.

(2) The exact value of (a) the Sidereal year,
       (b) the Anomalistic year, and
       (c) the Solar year.

(3) The definition of the exact mean sun distance.

(4) The definition of the Law of Gravitation in relation to the earth and its orbit.

(5) The definition of the exact limits of variation in the eccentricity of the earth's orbit.

(6) The definition of the exact formula for the annual rate of Precession.

(7) The definition of the exact formula for the annual rate of change in the Longitude of Perihelion, by means of the Gallery height, defined by the displacement factor.

(8) The definition of the year beginning the period for which the Pyramid's message was designed to apply, and in which the knowledge necessary for its understanding would be discovered; also thus defining the beginning of Daniel's "time of the end."

(9) The definition of the same year as ending a period of 1530 years—numerically ten times the number of fishes symbolising the gathering in of mankind for the kingdom of Heaven—from the 1st year of the Messiah's ministry.[1]

---

[1] 153 in Scriptural symbology means his "elect."
"*Simon Peter . . . drew the net to land full of great fishes, an hundred and fifty and three*" (John xxi, 11; also Matt. iv, 19; xiii, 47; Luke v, 10).
10 is symbolic of worldly completion and worldly power without Christ (153 × 10)—
A.D. 27 + 1530 = A.D. 1557, so that after 1557 His influence might be expected to be reasserted (see p. 302).
Christ's ministry on earth was for 918 days (6 × 153, 6 representing

ITS DIVINE MESSAGE

(10) The definition of the apex Pyramid—symbolic of the Messiah in Old and New Testament symbolism, as related to:
   (a) the Pyramid's external orbital features,
   (b) the Pyramid's internal passage chronology—astronomical and prophetical.
   (c) the Great Step dating and the Rhomboid of Displacement (p. 367).

The symbolism clearly conveyed by the ten categories is:

(1) By (10) in relation to the previous nine, that "*In Christ are hid all the treasures of wisdom and knowledge*" (Col. ii, 3), and that by Christ God "*created all things*" (Eph. iii, 9); "*the world was made by Him, and knew Him not*" (John i, 10).

(2) By (1) in relation to the succeeding nine, that "*God is Light*" (1 John i, 5) and "*God is Love*" (iv, 8); that Christ is "*the Light of the World*," Who, by His displacement—Daniel's "cutting-off" of the Messiah, "*the Light shining in Darkness*" (John i, 5)—neutralised in Himself the effect of sin (the breaking of Divine Law) by restoring the Light, which to man is the essence of Divine Law; the stabilising of the refracted rays.

The displacement therefore symbolises God's special sacrifice for mankind.

"*Herein is love, not that we loved Him, but that He loved us, and sent His Son to be the propitiation for our sins*" (1 John iv, 10).[1]

The graphical representation of this displacement factor, and of all it signifies, has been to constitute the one essential theme of the message of the Great Pyramid (p. 434).

To proclaim this same Jesus as the Deliverer and the

---

periods of "work")—from the Feast of Tabernacles in A.D. 27, which fell that year on Sunday, 3rd October (Greg.), the eve of His "thirty years of age" (Luke iii, 23), to His Resurrection on 7th April (Greg.) A.D. 30.

In the same symbology, the Grand Gallery is raised in height by 286·1" in 6 overlaps. The length of the roof of the Grand Gallery is 1836" (12×153, 12 representing governmental perfection).

Again from 0 A.K. to $2513\frac{1}{2}$ A.K. (17th Nizan, 6th April (Greg.))—4000–1486 B.C.—the deliverance of Israel at the crossing of the Sea of Reeds =918050 days, 1/1000th part =918.

Again the breaking of Ephraim (784–719 B.C., mid-date of siege of Samaria)	= 65 years
Interval	=788 ,,
The breaking of Judah (A.D. $70\frac{3}{4}$, siege of Titus, to $135\frac{3}{4}$, failure of Bar Cocheba's rebellion)	= 65 ,,
	918 years

[1] Davidson, *The Great Pyramid*, pp. 367, 368.

Saviour of men, to announce the dated circumstances related to His coming, and to prepare men by means of its message (whether they believed it or not in the first instance) to adapt themselves spiritually to the circumstances of *His Second Coming* when the fact of the message becomes to them a matter of certainty (p. 458).

The displacement factor symbolises:

(a) Man's displacement, whereby man lost his inheritance, *i.e.* dominion over the Earth.

(b) Christ's displacement to restore the spiritual conditions whereby man should regain his inheritance.

(c) The displacement from the existing conditions to higher conditions to effect the complete "restitution of all things" (p. 381).

The last phase—concerned with the Scriptural period of "the consummation"—is related to current history and symbolised by the Antechamber and King's Chamber and their connecting passages (see Chapter XI).

¶ 9. This then is the message, which the House of Seth put into hard scientific symbolism, for our scientific civilisation to elucidate: the knowledge they had from Divine tuition in that landlocked Adamic world, passed directly on to their successors who will hold the same standards, spiritual and scientific, that

(i) their civilisation (as in Gen. vi) "*grieved Him in His heart*" (Gen. vi, 6) and would be destroyed by a Flood; so under Divine instruction they built this Pyramid enshrining their knowledge of Natural Law in symbolism of events which are foreordained.

(ii) to Noah and Abraham, in whom is found righteousness, He gives an everlasting Covenant; from their seed He had foreordained an "elect" people to eradicate polytheism, idolatry and all their sequelæ (which was being "sown" as tares in the world by the brother House of Cain and the Devil, Bel), to obey His Divine Laws, and form the nucleus of His Kingdom on earth.

(iii) the foreordained Messiah would appear and live on this earth at a certain date to restore man's displacement and redeem his soul from death and destruction.

(iv) there was coming "a break in the continuity of something essential"—direct Divine tuition: but this would be restored after certain events, symbolised in the internal passages and chambers, by the establishment of His Kingdom on earth, this as a warning and an appeal to their successors, this "elect" race, to be prepared spiritually to receive and welcome His Second Coming.

This they put down in geometrical symbolism of stone so as to be preserved for thousands of years, and not in words and writings, which might become destroyed or lost, or its meaning talked away as some abstract philosophy.

Thus over the course of years the message stands as if given only yesterday. And the message expressed in symbols is in line with all Reality, which cannot be expressed in words but in some sublime concept—as in the music of a composer, as Handel's *Messiah*; in a picture of the beauty of Nature by an artist; by mathematical equations, as matter and the ether by present-day science; all aspects of the Nature of God.

Words are for the earthy of the earth, so there comes a flow of words in the present day from politics, religion, philosophy, economics, which all bewilder the mind as to what is Reality.

Words are of use only to transfer a concept—it needs Christ to do it in the fewest words, "*God is Love.*" To the mind earthy just a mere three words; but taken as a symbol (—except a man's mind be born again, he cannot see) there is opened up a vista as mighty as the Universe, and at the limit of human intelligence.

Thus these descendants of Seth after building the Pyramid migrated along the coasts of the Mediterranean to the west, gradually losing contact with their parent civilisation (which was finally destroyed by the Flood 300 years later), perhaps with some idea of finding the "Isles of the West"[1]; and finally some 1000 years later

---

[1] Did they expect that here would be their successors some long day in the future, and here the origin of the later science?

Camden's *Britannia* (p. 963), quoting from Postellius' lectures on Pomponius Mela, a first-century writer: "The Jews, being most wise sages

erecting at Stonehenge a circle of stone which still enshrined the perimeter of 3652·42 P" (the day-year circle).

Thus we see that the primary system of measures and other related data—astronomy, geometry; building and architecture; the cuneiform alphabet; agriculture, with cereal and fruit cultivation, and system of irrigation; the use of bronze and iron; the origin of religion, monotheism by the House of Seth, polytheism by the House of Cain; the colonisation of seaports along the maritime trading routes; "originated with a lost civilisation of the Ancient East, identical with the civilisation of the Adamic race of the First Book of Genesis," that their colonies formed in these migrations "built up the basis of civilisation in the countries bordering on the Persian Gulf, Indian Ocean, Red Sea, Mediterranean," even to Britain in the West and China and the Pacific in the East, before the migratory spread from the East of the Noachic branch of the Adamic race began.

By the excessive stimulation of world-trade and industry . . . taught by the ancient white race . . . mushroom civilisations sprang up, the appetite for luxuries was acquired, cultivated and "whetted," greed grew, aggression spread, and "violence filled the earth"; and the seeds of decline were sown in the sphere of national intercourse, through racial inter-marriage, inter-racial claims and inter-national rivalry. Deprived of its "splendid isolation" by a too intimate intercourse with the primitive races, the power of the white race underwent moral decline.

As it was in the former age, so it is in the present age. . . . The theme of the Great Pyramid's prophecy relates to the deliverance of man's spirit from the Robot of man's own creation.[1]

And it will be found that wherever in these ancient and succeeding civilisations there is polytheism, with its later developments (as a reaction from crude polytheistic

and learned philosophers, and knowing that the empire of the world should be settled in the strongest angle, which lieth in the West, seized upon these parts at an early period, and Ireland, the first"; so "Ireland was called Jurin" (= Jewsland).
(Here Jews mean the Hebrews, ancestors of the Jews of the Captivity period.)

[1] Davidson, *The Great Pyramid's Prophecy*, pp. 10, 12.

mythology on the part of a higher and philosophic type of mind)—Atheism, Anti-God—there predominates the influence of Cain-Bel. Among heathen aborigine races polytheism spread like wildfire. And wherever monotheism—the idea of One God—there is found the influence of Seth—Israel (meaning not only "ruling with God" but "fighting with God").

The war between good and evil in the Garden of Eden has now spread to the whole of the Near East, and with the spread of the family of Noah (Japheth, Shem and Ham) becomes universal over the world. During Christ's life on earth, the contrast vividly marks the universal spread over mankind of the Cain-Bel influence.

Israel having come under its influence and idolatrous is scattered (Jer. xxxi, 10), *"cast afar off among the heathen"* (Ezekiel xi, 16). *"I will sift the House of Israel among all nations, like as corn is sifted in a sieve"* (Amos ix, 9).[1]

And this war in the present times is intensified, working to a climax.

This carries out Gen. iii, 15: *"I will put enmity between thee and the woman."*

"Woe to the inhabitants of the earth and of the sea! for the devil is come down unto you, having great wrath, because he knoweth that he hath but a short time. And when the dragon saw that he was cast unto the earth,[2] he persecuted the woman[3] which brought forth the man child. . . . And the dragon was wroth with the woman and went to make war with the remnant of her seed,[4] which keep the commandments of

[1] Yet after each quotation above—showing His Divine Love—"*and keep him as a shepherd does his flock.*" "*Yet will I be to them as a little sanctuary in the countries where they shall come.*" "*Yet shall not the least grain fall upon the earth.*"
Then our forefathers, as the scattered nation of Israel, were spiritually useless to support Him, at His First Coming, as the duty of an "elect" people entailed.
Yet now is Israel-Britain any more worthy to form the nucleus of His Kingdom on earth, or to realise enough the purpose of the Logos to welcome His Second Coming to rule it?

[2] The compact with Cain.

[3] Israel, from the Pharez-Judah line was born the Christ.

[4] Britain and her Christian Church (and it is up to both to realise that they are in this fight with spiritual weapons; yea, and with material weapons if called upon to defend the Faith. What is the Faith?—modern philosophy has so clouded the issue, hence the question).

God, and have the testimony of Jesus Christ" (Rev. xii, 12–17).

"*. . . a beast rose up . . . and the dragon gave him his power, and his seat and great authority . . . and all the world*[1] *wondered after the beast. And they worshipped the dragon, . . . and the beast . . . saying . . . who is able to make war with him?*" (Rev. xiii).

[1] The world still under the Cain-Bel influence (the significance will be seen later, in Chapter XI).

CHAPTER IV

¶ 1. Such then is the civilisation of the Adamic white race being founded in the world before the Flood among the pre-Adamite, aborigine races, which Keith has described as being evolved from an anthropoid stock in four races of man — African, Australian, Mongolian and Caucasian.

2343 B.C. Noah, his three sons, Japheth, Shem and Ham, and their families migrated from the "mountains of Chin" westwards.

"*As they journeyed from the east, that they found a plain in the land of Shinar* (=Sumer, *Ency. Brit.*), *and they dwelt there*" (Gen. xi, 2)—in contact with the Cainite-Sumerian civilisation.

"*These are the sons of Shem, by their tribes, and in their languages, in their countries among the heathen.*

"*The above were the families of the sons of Noah, and their descendants by tribes. From them they spread themselves amongst the nations on the earth after the Flood*" (Gen. x, 31, 32)—Farrar Fenton's translation.

Professor Sayce (quoted p. 37) states that Genesis x is the land of the then known earth overspread and occupied by the sons of Noah, "all the nations mentioned belonged *to the white race.*"

¶ 2.

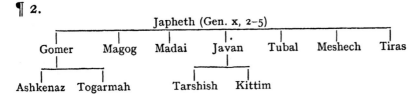

"*By these were the isles of the Gentiles divided in their lands*" (Gen. x, 5).

106 THE PILLAR IN THE WILDERNESS

"God shall enlarge Japheth, and he shall dwell in the tents of Shem, and Canaan shall be his servant" (Gen. ix, 27).

From some branch of Gomer with Madai came the Medes: Javan is identified with the Ionians, or Hellenes, of Greece; from Tiras, the Thracians; the modern name of Tubal is Tobolsk, and of Meshech, Moscow, these with Magog form the Slavs.

The Gomeri settled along the north and west shores of the Euxine (Black Sea), in Kimria (Crimea) and southwards to Deffrobanni (now Constantinople), and were known as Kimmerians and Kymri.[1]

A branch of this family migrated in early times to Britain under Hu Gadarn:

> Long before the Kymry came into Britain, the Llyn Llion, or Great Deep, broke up and inundated the whole earth. One vessel floated on the waters—this was the ship of Nevydd Nav Neivion (the Great Creator). In it were two persons preserved from the waters, Dwy Van (man of God) and Dwy Vach (woman of God). By the posterity of these two were the lands of the earth re-peopled.
>
> For a long time after the Deluge, the Kymry dwelt in the Summer Isle (Crimea) and land of Hav, called Deffrobanni. The land being exposed to sea floods, they resolved, under the guidance of Hu Gadarn to seek again the White Isle of the West (Albion). They journeyed westward towards the setting sun, being many in number and men of great heart and strength. . . . The Kymry still held onward until they saw the cliffs of the White Island. Then they built ships and in them passed over the Hazy Ocean. And they found no living creature in it but bisons, elks, bears, beavers, and water-monsters. And they took possession of it not by war, nor by conquest, nor by oppression, but by the right of man over nature. . . . And they called the island the White Island (Ynys Wen).[2]

Tribes of the Kimmerians round the Crimea were known as Brito-Legæ, Cumri, Kelts. Their journeyings westward

---

[1] Davies, *Celtic Researches*, derives Cymri, Kimmerii, from Gomeri.
[2] *The 300 Historic Triads of the Isle of Britain* (160 of these extant, referring to Bardic poems, Druidic teaching, laws and customs of the people. *Roll of Tradition and Chronology: The Traditional Annals of the Cymry*, p. 25.

CELTIC ORIGIN

were along two main lines, and later, in the seventh century B.C.,[1] when this land was invaded from the east by Scythians, the whole stock seems to have migrated. During the few centuries before, it is probable that with them were admixed some of Israel and the tribe of Dan, migrating before the Assyrian invasions.[2]

The more northerly line from the Crimea to the Baltic, up the riverways, was taken by the Kimmerians, displacing a sparsely populated country of Tartar races who were forced northwards and became the Esths, the Finns and the Lapps. In North-West Europe they occupied Scandinavia and Jutland.

The more southerly route along the Danube was taken by the Kymry and Kelts; when in sight of the Alps, part of their migration diverted southwards—the Veneti settled on the north Adriatic coast; the Umbri (before 1200 B.C.) and Brettani (towards the south) passed into Italy, the Ligures (in Greek, the Lygæ) into N.W. Italy. The main stock, after passing the Alps, left tribes, as the Lloegryws in the Riviera region, the Ambrones in N.E. of Spain and the Pyrenees, the Celts in Spain (coalescing with the

---

[1] Sharon Turner states: "Europe has been peopled by three great streams of population from the East, which have followed each other at intervals, so distant as to possess languages clearly separable from each other. The Kimmerian or Keltic nations must have first inhabited Europe, and reached the westerly portions. The Scythian or Gothic tribes, . . . from whom most of the modern nations of Continental Europe have descended, followed next. The third and most recent comprehends the Slavonian and Sarmatian nations . . . established themselves in Poland, Bohemia, Russia and their vicinities."

[2] *Roll of Tradition and Chronology*, Apud Iolo MSS., pp. 47, 426: "Rebelled against God and His fundamental truths: sinning and committing injustice with daring transgression: for which He poured upon them His retributive vengeance; whereupon dispersion and devastation ensued; upon which they became nearly extinct, having lost their territories and national rights. Then some betook to themselves their consciences; recovered to memory the name of the Deity and His Truth: and adhering to those principles, they conducted themselves under the influence of cautious reason in their sinking state. God now, out of His grace and unutterable love, imbued them with laudable intentions; placing among them wise and holy men, who, under the upholding of God and His peace, and in the refuge of His Truth and Justice, acquired a right knowledge of every superiority conducive to the well-being of the Cymry. Thus circumstanced they proceeded in their adopted course, admitting into their train all who would join them, from camp to camp; and in this manner retreated, until they escaped from the nations which had assailed them with devastation and plunder."

Iberians, as Celtiberians, and in West Spain as Ebruro-Brittium).

The main body, pursuing their course still farther, crossed the Rhône ("the river of eddies"), the Arar ("the slow river"), the Garonne ("the rough river"), the Loire ("the bright river"), leaving a large tribe in Gaul as the Bituriges, and seawards, the Pictones. They then turned northwards, where part of the Brython settled in Gwasgwyn and Armorica ("the expansion on the upper sea"), seven tribes of Brettani occupying the north coast as far as Jutland as the Brython, the Belgæ and the Kymry farthest east, uniting with the northern branch in Jutland as Cimbric Chersonese, and naming Denmark "Brittia."

Thence the migration to the Ynys Wen as the Cymri, who occupied Cambria (Wales), and later, by their invitation, came over their kin—the Lloegrians being given Kent to Cornwall, and the Brythons from Humber northwards.[1]

(The eastern parts and lands adjoining the great estuaries of Thames, Severn, Mersey, Humber, Trent, and the Fen countries were either submerged or mud swamps at this time.)

In the sixth century B.C. the Pictones migrated to S.W. Ireland as the Picts.

The religion of the Japhetidæ was Druidism, and Cambria being occupied by the senior tribe, the Cymry, became the seat and headquarters of the Druids, where the Druidic colleges were founded, and to which came the sons of all the Gallic nobility.

¶ 3.

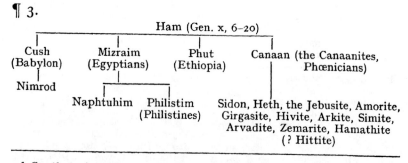

---

[1] So that Armorica was considered of Britain rather than of the Continent until A.D. 900.

Bishop Cumberland (in *Sanchoniathon's History*, p. 107) assumes Ham married Naamah, the only woman descendant of Cain mentioned in the Bible (Gen. iv, 22). By this right, Nimrod, "*the mighty hunter,*" may have succeeded to the Cainite throne in the Sumerian civilisation, for, Gen. x, 10, "*the beginning of his kingdom was Babel, and Erech, and Akkad, and Calneh, in the land of Shinar,*" and carried on the old Babylonian dynasty.

Thus the Noachic families become corrupt by promiscuous marriage with Cain-Turanian stock (as did the Adamic world) without regard to Divine command, nor recognising the deterioration brought about in the future race in moral and mental qualities.

The defeat of the Sumerians was brought about by Asshur, son of Shem, "*out of that land went forth Asshur and builded Nineveh,*" who founded the Assyrian Empire. Nimrod is said to have fled to his own race in Egypt—Mizraim—and "Nimrod appears to have been the human original of the Egyptian Osiris" (Garnier, *Worship of the Dead*, p. 36).

Mizraim is Egypt: Phut is Ethiopia, thither went other sons of Ham, and Ham becomes the mythological representative of the Egyptian god, Amon.

Ency. Brit., "Mexico," p. 330:

> that there are indications of some of Noah's family having travelled from Babylonia to Egypt, thence across Africa to America . . . that on account of these indications the first Spanish explorers of Mexico arrived at the curiously definite result that the Mexicans[1] were descended from Naphtuhim, son of Mizraim, who left Egypt for Mexico shortly after the confusion of tongues.

> Mexican belief in the stages of heaven and hell was apparently learnt from the Babylonian-Greek astronomical theory.

Thus Ham carries on the Cainite influence, with its polytheism and elaboration of Babylonian-Egyptian gods,

---

[1] *Discovery: a Monthly Journal of Knowledge*, June 1925: " . . . Dr H. J. Spinden, of Harvard, had discovered definite proofs of the exact year from which ancient Maya, builders of Central America, dated their time count, namely, 3373 B.C."

110 THE PILLAR IN THE WILDERNESS

human sacrifice and slavery, through S.W. Asia and N.E. Africa.

From his son, Canaan, also descended a mixture of peoples by intermarriage with Cain—aborigine stock, accursed and degraded.

¶ 4.

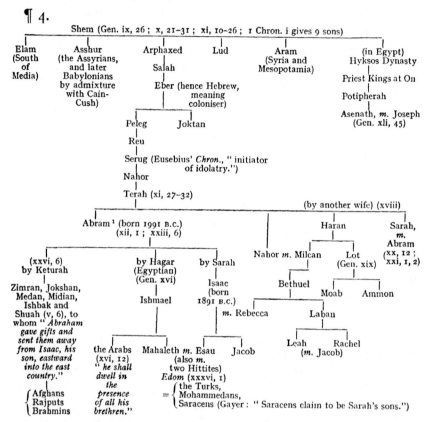

"When Noah came out of the Ark, he instructed Shem to arrange a fresh calendar, as everything had been altered by the Flood." [2] When these observations were finished, Shem left Babylon (2204 B.C.) and went to Egypt.

Here Shem is generally reputed to have been the second

---

[1] Called by bedouins, as founder of their race, "Ibrahim Khabil Abdurrahman," the Friend of God (C. L. Woolley, *Ur of the Chaldees*, p. 208).

[2] From very ancient writings called "The Chinese Shu King," known as *The Book of History*, which go back to the time of the Flood, and collected about the time of Confucius (500 B.C.).

king of the IVth Dynasty: his arrival in Egypt was the Hyksos (or Shepherd-King) invasion of the country—and was evidently intended as a clean-up of the accursed Cain-Ham rule.

M. H. Gayer, *The Heritage of the Anglo-Saxon Race*, quoting an old Egyptian papyrus about Ra and Osiris:

> During his stay in Egypt, Shem became the father of three sets of triplet sons. These sons ruled as priest-kings at Heliopolis or On, the City of Shem. In later years quarrels broke out between the brothers and they separated, and each set founded dynasties at Memphis, Coptos, and Elephantine. These Kings, reigning simultaneously, were the celebrated Hyksos (Shepherd Kings), worshipped one God under the name of Ra, and were of quite a different race to the Egyptians. Their High Priests were invariably princes of blood royal, thus was Potipherah, the father of Asenath, a Shemite princess, who married Joseph.

Manetho (a scholar of Egypt, 300 B.C.) says, concerning the Hyksos period, that after building the Pyramid, Shem and his people (numbering 240,000) quitted Egypt and went to Judea, where they built a city large enough to hold them, calling it Jerusalem.

"*Melchizedek, King of Salem* [Gen. xiv, 18–20] . . . *the priest of the most High God; without war God gave him this ascendancy: without descent . . . made like unto the Son of God:—to whom Abraham gave a tenth part of all, and was blessed*" (Hebrews vii, 1)—is believed to have been Shem.

In the Sumerian civilisation of Babylonia appears the first Semitic influence through Asshur, son of Shem.

Bristowe, *Sargon the Magnificent*, p. 115, states:

> Among thousands of polytheistic inscriptions, 3 or 4 tablets have been found . . . telling about One Supreme Being, the rebellious angels and the Fall of Adam . . . showing that the knowledge of God existed in Babylonia before the year 2200 B.C., when there is said to have been a literary revival, when older writings and traditions were reproduced and written in style more clear and lucid—proving Dr Kittel's theory that a common source existed for the Bible and the Babylonian inscriptions.

The House of Asshur seems to have quickly succumbed to the prevailing civilisation and adopted the gods of Bel, Marduk and others [1]; accepted the priesthood and their teaching, until the civilisations of Assyria and Babylon were nearly identical.

Circa 1900 B.C., King Hammurabi (thought to have been Abraham's contemporary, Amraphael) collected the laws and customs of the country, and promulgated his Code of Laws,[2] evidently to preserve some standard with the setting in of moral degeneration.

It differs from the Laws of Moses, which were

> a solemn reassertion of those Divine rules (laid down in the beginning for Adam and his descendants) made necessary by the falling away of the Israelites during their stay in Egypt.
> (1) Hammurabi's Code begins with what might be called a dedication to the gods Anu, Ea, Bel and Marduk; and
> (2) the greatest commandments—those against idolatry and murder—are omitted in the Babylonian code, while sorcery and witchcraft are encouraged.[3]

The Laws of Moses could not have been taken from such pagan code: the two had a common origin, whereas Cain had deliberately perverted the source with polytheism.

¶ 5. The reign of Hammurabi (1933–1890 B.C.) opens with eleven years of war with Elam, to be followed by nineteen years of peace between them, until in 1903 B.C. he overthrew the Elamite power and there began the Chaldean Empire.

The year 1921 B.C. still sees the supremacy of Kudur-laghamer, the Elamite (Chedorlaomer of Gen. xiv), reign-

---

[1] These are, however, inferior to the one supreme god that the Assyrians always recognised and named as Asshur—"Asshur, the great lord, who rules the host of the gods" on the inscription of Tiglath-Pileser I: see Ragozin, *Assyria*, pp. 5–18.

[2] *One-Volume Bible Commentary*, p. 35: "This is the oldest known code in the world . . . the laws themselves must have been in operation long before their codification and promulgation by Hammurabi." *Religions of Egypt and Babylonia*: "Fragments of law exist which antedate Hammurabi's age, which reveal an organised life not inferior in its cultural developments to that attested by his code. . . . We must not forget that previous to Hammurabi there existed a high culture and social developments underestimated."

[3] Bristowe, *Sargon the Magnificent*, pp. 124–125.

THE HYKSOS

ing at Larsa over the land of Sumer and ruling over Canaan for twelve years (1921-1909 B.C.).

Then the five Canaanite kings revolted, and in the spring of 1907 B.C. occurred *"the battle of four kings with five"* (xiv, 9), when they were defeated by Chedorlaomer in alliance with Amraphael (Hammurabi), King of Chaldea, Arioch, and Tidal, king of the Hittites.

Then followed the Hittite pressure on South Canaan (Gen. xxiii).

This pressure, with the Elamite invasion of Sumer in 1935 B.C., and the periodic famine which occurred in 1917 B.C.,[1] caused the Hyksos migration into Egypt.

The Elamite and Kassite pressure from the East caused an emigration from Babylonia and her dependencies westwards and southwards, and the people thus dispossessed drove before them the nomadic tribes on the north-east of Egypt. . . .

The dwellers in Syria and Palestine joined with the nomadic tribes of the Eastern Desert, and fled to Egypt for safety, and it needed little foresight to see that they might easily be pursued thither by the victorious armies of Chaldea and Babylon (Budge, *Hist. Egypt*, vol. iii, pp. 135-137).

So they immediately fortified the eastern frontier of the Delta.

The Hyksos entry first began in 1937 B.C., when a deputation of thirty-seven Aamu of Shu was received by Senusert II (of XIIth Dynasty), and granted land in the Delta. Their first king was Apepa I, being given the title of "Heq" (prince),[2] and a record states that they "had destroyed the ancient works; they reigned ignorant of the god, Ra."[3]

One might here see in history some purer element of Shem, trying to get away from the polytheism of Sumer, and perhaps to purify the land of Ham, acquired land in the Delta, to be afterwards joined by a later migration for the reasons given above.

[1] Davidson (p. 292) verifies that the famine conditions led to dredging operations in the rocky channel at the First Cataract in the eighth year of Senusert III (1917-1916 B.C.): that Abraham was driven by famine in Canaan into Egypt in 1916 B.C.
[2] Hyksos thus derived from Heq-shu, shu = shepherd.
[3] Davidson, pp. 310-320.

After the war, 1855–1830 B.C., the Hyksos were expelled.

1770 B.C. saw the Egyptian power under Tothmes II extending, until under Tothmes III it reached to the Euphrates.

In 1750 B.C. the Hittites overthrew Babylon, but were pushed back by the Kassites, in 1726 B.C., who remained supreme in Sumer until 1458 B.C.

1458 B.C. saw the rise of the Aramaic [1] kingdom in Mesopotamia; in 1345 B.C. they swarmed into Babylon and Assyria, obliterating all trace of Assyrian dynasty between 1284–990 B.C.

"*Babylon hath been a golden cup in the Lord's hand, that made all the earth drunken; the nations have drunken of her wine: therefore the nations are mad.*"

"*We would have healed Babylon, but she is not healed*" (Jer. li, 7, 9).

Thus it is written of this chaotic assemblage of mingled peoples—Cain, Ham, and degenerated parts of Shem intermarried with aborigine stock; all warring between themselves, seeking power and wealth out of conquered nations, worshipping that polytheistic galaxy of Cain-Bel brewing —Babylon indeed a cup in which Cain-Bel have brewed a wine, of which all nations have drunken—"*therefore the nations are mad.*"

With His Divine Tuition forgotten, His Divine Laws disobeyed—as the Adamic race except Noah had become "*corrupt,*" so now also was the Noachic race: how was His Divine Purpose to be carried out?

¶ 6. Out of Ur of the Chaldees, He called Abram (1916 B.C.).

"*Get thee out of thy country, and from thy kindred, and from thy father's house, unto a land that I will shew thee; and I will make of thee a great nation, and make thy name great, . . . and in thee shall all the families of the earth be blessed*" (Gen. xii, 1–3).

"*And Abram went forth into the land of Canaan.*

[1] Aram, son of Shem, settled in Syria and Mesopotamia.

ABRAHAM

"And the Lord appeared unto Abram and said, Unto thy seed will I give this land."

Three years later, after his sojourn in Egypt owing to the famine in Canaan (1913 B.C., dating in Pyramid ascending passage produced to rock-level):

"I will make thy seed as the dust of the earth [xiii, 16], tell the stars, if thou be able to number them, . . . so shall thy seed be [xv, 5], unto thy seed have I given this land from the river of Egypt and the great river Euphrates, and the Lord made a covenant with Abram" (xv, 18).

So far Abram's Freewill had operated in faith,[1] according to Divine command: then came his test.

When Abram, aged one hundred, and Sarah, his wife, was ninety, so that *"she laughed,"* a son was promised. *"I will make My covenant between Me and thee . . . for an everlasting covenant . . . thou shalt be a father of many nations . . . kings shall come out of thee . . . all the land of Canaan for an everlasting possession, and I will be their God [xvii, 1–8] . . . thou shalt call his name Isaac, and I will establish My covenant with him for an everlasting covenant, and with his seed after him"* (xvii, 19).

Isaac was born, through whom was His Covenant to be fulfilled, *"for in Isaac shall thy seed be called"* (xxi, 12).[2]

When Isaac was aged fourteen, to Abraham, *"Take now thy son, thine only son Isaac, whom thou lovest, and get thee into the land of Moriah, and offer him there for a burnt offering upon one of the mountains which I will tell thee of"* (Gen. xxii).

"And Isaac spake unto his father . . . Where is the lamb for a burnt-offering? And Abraham said, My son, God will provide Himself a lamb for a burnt-offering" (8).

"And Abraham stretched forth his hand, and took the knife to slay his son" (10). And he was stopped.

"By Myself have I sworn, saith the Lord, for because thou hast done this thing, and hast not withheld thy son, thine only son. . . ."

---

[1] Freewill operating in faith, without any quibbling: whereas Adam failed when tested.

[2] Isaac = Tsak (Hebrew); Saca-sena, Sak-sons = Saxons.

In effect, the Most High says, The sacrifice is postponed, for the sacrifice is of Me, the Lamb has been provided since the foundation of the world, Mine only Son.[1]

"That in blessing I will bless thee, and in multiplying I will multiply thy seed as the stars of heaven, and as the sand which is upon the sea-shore: and thy seed shall possess the gate of his enemies: And in thy seed shall all the nations of the earth be blessed, because thou hast obeyed My voice" (xxii, 16-18).

His Covenant is confirmed to Isaac (Gen. xxvi, 3-5); and to Jacob (xxviii, 13-15),—

"Thy seed . . . shall spread abroad to the west, and to the east, and to the north, and to the south,[2] . . . I am with thee in all places whither thou goest and will bring thee again into this land; for I will not leave thee until I have done that which I have spoken to thee of.

"A nation, a company of nations shall be of thee, and kings shall come out of thy loins" (Gen. xxxv, 10-14),

—was said after Jacob had wrestled all night, purified, so that his name of usurper was cast away, and substituted—Israel, "ruling with God" (xxxii, 28).

And the birthright passes through Joseph to Ephraim before the elder son, Manasseh (Gen. xlviii, 20).

Jacob,[3] blessing his sons, *"that I may tell you that which shall befall you in the last days,"* gives to Joseph the birth-

---

[1] Davidson, p. 419:

Twelve times 153—the length of Grand Gallery	1836 A.K.
The Displacement defining God's sacrifice	286
Isaac on the altar when 14 years old	2122 A.K. (1878 B.C.).
Interval	1906½ yrs.
Christ on the Cross (Commencement of Grand Gallery)	4028½ A.K. (=A.D. 30).
Same Interval	1906½
Threshold of King's Chamber	5935 A.K. (=A.D. 1936).

When Judah looks upon "*Me whom they have pierced*" (Zech., xii, 10).

[2] The order is significant, to W., E., N., S. (for such the direction his seed spread).

[3] Jacob died in Egypt, no kings from his loins as yet, and his people in exile: thus the promise applies to his seed, "*I will not leave thee.*"

right and the blessings threefold, "from whom (Jacob) is Israel's guardian stone"[1] (the stone he had raised at Bethel).

Repeated by Moses: "*His glory is like the firstling of his bullock, and his loins are like the horns of unicorns: with them he shall push the people together to the ends of the earth.*"

And He had already fore-ordained the boundaries of the nations:

"*When the Most High divided to the nations their inheritance, when He separated the sons of Adam, He set the bounds of the people according to the number of the Children of Israel. For the Lord's portion is His people, Jacob is the lot of His inheritance*" (Deut. xxxii, 8, 9).

¶ 7. The Mosaic Covenant (Deut. xxviii, Lev. xxvi) was conditioned by obedience to the Divine laws of righteousness, summed up in His first law of Love, that Harmony should reign throughout the world and the whole of His creation.

After contact with the mixed civilisations of Cain and Ham in Babylon and Egypt, with their impurity, polytheism and idolatry, His servant nation was isolated in the Wilderness during forty years for purification, to be trained in His Law, restated by Moses, to become a holy people, His elect.[2]

His Purpose for them was that when they entered the promised land of Canaan, still obedient, they were first to clear out the races of Ham without coalescing by intermarriage, so degrading their stock; then Joshua's campaigns would have been extended to the Euphrates, to the Nile,[3] until from the world was eradicated the Cain-Bel influence. Then there would have been a different history,

---

[1] Ferrar Fenton's translation.

[2] "*Israel is My son, My firstborn*" (Exodus iv, 22, 23). "*Thy God hath chosen thee to be a special people unto Himself above all people*" (Deut. vii, 6). "*This day thou art become the people of the Lord thy God*" (xxvii, 9). "*The Lord shall establish thee an holy people unto himself*" (xxviii, 9). "*Thou art mine*" (Isa. xliii, 1). "*My people, My chosen*" (8). "*This people have I found for Myself*" (21). "*Mine elect*" (lxv, 9).

[3] "*Then the Lord shall drive out all these nations from before you, and ye shall possess greater nations and mightier than yourselves*" (Deut. xi, 23). "*Thou art My battle-axe and weapons of War, for with thee will I break in pieces the nations*" (Jer. li, 20).

no succession to the Nebuchadnezzar kingdoms—Babylon, Persia, Greece, Rome: no Græco-Roman mythology, and their dried get-nowhere philosophy: but the universal worship of the One God.

Thus the punishments for disobedience, the standards to be maintained, the detail as to the Law and the Worship were so minutely stressed and put out in Leviticus, Numbers and Deuteronomy. In His foreknowledge of His Israel, He puts it to their Freewill as to how they should operate, in words which cannot be mistaken.

Yet the punishments are for a limited number of years, that during the years of Captivity and Exile they might learn repentance. Such does not annul His everlasting Covenant to their forefathers, to Abraham *"because thou hast obeyed my voice,"* to Isaac, and to Jacob. His everlasting Covenant is literally meant, His people may even be lost, but never destroyed.

In sublime language (and read with it the history of Britain [1] and our people):

"And it shall come to pass, if thou shalt hearken diligently unto the Voice of the Lord thy God, and to do all His commandments . . . that the Lord thy God shall set thee on high above all nations of the earth,[2] *and all these blessings shall come on thee and overtake thee."*

v. 7. *"The Lord shall cause thine enemies that rise up against thee to be smitten before thy face: they shall come against thee one way, and flee before thee seven ways.*[3]

"The Lord shall command the blessing upon thee in thy storehouses,[4] *and all that thou settest thine hand unto, and He shall bless thee in the land which the Lord thy God giveth thee.*[5] *The Lord shall establish thee an holy people unto Himself, as He hath sworn unto thee*[6] *. . . and all the people*

---

[1] Since 1804. ("Seven times" punishment = 7 × 360 = 2520 years from 717 B.C., the last Assyrian captivity.)

[2] Such has become the position of Britain.

[3] At Trafalgar, Waterloo: in the Crimea, the Indian Mutiny: the Boer War: in the Great War, more than ever before, such is so.

[4] Britain sells to the world from her storehouse.

[5] The prosperity of herself and her Colonies; the privilege of being born British.

[6] The British Church has realised this; its missionary societies, etc.

THE MOSAIC COVENANT

shall see that thou art called by the name of the Lord, and they shall be afraid of thee.[1]

"And the Lord shall make thee plenteous in goods, in the fruit of thy body,[2] *and in the fruit of thy cattle, and in the fruit of thy ground*[3] *... the Lord shall open unto thee His good treasure, the heaven to give the rain unto thy land in His season,*[4] *and to bless all the work of thine hand; and thou shalt lend unto many nations, but thou shalt not borrow.*[5] *And the Lord shall make thee the head and not the tail; and thou shalt be above only, and thou shalt not be beneath.*[6] *... And thou shalt not go aside from any of the words which I command thee*[7]*—to go after other gods to serve them."*[8]

v. 15. *"But it shall come to pass, if thou wilt not hearken unto the voice of the Lord thy God, to observe to do all his commandments . . . that these curses shall come upon thee, and overtake thee.*[9]

"The Lord shall send upon thee cursing, vexation and rebuke, in all that thou settest thine hand unto for to do . . . the Lord shall make pestilence cleave unto thee, until He has consumed thee from off the land whither thou goest to possess it."

v. 22. *"The Lord shall smite thee with a consumption, and with a fever, and with an inflammation, and with an extreme burning*[10]*; and with the sword, and with blasting, and with mildew; and they shall pursue thee until thou perish."*

v. 24. *"The Lord shall make the rain of thy land powder and dust*[11] *... the Lord shall cause thee to be smitten before*

---

[1] Read *England*, by Wilhelm Dibelius, Professor of English at Berlin University, and other Continental writers.

[2] The population in days of Queen Elizabeth, 6 millions; in 1800, 16 millions; in 1933, 160 millions (ten times in a century and a third).

[3] Our cattle the finest, so our farming; and as to goods, see the Savings Banks, and the household goods of the average family.

[4] No prolonged drought, nor famine.

[5] Britain lent everywhere, especially in the War, and borrowed only from its own stock, the U.S.A.

[6] Britain holds the League of Nations and the balance of power.

[7] Our English Prayer Book is unique; the Bible; the daily broadcast service.

[8] Meaning Communism, and the Anti-God attitude of Russia.

[9] Britain—her history before 1557 especially: and still while not realising that she is His servant Nation; for each individual can be picked out the punishment which applies.

[10] Tuberculosis, influenza, venereal disease, cancer.

[11] Canaan, a land flowing with milk and honey—now Palestine, parched, sandy, lacking water.

thy enemies; thou shalt go one way before them and flee seven ways before them: and shalt be removed into all the kingdoms of the earth.[1] And thy carcase shall be meat unto all the fowls of the air, and unto the beasts of the earth."[2]

v. 28. "The Lord shall smite thee with madness, and blindness and astonishment of heart; and thou shalt grope at noonday as the blind gropeth in darkness, and thou shalt not prosper in thy ways: and thou shalt be only oppressed and spoiled evermore, and no man shall save thee."

v. 30. "Thou shalt betroth a wife and another man shall lie with her: thou shalt build an house, and thou shalt not dwell therein; thou shalt build a vineyard, and shalt not gather the grapes thereof. Thine ox shall be slain before thine eyes, and thou shalt not eat thereof; thine ass shall be violently taken away from before thy face, and shall not be restored to thee: thy sheep shall be given unto thine enemies, and thou shalt have none to rescue them."

v. 32. "Thy sons and daughters shall be given unto another people, and thine eyes shall look and fail with longing for them all the day long, and there shall be no might in thy hand. The fruit of thy land and all thy labours shall a nation which thou knowest not eat up. . . . The Lord shall smite thee in the knees, and in the legs, with a sore botch that cannot be healed, from the sole of thy foot to the top of thy head."

v. 36. "The Lord shall bring thee and thy king which thou shalt set over thee, unto a nation which neither thou nor thy fathers have known; and there shalt thou serve other gods, wood and stone[3]: and thou shalt become an astonishment, a proverb and a byword among all nations whither the Lord shall send thee."[4]

v. 38. "Thou shalt carry much seed into the field, and shalt gather but little in, for the locust shall consume it: thou shalt plant vineyards and dress them, but shalt neither drink of the

---

[1] Thus Israel, lost to history, but not destroyed. If the Jews survive after A.D. 70, when by all historical laws they should have disappeared, then why not Israel?

[2] So in the migration of our race through Europe; in innumerable wars; of our pioneers and adventurers.

[3] Israel, in Scythia.

[4] The Jews.

THE MOSAIC COVENANT

wine, nor gather the grapes, for the worms shall eat them. Thou shalt have olive trees throughout all thy coasts, but thou shalt not anoint thyself with oil, for thine olive shall cast his fruit. All thy trees and fruit of thy land shall the locust consume."

v. 41. *"Thou shalt beget sons and daughters, but thou shalt not enjoy them, for they shall go into captivity."*

v. 43. *"The stranger that is within thee shall get up above thee very high, and thou shalt come down very low. He shall lend to thee, and thou shalt not lend to him: he shall be the head, and thou shalt be the tail."*

v. 45. *"And these curses shall be upon thee for a sign and for a wonder, and upon thy seed for ever."*

v. 48. *"Therefore shalt thou serve thine enemies which the Lord shall send against thee, in hunger, and in thirst, and in nakedness,[1] and in want of all things; and he shall put a yoke of iron upon thy neck, until he hath destroyed thee."*

v. 49. *"The Lord shall bring a nation against thee from far, from the end of the earth, as swift as the eagle flieth, a nation whose tongue thou shalt not understand, a nation fierce of countenance, which shall not regard the person of the old, nor show favour to the young, and he shall eat of the fruit of thy cattle, and the fruit of thy land, until thou be destroyed."*

v. 52. *"And he shall besiege thee in all thy gates, until thy high and fenced walls come down, wherein thou trustedst, throughout all thy land [2] . . . which the Lord thy God hath given thee. And thou shalt eat the fruit of thine own body, the flesh of thy sons and thy daughters . . . in the siege,[3] and in the straitness wherewith thine enemies shall distress thee."*

v. 59. *"And the Lord will make thy plagues wonderful, and the plagues of thy seed, even great plagues, and of long continuance, and sore sicknesses . . . also every sickness and every plague which is not written in the book of this law."*

v. 62. *"And ye shall be left few in number, whereas ye were as the stars of heaven for multitude."*

---

[1] *Cf.* Matthew xxv, 35–45: the parable of His judgment of the nations at His Second Coming.

[2] Assyria and Babylon.   [3] A.D. 70, by the Romans.

v. 63. "And it shall come to pass that as the Lord rejoiced over you to do you good, and to multiply you; so the Lord will rejoice over you to destroy you, and to bring you to nought; and ye shall be plucked from off the land whither thou goest to possess it. And the Lord shall scatter thee among all people, from the end of the earth even unto the other [1] *: and there thou shalt serve other gods which neither thou nor thy fathers have known, even wood and stone."*

v. 65. "And among the nations shalt thou find no ease, neither shall the sole of thy foot have rest: but the Lord shall give thee there a trembling heart, and failing of eyes, and sorrow of mind: and thy life shall hang in doubt before thee, and thou shalt fear day and night, and shalt have none assurance of thy life."

v. 67. "In the morning thou shalt say, Would God it were even! and at even thou shalt say, Would God it were morning! [2] *for the fear of thy heart wherewith thou shalt fear, and for the sight of thine eyes which thou shalt see."*

Also the parallel chapter in Leviticus xxvi:

v. 17. "And ye shall flee when none pursueth." [2]

v. 32. "And I will bring the land into desolation,[3] *and your enemies which dwell therein shall be astonished at it."*

v. 36. "And upon them that are left alive of you I will send a faintness into their hearts in the lands of their enemies, and the sound of a shaken leaf shall chase them." [2]

In verses 18, 21, 24, 28 is regularly set down, "*I will punish you seven times.*" "A time" in Scriptural phraseology means 360 years. Seven times = 2520 years.[4] This period is also verified in the Pyramid geometry, as also the half-period, 1260 years, which appears scripturally as "*a time, times and a half.*"

Yet follows, Lev. xxvi, 40–42: "*If they shall confess their iniquity, and the iniquity of their fathers, . . . if then their uncircumcised hearts be humbled, and they then accept of the punishment of their iniquity, then will I remember My*

---

[1] This must have been the fate of Israel after the Assyrian Captivity.

[2] Is there a better description for the beginning of neurasthenia (or "nerves")?

[3] So did Canaan become after the captivities.

[4] The lowest number to be divided by factors, 1–10.

THE MOSAIC COVENANT

covenant with Jacob . . . with Isaac, . . . and with Abraham will I remember, and I will remember the land."

v. 44. *"And yet for all that, when they be in the land of their enemies, I will not cast them away, neither will I abhor them, to destroy them utterly, and to break My covenant with them, for I am the Lord their God. But I will for their sakes remember the covenant of their ancestors . . . that I might be their God; I am the Lord."* [1]

And in Deut. xxxiii, 1–6: *"And it shall come to pass that when all these things are come upon thee . . . and thou shalt call them to mind among all the nations, whither the Lord thy God hath driven thee, and shalt return unto the Lord thy God, and shalt obey His voice . . . that then the Lord thy God will turn thy captivity, and have compassion upon thee, and will return and gather thee from all the nations, whither the Lord thy God hath scattered thee. If any of thine be driven out unto the utmost parts of heaven, from these will the Lord thy God gather thee, and from thence will He fetch thee; and the Lord thy God will bring thee into the land which thy fathers possessed, and thou shalt possess it. . . . And the Lord thy God will circumcise thine heart, and the heart of thy seed, to love the Lord thy God with all thine heart, and with all thy soul, that thou mayest live."*

Deut. xxxiii, 27: *"The eternal God is thy refuge, and underneath are the everlasting arms, and He shall thrust out the enemy before thee. . . . [v. 29] Happy art thou, O Israel, . . . O people, saved by the Lord . . . thou shalt tread upon their high places."*

Moses in his final exhortation to Israel puts the issue of whether their Freewill will operate in accord or deviate:

Deut. xxix, 29: *"The secret things belong unto the Lord our God: but those things which are revealed belong unto us and to our children for ever, that we may do all the words of this law."*

Deut. xxx, 19, 20; also Deut. iv, 26–40: *"I call heaven and earth to record this day against you, that I have set before you life and death, blessing and cursing: therefore choose life, that both thou and thy seed may live, that thou mayest*

---

[1] Therefore, Israel was not destroyed in Assyria.

love the Lord thy God, and that thou mayest obey His voice . . . for He is thy life, and the length of thy days, that thou mayest dwell in the land."

8. ¶ God, in his foreknowledge, knew that the Freewill of Israel would deviate, as Adam's had done: so that the time of their punishment and the means of their redemption are prepared. This time coincides with the grant to Nebuchadnezzar as *"the head of gold,"* the rule of the Gentiles with the successive kingdoms of Babylon, Persia, Greece and Rome, for 2520 years: during which time the name of Israel disappears from the ken of man, but not the nation, His people, from before His eyes. In both the above chapters, and later, by the prophets, is the expectation of their repentance, of the fulfilling of the promises in the Covenant, also *"for all that, I will not cast them away to destroy them utterly."* For His Covenant with Abraham, with Isaac and with Jacob is everlasting, Israel as a nation survives to this day.

If not, you doubt His power to eradicate the sin of idolatry from His chosen people, and you will doubt His power to eradicate the Cain-Bel influence from the world; you will make of him a Jahwe, some local god of a wandering Bedouin tribe, on a level with Anu, Bel, or of some lesser category: you will break the continuity of His-Story; and you will presuppose, with the failure of Israel, he most needs introduce a new religion with the birth of Christ, not as God, but as the perfect example of the Good and the Beautiful in Man to fit the "run-to-and-fro" philosophy of Socrates, Plato and Aristotle.

Davidson (pp. 490–507) shows, in comparison of the Passover at the Exodus and at the Crucifixion in direct parallel, that Israel is bound to Christ in a wonderful way. He also shows the Day of Pentecost occurred on the exact anniversary of the giving of the Mosaic Law on Mount Sinai.

Christ based the last acts of His life on earth for the salvation of mankind (such foreordained *"in the beginning"* before the Creation) on the salvation of Israel at the Exodus,

Crucifixion Year A.D. 30.	The day and date corresponding with		Exodus Year 1486 B.C.
(Prior to 6 A.M.) The Compact of Judas with the Priests. (Luke xxii, 1–7): "*I am the Bread of Life.*"	14th Nisan, 5th day. 6 A.M.		The Sacrifice of the Lord's Passover.
The first day of unleavened bread began. The Killing of the Passover.	Noon	Moses' demand of Pharaoh—the "three days' journey into the wilderness to sacrifice."	The first day of unleavened bread. The Killing of the Lamb of the Passover.
Our Lord's Last Supper: "*This is My Body.*" (The Lamb giving of Himself; the blood a token.)	6 P.M. 15th Nisan, 6th day.	Christ for "three days and three nights in the heart of the earth."	The Eating of the Lamb; the blood a token.
Gethsemane. Midnight Crisis and Victory. "Passing over of Death from Mankind."	12.0		The Smiting of the Firstborn of Egypt: "*I will pass over you.*"
The Betrayal by Judas before cockcrow.	3 A.M.		
	6 A.M.		The Flight of Israel began.
Our Lord crucified — the 3rd hour	9 A.M.		
Our Lord crucified — the 6th hour	Noon		The first Feast of baked unleavened bread.
The Death of our Lord — the 9th hour	3 P.M.		The Journey into the Wilderness began (diverted from ordinary caravan route which would have meant the 26° 18' 10" descent to destruction by the Egyptian).
Burial before Sabbath began — the 12th hour	6 P.M.		
(The means of Ascent, symbolised by ascending passage angle of 26° 18' 10", of Mankind. The raising of the Grand Gallery height by 286·1" in 6 overlaps signifying work accomplished, *i.e.* from stooping posture to upright.)	16th Nisan, 7th day.		
In the Sepulchre, prefigured by Israel in the *cul-de-sac* in the Wilderness.			Into the wilderness of "*the Sea of Reeds*": Israel in the apparent *cul-de-sac*, which becomes the highway of
	17th Nisan, 12.0		
(Sunday, 7th April, Greg.) (Prior to 6 A.M.) The Resurrection: the Salvation of Mankind, redemption from Satan (Bel) and death. = "The Crossing of the Pure Waters of Life" (Egyptian).	1st day. 6 A.M. EXODUS SYMBOLISM.		The Salvation of Israel at the crossing of "*the Sea of Reeds*" (the destruction of the pursuing Egyptians). Israel redeemed from Cain-Ham.

Davidson, Table XXXVIII, adapted.

to the very day and hour, and Himself as the Passover Lamb—not mere words, but in symbols to reach man's concept of Reality. With same symbols—man is redeemed and survives death: then the nation Israel [1] is redeemed and survives, though now dead to man's knowledge.

Paul (Romans xi, 1, 2)—"*Hath God cast away his people? God forbid. For I also am an Israelite, of the seed of Abraham, of the tribe of Benjamin. God hath not cast away His people whom he foreknew*"—means this literally. If here some theological explanation that Israel being "lost" for 750 years and Jews recently rejecting him, *His people* means the Christian Church which is just being formed; then why the insistence on his tribe, what meaning to *cast away* and *foreknow*?

Paul continues (Rom. xi, 25): "*Blindness in part has happened to Israel, until the fullness of the Gentiles be come in.*"

¶ 9. Among nations drunken of the golden cup of Babylon, a Cain-Bel brewing of polytheism and man's natural ego unleashed—to each its own special god devised from the Sumerian official pantheon, and where the free-will of each king sought only the aggrandisement of himself and his race in war, towards more and ever more power, wealth and women, with cruelty to and slavery of the vanquished—the preparation of Israel to be His "elect" nation, to fulfil His Purpose of becoming "*a blessing to all the families of the earth*" is seen in

(i) forty years of isolation, from such contamination, in the Wilderness, to be trained in His commands as given in Exodus (the Nation and the Law), Leviticus (the Worship), Numbers (the National Services), and Deuteronomy (the Constitution); and

(ii) the birthright to leadership was given to the House of Joseph. His training for such is given in Gen. xxxvii–xlvii, being separated by Divine Purpose from Israel.

---

[1] Deut. vii, 6.

THE RULE OF JOSEPH

The effect is seen historically:

Tothmes II and III, of the XVIIIth Dynasty, had extended the Egyptian Empire by conquest from the Nile to the Euphrates.

Whereas all kings rule by force of arms and show of might (*vide*, even in the civilisation of to-day, the rule of the Totalitarian States), Egyptian rule at this time is well summed up in

> for over 200 years[1] their sovereign power over distant lands and peoples which were still half-barbaric, without an army of occupation, fussy officials or abuses of authority ... a wise policy and a skilful diplomacy, based on a national opinion proud of its intellectual superiority and economic power.
>
> Men would have to stride over the centuries down to the modern British Empire before they would again find the wise principles of Egyptian colonisation.[2]

Why the similarity and who the author of this policy?

Joseph, Governor of Egypt, as Zaphnath-Paaneah, 1710–1630 B.C., who thus by experiment and experience on the Egyptians transmitted the type of rule *via* Ephraim[3] and his son's sons.

And what his influence on religion?[4]

Monotheism was first begun by Amenhotep III (1656 B.C.), but his son Amenhotep IV (1625 B.C.) introduced the worship of one god for the whole of the races of his empire, a revolutionary idea and bitterly opposed by the priests.

He chose the Sun as common and visible to all men, and in introducing the worship of Aten, the Solar Disc, he may have been influenced by his wife, Nefertiti, daughter of the king of Mitanni, whose worship of Ishtar (derived from Ea) was similar to that of the Sun (Marduk, son of Ea).

Amenhotep IV changed his name to Akhenaten ("the

---

[1] 1768 B.C., Tothmes II begins his reign, to 1570, end of XVIIIth Dynasty.
[2] Madame G. R. Tabouis, *The Private Life of Tutankhamen*, p. 84.
[3] England—John (Jack, Jacob) Bull (Joseph, Ephraim, whose sign was the unicorn or wild bull).
[4] Neither influence, of course, would be mentioned by the priests of Amen-Ra in their records, being of Semite origin and contrary to their own.

glory of Aten") and founded a new capital, now known as Tell el-Amarna.

On his death, Tutankhaten (son by a concubine, and married to the third of six royal daughters, Ankhsenpaaten, Nefertiti having no son), then quite a boy, for four years held out, but by priestly intrigue was forced back to Thebes, and to change his name to Tutankhamen.

And with the decease of Horemheb, Tutankhaten's adviser, the aged and mightiest of Egypt's generals, who protected the Semites and became Pharaoh for four years, the XVIIIth Dynasty ended (1570 B.C.).

With the coming of the XIXth Dynasty, which restored Amen worship, the might of Egypt began to decline; then arose Ramessu II (1558–1491 B.C.), "*a new king which knew not Joseph*" (Exod. i, 8).

In 1447 B.C., as Israel under Joshua is about to enter Canaan, Balaam sees them prophetically in Num. xxiii and xxiv, as the lion and the unicorn, "*with seed in many waters,*" "*an exalted kingdom.*"

The Israelitish Empire would have extended from the Nile to the Euphrates [1] on the model of Joseph's rule, and founded on His commandments would have eradicated the Cain-Bel supremacy of nations and all their influence on men's thought.

But owing to Israel's divergence from His Covenant, Balaam's prophecy did not begin to be fulfilled for 2520

---

[1] This time was the psychological moment for a general of Joshua's ability to strike:

The Hittites had retired north of Syria into Asia Minor; they had been fought to a standstill by Ramessu II.

The power of the Egyptians after the Exodus faded out for 260 years, so that it is recorded that even "*Assu the Syrian made himself prince over them.*"

1458 B.C. saw the fall of Kassite supremacy in Sumer, and the beginning of the Aramaic kingdom (Judges iii, 8), but still weak and checked by Tiglath-Peleser I of Assyria in 1349 B.C.

Israel could have possessed the land from the Nile to the Euphrates, as promised to Abraham, Gen. xv, 18.

The moment passed. The IInd Dynasty of Isin in Babylonia (1458 B.C.) was ended in 1326 B.C. by the Arameans, who swarmed into Assyria and Babylon, and their supremacy lasted till 1077 B.C. All traces of Assyrian records were wiped out from 1284–990 B.C.

Thus was allowed to arise, through Israelite material contentment and idolatry, the powers of Assyria and Babylon, to which Israel and Judah eventually succumbed.

CANAAN

years (see p. 313); and then, with the white or Adamic race drunken of Babylon's cup and spread all over the earth, the foundation of a world-wide British Empire was laid down.

¶ 10. On the conquest of Canaan, Israel settled down to comfortable life after wilderness wandering, and the coastal tribes coalesced and intermarried with the local inhabitants, against Divine command.[1]

The command to expel or exterminate the race of Ham from the land of Canaan has been interpreted as too bloodthirsty on the part of Jehovah.

Before you judge: as a father, you wish your son to marry a good woman and not a harlot; and if he marries a harlot, ruins his life in misery and treachery, and the business you have spent your life to build, his children tainted, you would wish he had never been born—

So in a human way, God as a Father watching His son, Israel, "*My firstborn,*" deviating from "*the image*" set (p. 23).

They adopted the gods of Cain-Bel-Ham, and forgot His Laws; and the conquest[2] to the Euphrates and Nile by their own freewill choice sank into oblivion. Having ease and material content,[3] Israel rested, with only an occasional spiritual endeavour.

They elected a king of their own, to be like unto other nations,[4] Saul (1090 B.C.):

1 Sam. x, 24. "*And all the people shouted, and said, God save the King.*"

When he failed, God provided one for them from the Judah-Pharez[5] line, and to David (1050-1010 B.C.) He gave His Covenant—

2 Sam. vii, 16. "*And thine house and thy kingdom shall be established for ever before thee: thy throne shall be established for ever.*"

---

[1] Deut. vii, 1-9; xx, 16-18; and Judges iii, 2: "*to teach them war*"; "*the children of Israel dwelt among the Canaanites . . . and they took their daughters to be their wives*" (iii, 5, 6).   [2] Deut. xi, 22-25.
[3] As Goethe says of man, "Why do the people so strive and cry? They will have food, and they will have children." And so do all animals.
[4] Deut. xvii, 14, 20.
[5] 1 Chron. v, 2: "*For Judah prevailed above his brethren, and of him came the chief ruler; but the birthright was Joseph's.*"

This Covenant is repeated to Solomon, his son and successor—

1 Kings ix, 5. "*I will establish the throne of thy kingdom upon Israel for ever, as I promised to David, thy father, saying, There shall not fail thee a man upon the throne of Israel.*"

Even Jeremiah, inveighing against the last Judah-Pharez king, Zedekiah, and prophesying his fall, says: "*For thus saith the Lord, David shall never want a man to sit upon the throne of the House of Israel.*"

At the accession of Joash, a boy seven years old—

2 Kings xi, 12–14. "*They clapped their hands, and said, God save the King . . . as he stood by a pillar, as the manner was.*"

The custom is referred to when they "*made Abimelech King by the plain of the pillar that was in Shechem*" (Judges ix, 6), going back to the time of Joshua.

Joshua xxiv, 25–27. "*Joshua . . . set them a statute and an ordinance in Shechem . . . and took the great stone, and set it up there under an oak, that was by the sanctuary of the Lord, . . . and said, Behold, this stone shall be a witness unto us. Israel's guardian stone*" that Jacob had entrusted to Joseph (Gen. xlix, 24) and which had been carried throughout the wilderness wandering.

The signs of cleavage in the nation appeared early, and on the death of Solomon (970 B.C.) it split into two kingdoms, the northerly of ten tribes, and the southerly of Judah, to whom the tribe of Benjamin was added.

1 Kings xi, 36. "*. . . will I give one tribe, that David My servant may have a light alway before Me in Jerusalem, the City which I have chosen Me to put My name there.*"

At the beginning, when Rehoboam intended to make war on Israel, civil war was diverted by Divine order, for the separation fulfilled His purpose.

1 Kings xii, 24. "*Thus saith the Lord, Ye shall not go up, nor fight against your brethren the children of Israel; return every man to his house: for this thing is from Me.*"

Though their Freewill deviates, His guidance controls each vital step: they cast off theocracy, then their line of kings appointed; they cause a division, then civil war is prevented and Jerusalem preserved for a "light."

"For the next 254 years, the Northern Kingdom, ruled over by a series of military adventurers, falls away into idolatry" (Gayer)—initiated by their first king, Jeroboam, who set up two golden calves, one in Bethel and one in Dan (1 Kings xii, 28, 29).

Dan coalesced with the Phœnicians, and with them was carrying on trade with the western isles; was rebuked by Deborah for "*abiding in his ships*" (Judges v, 17), *circa* 1285 B.C.: and the whole tribe seems to have migrated altogether before the Assyrian invasions, as they are not mentioned in the genealogy of 1 Chron.

Ahab married Jezebel, daughter of Ithi-bel, King of Tyre, and under her influence introduced the Phœnician gods, Baal and Ashtoreth (derived from the Babylonian Bel and Ishtar).[1] Athaliah, their daughter, married Jehoram, King of Judah, and attempted to murder the whole seed royal of David (2 Kings xi, 1), only baby Joash being rescued, later to be anointed king.

¶ 11. And so it went on until the Assyrian invasion, when the punishment fell:

1 Chron. v, 26 (771 B.C.). "*The God of Israel stirred up the spirit of Pul, King of Assyria, . . . and he carried them away, even the Reubenites, and the Gadites and the half tribe of Manasseh,*[2] *and brought them unto Halah, Habor and Hara, and to the River Gozan,*[3] *unto this day.*"

2 Kings xv, 29 (740 B.C.). "*In the days of Pekah, King of Israel, came Tiglath-Pileser, King of Assyria, and took . . . Galilee, all the land of Naphtali and carried them captive to Assyria.*"[4]

---

[1] 1 Kings xviii, 21. "*And Elijah came unto all the people, and said, How long halt ye between two opinions? if Jehovah be God, follow him; but if Bel, then follow him,*" and there followed the scene on Mount Carmel.

[2] These lying on the east side of the Jordan.

[3] This region was called Gutium, the ancient name for the whole of Assyria.

[4] Placed in mountainous regions of Media, called 'Ara: these Israelites were called "Yasubi-Galileans," who in 705 B.C. rebelled and fought Sennacherib, who succeeded Sargon, possibly in revolt, because in 713 B.C. Sennacherib had captured from Judah 200,150 of their race.

132 THE PILLAR IN THE WILDERNESS

2 Kings xvii, 6 (721–717 B.C.). "*In the ninth year of Hoshea, Shalmaneser, King of Assyria, took Samaria, and carried Israel away into Assyria,*[1] *and placed them in Halah, and in Habor, by the river of Gozan, and in the cities of the Medes.*"

2 Kings xvii, 23. "*So was Israel carried away out of their own land to Assyria.*"

2 Kings xviii, 12. "*Because they obeyed not the voice of the Lord their God, but transgressed His Covenant, and all that Moses the servant of the Lord commanded.*"

2 Kings xvii, 24. "*And the King of Assyria brought men from Babylon . . . and placed them in the cities of Samaria instead of the children of Israel, and they possessed Samaria.*"

2 Kings xviii, 13 (713 B.C.). "*In the fourteenth year of King Hezekiah did Sennacherib come up against all the fenced cities of Judah, and took them.*"

But his attempt against Jerusalem failed by direct Divine intervention, because the Divine Time record for Jerusalem was not to open until 604 B.C.—"*the time of the Gentiles*" of Christ's prophecy in Luke xxi, 24, fulfilled by a dating 2520 years later (A.D. 1917).

2 Kings xix, 35. "*And it came to pass in that night that the angel of the Lord went out, and smote in the camp of the Assyrians an hundred and fourscore and five thousand and . . . in the morning, behold, they were all dead corpses.*"

Thus Israel [2] passed into captivity into the mountainous region ("'Ara") south of the Caspian Sea, into which flows the River Gozan, called Gutium (or land of Guta), the ancient name for the whole of Assyria. Here their labour was no doubt utilised by the Medes to build their capital, Ecbatana (=Khanikan, 709 B.C.).

---

[1] These were called Beth-Kumri ("*children of Omri,*" Omri being the King of Israel who built Samaria as his capital).

[2] The separation of Israel and the Jews is completed—(Zech. xi, 14): "*Then I cut asunder mine other staff, even Bands, that I might break the brotherhood between Judah and Israel.*"

After the release of Babylon, the Jews to Israel (Ezek. xi, 15): "*All the House of Israel . . . unto whom the inhabitants of Jerusalem have said, Get you far from the Lord; unto us is this land given in possession.*" (And to-day the Jews still claim Palestine—when the birthright is Ephraim's.)

THE CAPTIVITIES

710 B.C. Sargon put down a revolt of the Medes.

691 B.C. Sennacherib sacked Babylon, which had revolted.

674–670 B.C. Esarhaddon conquers Egypt.

660 B.C. Egypt had revolted and gained independence.

(*circa*) 650–610 B.C. The Scythians (Turanians, hordes north of the Caspian), having attacked the Kimmerians round the Black Sea, pursued the more southerly of them over the Caucasus into Asia Minor, where, missing their prey, they swung eastwards and ravaged Assyria and contiguous lands, even as far south as Egypt, for twenty-eight years.

625 B.C. The Medes and Babylon unite in revolt for independence against Assyria, aided by the Scythians.

612 B.C. Pharaoh Necho, of Egypt, desiring some of the spoils of Assyria, asks permission for passage of Josiah, King of Judah, who opposes in spite of Jeremiah's advice, and falls mortally wounded in the battle "*in the valley of Megiddo.*"[1] The Pharaoh is opposed by Nebuchadnezzar, who defeats the Egyptians at Carchemish, on the Euphrates, 608 B.C.

The power of Nebuchadnezzar, of Babylon, is now in the ascendant.

606 B.C. After a three years' siege, Nineveh falls, and the Assyrian Empire splits up into Babylon, Media, Egypt, Cilicia, Lydia.

Nebuchadnezzar, fearing his former allies, the Medes, fortifies Babylon, and, turning on the Scythians, defeats them at Askelon, who retire over the Caucasus. He is now free to open his campaign against Egypt and Judah.

The Captivity of Judah to Babylon:—

604 B.C.[2] (First). "*Against Jehoiakim came up Nebuchadnezzar, and bound him in fetters, to carry him to Babylon . . . also carried off the vessels of the house of the Lord . . . and put them in his temple at Babylon*" (2 Chron. xxxvi, 6, 7; 2 Kings xxiv, 1).

---

[1] 2 Chron. xxxv, 20–27.
[2] From this, dates the seventy years' *desolation* of Jer. xxv, 11, to 534 B.C.—"*first year of Cyrus the Persian*," with the fall of Babylon—ended by decree of Cyrus (2 Chron. xxxvi, 22, 23). Daniel is taken captive.

595 B.C.[1] (Second). "*The King of Babylon took Jehoiachin . . . and he carried thence all the treasures of the house of the Lord . . . and he carried away all Jerusalem, and all the princes and all the mighty men of valour, even ten thousand captives and all the craftsmen and smiths, none remained, save the poorest sort of people of the land . . . and made Mattaniah, his father's brother, king in his stead, and changed his name to Zedekiah*" (2 Kings xxiv, 10–17; 2 Chron. xxxvi, 10).

584 B.C.[2] (Third and Final). "*Zedekiah rebelled against the King of Babylon . . . and Nebuchadnezzar . . . came against Jerusalem . . . and they slew the sons of Zedekiah before his eyes and put out the eyes of Zedekiah . . . and carried him to Babylon. But the captain of the guard left of the poor of the land to be vinedressers and husbandmen.*" Then follows in detail the vessels removed from the Temple (2 Kings xxv, 1–17; 2 Chron. xxxvi, 13–20; Jer. xxxix, 1–12). Nebuchadnezzar orders the captain of the guard to remove Jeremiah from prison, "*Take him and look well to him, and do him no harm, but do unto him even as he shall say unto thee*" (Jer. xxxix, 12).

Jer. xl–xlv is the subsequent story of Jeremiah, Simon Baruch his scribe, and the two "King's daughters," with a remnant, abducted by Ishmael, rescued by Johanan,

---

[1] Ezekiel is taken captive. 590 B.C.—Ezekiel viii, 1; ix, 3—the departure of the Shekinah glory from the Temple. From this, dates the seventy years' indignation of Zech. i, 12—to 520 B.C., the founding of the Temple in the second year of Darius.

[2] After a siege lasting eighteen months, the city and Temple are destroyed. Jeremiah is saved.

From Basil Stewart, *The Times of the Gentiles*: "The decree of Cyrus to rebuild the Temple at Jerusalem was lost and forgotten until found in Ecbatana in the palace of Darius the Mede (Darius Hystaspes), who confirmed by a new decree, and the building of the Temple completed in the sixth year of his reign (515 B.C.). But with opposition from the people of Judea, the city was not rebuilt until 444 B.C. by the decree of Artaxerxes Longimanus, and completed in the thirty-second year of his reign, 432 B.C. (Nehemiah returns to Jerusalem, Neh. xiii, 6)."

So Daniel's prophecy of 69 weeks (=483 years lunar =470 years solar) dates from 444 B.C.–A.D. 27—"*First year of Christ's Ministry*" (Mark i, 14–15; Luke iii, 23).

In the seventieth week "*the Messiah shall be cut off*" (Dan. ix, 26)—*i.e.* A.D. 30, His death on the Cross.

A.D. 34 (end of the week), Acts xiii, 44–47. Then Paul and Barnabas said to the Jews, "*Lo, we turn to the Gentiles.*"

THE TIMES OF THE GENTILES BEGIN

taken to Egypt against Divine command, and as "*a small number that escape . . . return out of the land of Egypt into the land of Judah*" (Jer. xliv, 28).

¶ 12. A wonderful picture of these times is given by Madame Tabouis' *Nebuchadnezzar*.

When reigned polytheism and superstition and debauchery, wealth and power and war, intrigue and cruelty of man to man. Perhaps in no other age has the contrast in man's condition been greater from absolute power to serfdom: in which nothing can be seen of His Divine Purpose—Israel exiled, Judah in captivity.

"*Babylon*[1] *hath been a golden cup . . . the nations have drunken of her wine, therefore the nations are mad.*"

Indeed the wine of Cain-Bel's distilling had filled the minds of men, henceforth religious and philosophic ideas of life will be soaked in the magnificence of Babylon, its perversion of His-Story, its mythology of gods later adopted by Greece and Rome, in fact totally submerging the true His-Story carried by the House of Seth, even in this our Britain.

And to this civilisation of the eldest son of Adam (His Israel, having failed, through the Seven Times' punishment were to seek their God) the Divine edict is given in 603 B.C., the rule of the nations for 2520 years, "*The Times of the Gentiles.*"

Dan. iv, 16. "*Let his heart be changed from man's, and let a beast's heart be given unto him, and let seven times . . .*"

Dan. iv, 25. "*Seven times shall pass over thee till thou know that the Most High ruleth in the kingdom of men, and giveth it to whomsoever He will.*"

Dan. vii, 3. "*And four great beasts came up from the sea*"—*Lion*=Babylon; *Bear*=Persia; *Leopard*=Greece.

Dan. vii, 7. "*A fourth beast, dreadful and terrible, . . . and it had great iron teeth . . . and ten horns*"=Rome.

Rev. xiii, 1. "*I . . . saw a beast rise up out of the sea,*

---

[1] *Ency. Brit.*, "Marduk": "The history of the City of Babylon can now be traced back to the days of Sargon of Agade (before 3000 B.C.), who appears to have given the City its name."
Ibid., "Babylon": Sargon of Agade (3800 B.C.), who is stated to have built sanctuaries there to Anu and Ea.

having seven heads and ten horns, and upon his horns, ten crowns, and upon his heads the name of blasphemy."

Rev. xiii, 2. "*And the beast was like unto a leopard, and his feet were as the feet of a bear, and his mouth as the mouth of a lion: and the dragon* [Bel] *gave him his power, and his seat and great authority . . . and all the world wondered after the beast. And they worshipped Bel who gave power unto the beast, and they worshipped the beast, saying, who is like unto the beast? who is able to make war with him? And there was given unto him a mouth speaking great things and blasphemies, . . . and power was given unto him to continue forty and two months.*[1] *And it was given unto him to make war with the Saints and to overcome them. . . . And all that dwell upon the earth shall worship him whose names are not written in the book of life of the Lamb slain from the foundation of the world.*"

Rev. xiii, 9, 10. "*If any man have an ear, let him hear. He that leadeth into captivity shall go into captivity, he that killed with the sword must be killed by the sword.*

"*Here is the patience and the faith of the Saints*" (= His Israel).

Rev. xiii, 11. "*And I beheld another beast coming up out of the earth, and he had two horns like a lamb, and he spake as a dragon. And he exerciseth all the power of the first beast before him, and causeth the earth and them which dwell therein to worship the first beast.*"

Seven times (= 2520 years) 603 B.C. = A.D. 1918.

And the characteristics of the rulers of these empires are:

(1) Idolatry (the image of gold).
(2) Persecution of God's people (as the three faithful into the "*burning fiery furnace*").

The heart of a beast typifies self-exaltation, and moral degradation" (Basil Stewart, *The Times of the Gentiles*).

---

[1] Forty-two months = $3\frac{1}{2}$ times = $3\frac{1}{2} \times 360$ = 1260 years from 603 B.C.
A.D. 652, Penda, last heathen king in Britain, slain, and England reckoned Christian.
A.D. 657, Fall of the Roman Empire; subsequent power taken by the two feet, Western and Eastern, Roman Catholic and Greek Orthodox Churches.

THE TIMES OF THE GENTILES BEGIN 137

That Nebuchadnezzar, as representative of his nation ("*Thou art this head of gold*") and that of the symbolic tree, when cut down, being bound with a band of brass and iron (Dan. iv, 33), the metals significant (in the fourfold image) of the Greek and Roman Empires, which did for ages hold sway over the prostrate region of Babylon; all these considerations induce me to believe that the seven times (7 × 360) that passed over Nebuchadnezzar in his madness represent the 2520 years of the *times of the Gentiles* (Rev. E. B. Elliot, *Horæ Apocalypticæ*).

So to Nebuchadnezzar, as the first of its kings, "*the God of Heaven hath given thee a kingdom, power, and strength and glory. . . . Thou are this head of gold*" (Dan. ii, 37, 38).

And Nebuchadnezzar, puffed with pride with the defeat of Judah, insisting that his gods, Bel and Marduk, had overruled Jehovah, erected "*an image of gold, whose height was threescore cubits, and the breadth thereof six cubits*" (Dan. iii, 1), 590 B.C.

This edict replaced the Divine economic system—"produce goods, give service," gold as money a mere matter of exchange;—and on it has been built the man-made edifice of money, insurance, bills of exchange, interest and dividends, banks, the "gold standard."[1]

As man evolved with Body-Mind from protoplasm, and as protoplasm formed from the affinity of the carbon atom (element No. 6) to string itself in chains (pp. 6, 13, 327)—so Nebuchadnezzar's image, 60 × 6, shows the acme of man's Body-Mind Stature, and

666—the completeness of the string of carbon atoms that formed protoplasm, and evolved life and man—"*is the measure of a man*"—his earthly ambition and desire; "*the imagination of the thoughts of his heart*" can be summed up from history as three:

(1) Power (as that that was Lenin's—"*for it is the number of a man, and his number, and his number is six hundred threescore and six.*"—Rev. xiii, 18).

---

[1] Nebuchadnezzar's image set up 590 B.C., +2520 years, = A.D. 1931. In 1933 the World Economic Conference sat in London, of sixty nations to bolster up the gold standard, and failed (see pp. 315, 338).

(2) Wealth (as that that was Solomon's—"*now the weight of gold that came to Solomon in one year was six hundred threescore and six talents of gold.*"—1 Kings x, 14; 2 Chron. ix, 13).

(3) Woman (disobeying Seventh Commandment—Lev. xx, 14—with Matt. v, 28—"*Thou shalt not look on a woman to lust after her*"—and Gen. ii, 18, "*I will make him an help, meet for him*").[1]

(How many women superstitiously have 6 as their lucky number, that it will bring to them their desire of the whole three? How many the wreck of marriages on (*a*) his pursuit of the three; (*b*) her putting of her body, its needs and its trappings before the work that he feels to be his own job?)

"*For what shall it profit a man if he gain the whole world*" (power, wealth, women) "*and lose his own soul?*"

Thus have been the Times of the Gentiles, the succession of Kingdoms being given in the image of Nebuchadnezzar's dream, Dan. ii, 27–45, and in the later visions amplified (Dan. vii, viii, x):

"*Thou, O King, sawest a great image . . . the form thereof was terrible. This image's head was of fine gold, his breast and his arms of silver, his belly and his thighs of brass, his legs of iron, his feet part of iron and part of clay. Thou sawest; till a stone cut out without hands, which smote the image upon his feet that were of iron and clay, and brake them to pieces.*"

These have been interpreted as four succeeding Kingdoms (later visions verify as Babylon, Medo-Persia, Greece, and Rome with its two divisions, Western and Eastern, "*they shall mingle themselves with the seed of men, but they shall not cleave one to another, even as iron is not mixed with clay,*" ii. 43).

"*And in the days of these kings shall the God of Heaven set up a kingdom which shall never be destroyed: and the kingdom shall not be left to other people, but it shall break in pieces and consume all those kingdoms, and it shall stand for*

---

[1] A help, worthy of his spirit: how often are the two words run together as "help-mate," suggestive of the domestic, who merely looks after his bodily needs.

ever. Forasmuch as thou sawest that stone was cut out of the mountain without hands, and that it brake in pieces the iron, the brass, the clay, the silver, and the gold; the Great God hath made known to the king what shall come to pass hereafter; and the dream is certain, and the interpretation thereof, sure."

Emphasis is laid on a *stone* Kingdom [1] "*cut out without hands,*" set up during the times of these four Kingdoms, maintained by one people (not left to other people), that will "*break in pieces all those kingdoms,*" and "*which shall never be destroyed, . . . it shall stand for ever.*"

And Christ:

"*Therefore say I unto you* [the Jews], *The Kingdom of God shall be taken from you and given to a **nation** bringing forth the fruits thereof. And whosoever shall fall on this stone shall be broken; but on whomsoever it shall fall, it will grind him to powder*" (Matt. xxi, 43, 44).

Now seek it.

[1] The usual interpretation of the "*Stone Kingdom*" is the Christian Church. But this was formed centuries later, not during the times of these kings; it is universal to include all peoples, and was not meant to break kingdoms, but to convert them: also, there is the simile of four kingdoms as actual nations, therefore so must be the fifth.

In Britain, Druidism accepted Christianity, from the first coming of Joseph of Arimathea in A.D. 35, and in A.D. 155 King Lucius established Christianity as the national religion of the whole country in place of Druidism.

And this island of Britain, with its rocky hill ranges, was literally "*cut out*" from the continent of Europe "*without hands,*" by the eroding of the sea into the marshlands, which are now the Straits of Dover.

CHAPTER V

¶ 1. WHAT say the prophets?

Israel survives; they all start by cursing her, then plead with her, then bless her. Even before the Israelite entry into Canaan, the first, Balaam, was sent for by the king of Moab from distant Pethor on the Euphrates, to curse, and he came to bless.[1]

"*Lo, the people shall dwell alone, and shall not be reckoned among the nations*" (Num. xxiii, 9).

"*He hath as it were the strength of an unicorn. . . . Behold, the people shall rise up as a great lion, and lift himself up as a young lion*" (22).

"*His seed shall be in many waters . . . his kingdom shall be exalted*" (xxiv, 7).

"*He hath as it were the strength of an unicorn, he shall eat up the nations his enemies. . . .* [8]. *He couched, he lay down as a lion and as a great lion*" (9).

"*There shall come a Star out of Jacob, and a Sceptre shall rise out of Israel* [17]. *Israel shall do valiantly* [18]. *Out of Jacob shall come He that shall have dominion, and shall destroy him that remaineth of the city*" (19).

"*I will advertise thee what this people shall do to thy people in the latter days*" (xxiv, 14).

"*God is not a man that He should lie, neither the son of man that He should repent, hath He said, and shall He not do it? or hath He spoken, and shall He not make it good?*" (xxiii, 19).

Concerning Babylon:

"*Babylon, the glory of kingdoms, the beauty of the Chaldees' excellency . . . her time is near to come, and her days shall not be prolonged*" (Isa. xiii, 19–22).

[1] See p. 313.

ISRAEL IN PROPHECY

"*I was wroth with My people . . . and given them into thine hand : thou didst shew them no mercy*" (Isa. xlvii, 6).

"*Is this the man* [1] *who made the earth to tremble, that did shake kingdoms : that made the world a wilderness . . . that opened not the houses of his prisoners ? all the kings of the nations, even all of them, lie in glory, every one in his own house*" (xiv, 16–18).

"*For I will rise up against them, saith the Lord of Hosts, and cut off from Babylon the name and remnant*" (22).

"*I will break the Assyrian in my land . . . then shall his yoke depart from off them, and his burden depart from off their shoulders*" (25).

"*This is the purpose which is purposed upon the whole nation : and this is the hand stretched out upon all the nations. For the Lord of Hosts hath purposed, and who shall disannul ? and His hand is stretched out, and who shall turn it back ?*" (Isa. xiv, 26–27).

"*Thus saith the Lord, your redeemer, the Holy One of Israel ; For your sake I have sent to Babylon. . . . I am the Lord, your Holy One, the Creator of Israel, your King*" (xliii, 14–15).

"*Thus saith the Lord of Israel, and his redeemer, the Lord of Hosts. . . . I am the first and I am the last, and beside Me there is no God. . . . I appointed the ancient people, and the things that are coming*" (Isa. xliv, 6–7).

Then concerning Israel:

"*The remnant shall return, even the remnant of Jacob, unto the Mighty God. For though thy people Israel be as the sand of the sea, yet a remnant of them shall return*" (Isa. x, 21, 22).

"*And it shall come to pass in that day that the Lord shall set His hand **again the second time** to recover the remnant of His people which shall be left from Assyria . . . and **from the islands of the sea**" (Isa. xi, 11).

"*He shall assemble the outcasts of Israel and gather together the dispersed of Judah **from the four corners of the earth**" (Isa. xi, 12).

"*For the Lord shall have mercy on Jacob, and will yet*

---

[1] Referring to Lucifer (v. 12): meaning Cain-Bel, that founded the Babylon system among men.

choose Israel, and set them in their own land: and the strangers shall be joined with them and they shall cleave to the House of Jacob" (xiv, 1).

"*My servant whom I uphold, Mine elect in whom My soul delighteth, I have put My spirit upon him, he shall bring forth judgment to the Gentiles*" (xlii, 1).

"*Yet now hear, O Jacob My servant, and Israel, whom I have chosen: thus saith the Lord who made thee . . . I will pour My spirit upon thy seed, and My blessing upon thy offspring*" (xliv, 1-3).

"*Thou art My servant, I have formed thee; O Israel, thou shalt not be forgotten of Me*" (21).

"*I have blotted out as a thick cloud thy transgressions, and as a cloud, thy sins; return unto Me, for I have redeemed thee. . . .* [22]. *The Lord hath redeemed Jacob and glorified Himself in Israel*" (23).

"*But Israel shall be saved in the Lord with an everlasting salvation*" (xlv, 17).

"*I have chosen thee in the furnace of affliction*" (xlviii, 10).

"*I will bring forth a seed out of Jacob, and out of Judah an inheritor of My mountain; and Mine elect shall inherit it, and My servant shall dwell therein*" (lxv, 9).

"*Shall the earth be made to bring forth in one day? or shall a nation be born at once?*" (lxvi, 8).

"*I will set a sign among them and I will send those that escape of them, unto the nations, to Tarshish . . . to the isles afar off that have not heard My fame, neither have seen My glory, and shall declare My glory among the Gentiles*" (19).

"*O Lord, return for thy servants' sake, the tribes of thine inheritance. . . . We are thine*" (Isa. lxiii, 17-19).

"*O Lord, thou art our father, we are the clay, and thou our potter, and we all are the work of thy hand. Be not wroth very sore, O Lord, neither remember iniquity for ever; behold, see, we beseech thee, we are all thy people*" (Isa. lxiv, 8-9).

"*Behold we come unto thee, for thou art the Lord, our God, . . . we and our fathers, from our youth even unto this day, have not obeyed the voice of the Lord our God*" (Jer. iii, 22, 25).

ISRAEL IN PROPHECY

¶ 2. What of the Isles? [1]

"*Wherefore glorify ye the Lord . . . God of Israel, in the Isles of the Sea*" (Isa. xxiv, 15).

"*O islands*" (xli, 1).

"*And the isles shall wait for His law, ye that go down to the sea, and all that is therein ; the isles and the inhabitants thereof*" (xlii, 4, 10).

"*I the Lord have called thee in righteousness, and will hold thine hand and will keep thee, and give thee for a covenant of the people, for a light of the Gentiles ; to open the blind eyes, to bring out the prisoners from the prison, and them that sit in darkness out of the prison house* [2] [xlii, 6–7]. *I am the Lord . . . and My glory will I not give to another*" (8).

"*Behold the former things are come to pass, and new things do I declare: before they spring forth I tell you of them*" (9).

"*Listen, O isles, unto Me, and hearken, ye people from afar*" (xlix, 1).

"*Thou art My servant, O Israel, in whom I will be glorified,* [3] *to raise up the tribes of Jacob and to restore the preserved of Israel ; I will also give thee for a light to the Gentiles, that thou mayest be My salvation unto the end of the earth*" (6).

"*I will preserve thee, and give thee for a covenant of the people to establish the earth, to cause to inherit the desolate heritages*" [3] (8).

"*That thou mayest say to the prisoners, Go forth, to them that are in darkness, Shew yourselves*" (9).

"*These shall come from afar . . . from the north and from the west*" (12).

"*Can a woman forget her sucking child . . . yea. . . . Yet will I not forget thee. Behold, I have graven thee upon the palms of My Hand, thy walls are continually before Me*" (15–16).

"*The children which thou shalt have, after thou hast lost*

---

[1] Why isles?—none near Canaan and Babylon, except of Greece; therefore these British Isles: and watch its mission to the earth breaking forth in the words that follow.

[2] The crusade against the slave trade and the prison reform in the eighteenth century.

[3] Canada, Australia, etc.

*the other, shall say **again** in thine ears, The place is too strait for me, give place to me that I may dwell"* [1] (20).

"The isles shall wait upon me, and on Mine arm shall they trust" (li, 5).

"For thou shalt break forth on the right hand and on the left: and thy seed shall inherit the Gentiles, and make the desolate cities to be inhabited" (liv, 3).

"In righteousness shalt thou be established; thou shalt be far from oppression; for thou shalt not fear; and from terror, it shall not come near thee [14]. *Behold, they shall surely gather together, but not by Me; whosoever shall gather together against thee shall fall for thy sake"* (15).

"No weapon that is formed against thee shall prosper" [2] (17).

"If thou turn away thy foot from the Sabbath, from doing thy pleasure on My holy day . . . I will cause thee to ride upon the high places of the earth" [3] (lviii, 13, 14).

"Surely the isles shall wait for Me, and the ships of Tarshish first, to bring thy sons from far, their silver and their gold with them . . . to the Holy One of Israel, because He hath glorified thee" (lx, 9).

"Therefore thy gates shall be open continually, they shall not be shut day nor night, that men may bring unto thee the wealth of the Gentiles" [4] (11).

"For the nation and kingdom that will not serve thee shall perish: yea, those nations shall be utterly wasted" (12).

"A little one shall become a thousand, and a small one a strong nation, I the Lord will hasten it in His Time" [5] (22).

"And they shall build up the old wastes, they shall raise up the former desolations, and they shall repair the waste cities, the desolations of many generations" (lxi, 4).

"But ye shall be named the Priests of the Lord, men shall call you the Ministers of our God; ye shall eat the riches of the Gentiles" (6).

"And their seed shall be known among the Gentiles, and their

---

[1] Hence, the colonies: as first said to Joshua (Josh. xvii, 14, 16).
[2] The Armada, the submarine, the Zeppelin.
[3] So it was in the Victorian era.
[4] Our trade before "the tribulation" beginning 1914.
[5] See Davidson on "The Rhomboid of Displacement," 1558–1844, pp. 302, 309.

offspring among the people; all that see them shall acknowledge them, that they are the seed which the Lord hath blessed" (9).

"And thou shalt be called by a new name, which the mouth of the Lord shall name" [1] (lxii, 2).

"And Mine elect shall long enjoy the work of their hands. ... And it shall come to pass, that before they call, I will answer; and while they are yet speaking, I will hear" (lxv, 21–24).

¶ 3. Jeremiah to Babylon:

"I will punish the King of Babylon, and that nation, saith the Lord, for their iniquity, and the land of the Chaldeans, and will make it perpetual desolation" (Jer. xxv, 12).

"Take the wine cup of this fury at My hand, and cause all the nations, to whom I send thee, to drink it. And they shall drink, and be moved, and be mad" [2] (15, 16).

"Thus saith the Lord, Behold I will raise up against Babylon ... a destroying wind. For Israel hath not been forsaken, nor Judah of his God, of the Lord of Hosts, though their land was full of sin against the Holy One of Israel" (li, 1, 5).

"Flee out of the midst of Babylon and deliver every man his soul. Babylon hath been a golden cup that made all the earth drunken, the nations have drunken of her wine, therefore the nations are mad" (6–7).

"Forsake her, and let us go every one into his own country" (9).

To Israel:

"I will scatter them among the heathen, whom neither they nor their fathers have known; and I will send a sword after them till I have consumed them" (Jer. ix, 16).

"They have filled Mine inheritance with the carcases of their detestable and abominable things" (xvi, 18).

"I will scatter them as with an east wind before the enemy" (therefore to the west) (xviii, 17).

---

[1] Brith = covenant, ish = man. Saxons = Sak-seni (sons of Isaac). *Cf.* Deut. xxviii.

[2] For 2520 years, the Times of the Gentiles.

"O House of Israel, cannot I do with you as this potter? saith the Lord. Behold, as the clay is in the potter's hand, so are ye in Mine hand, O House of Israel" (xviii, 6).

"The Lord said unto me, The backsliding Israel hath justified herself more than treacherous Judah" [1] (iii, 11, 12).

"Only acknowledge thine iniquity [13]. *Turn, O backsliding children, saith the Lord, for I am married unto you"* (14).

*"Lo, the days come, that I will bring **again** the captivity of My people, Israel and Judah* [2] ; *and I will cause them to **return to the land that I gave to their fathers, and they shall possess it***"* (Jer. xxx, 3–11).

"Fear thou not, O My servant Jacob, neither be dismayed, O Israel: for lo, I will save thee from afar, and thy seed from the land of their captivity" (10).

"For I am with thee, saith the Lord, to save thee: though I make a full end of all nations whither I have scattered thee; yet I will not make a full end of thee; but I will correct thee in measure, and will not leave thee altogether unpunished" (11).

"Saith the Lord, I will be the God of all the families of Israel, and they shall be My people" (xxxi, 1).

"The people which were left of the sword found grace in the wilderness, even Israel, when I went to cause him to rest . . ." (2).

"Yea I have loved thee with an everlasting love" (3).

"Again I will build thee" (4).

*"Behold, I will bring them **from the north country** and gather them **from the coasts of the earth** . . . great company shall return thither"* (8).

". . . for I am a father to Israel, and Ephraim is My firstborn" (9).

*"Hear the word of the Lord, O ye nations, and declare it **in the isles afar off** and say, He that scattered Israel will gather him, and keep him as a shepherd doth his flock"* (10).

"For the Lord hath redeemed Jacob" (11).

---

[1] Yet Israel "lost," and Judah survives as the Jews! *Vide* Matt. xxi, 43.

[2] If Judah as the Jews returned to Palestine, then where the Israel associated?

"*They shall come again from the land of the enemy*" (16).

"*I have surely heard Ephraim bemoaning himself thus, Thou hast chastised me, and I was chastised, as a bullock unaccustomed to the yoke*" (18).

"*Is Ephraim My dear Son? . . . for since I spake against him, I do earnestly remember him still. . . . I will surely have mercy upon him, saith the Lord*" (20).

"*Set up thy waymarks, make thee high heaps . . . even the way which thou wentest: turn again, O virgin of Israel, turn again to these thy cities*" (21).

"*Saith the Lord, I will sow the House of Israel and the House of Judah with the seed of man*" (27).

"*I will gather them out of all countries, whither I have driven them in Mine anger, in My fury, and in My great wrath, and I will bring them again unto this place, and I will cause them to dwell safely, and they shall be My people, and I will be their God*" (xxxii, 37–38).

"*And I will make an everlasting covenant with them*" (40).

"*Yea, I will rejoice over them . . . and I will plant them in this land. . . .*" (41).

"*Like as I have brought all this great evil upon this people, so will I bring upon them all the good that I have promised them*" (42).

"*For I will cause their captivity to return, saith the Lord*" (44).

"*Israel is the rod of His inheritance, the Lord of Hosts is His name. Thou art My battle-axe, and weapons of war: for with thee will I break in pieces the nations, and with thee will I destroy kingdoms*" (li, 19, 20).

¶ 4. Hosea.

"*Call his name Lo-ammi* [rejected]; *for ye are not My people, and I will not be your God. Yet the number of the children of Israel shall be as the sand of the sea, . . . and it shall come to pass that in the place where it was said unto them, Ye are not My people, there it shall be said unto them, Ye are the sons of the living God*" (Hosea i, 9, 10).

148 THE PILLAR IN THE WILDERNESS

"I will sow her unto Me in the earth, and I will have mercy upon her that had not obtained mercy, and I will say to them which were not my people, Thou art my people, and they shall say, Thou art my God" (ii, 23).[1]

"My people are destroyed for lack of knowledge; because thou hast rejected knowledge, I will also reject thee, that thou shalt be no priest to Me" (iv, 6).

"Ephraim is joined to idols" (iv, 17).

"I know Ephraim and Israel is not hid from Me" (v. 3).

"They shall go with their flocks and their herds to seek the Lord, but they shall not find Him, for He hath withdrawn Himself from them" (v. 6).

"For I will be unto Ephraim as a lion, and as a young lion to the House of Judah" (v. 14).

"After two days will He revive us, in the third day He will raise us up, and we shall live in His sight" (vi, 2).[2]

"They shall be wanderers among the nations" (ix, 17).

"I found Israel like grapes in the wilderness" (10).

"How shall I give thee up, Ephraim? how shall I deliver thee? . . . Mine heart is turned within Me" (xi, 8).

"O Israel, return unto the Lord thy God" (xiv, 1).

"I will heal their backsliding. I will love them freely, for Mine anger is turned away from him" (4).

"Ephraim shall say, What have I to do any more with idols?" (8).

Amos.

"O children of Israel . . . you only have I known of all the families of the earth" (Amos iii, 2).

"I will sift the House of Israel among all nations, like as

---

[1] "*. . . to the strangers scattered . . . elect according to the foreknowledge of God the Father who are kept by the power of God through faith unto salvation ready to be revealed in the last time . . .*" (1 Peter i, 1, 5).

"*Ye are a chosen generation, a royal priesthood, an holy nation, a peculiar people . . . which in time past were not a people, but are now the people of God, which had not obtained mercy, but now have obtained mercy*" (ii, 9, 10).

[2] A day as 1000 years: 717 B.C. + 2000 = A.D. 1284.

"The greatest achievement of the careful researches of the late Bishop of Oxford was the conclusive establishment of the fact that by the close of the thirteenth century the consciousness of any difference in ancestry had entirely disappeared. . . . Surely the fact that Englishmen became conscious of their common blood is a striking and important fact" (Professor R. G. Usher (Washington Univ.), *Woman's Magazine*, February 1915, "The Secret of England's Predominance in Europe").

corn is sifted in a sieve, yet shall not the least grain fall upon the earth" (ix, 9).

"*I will bring again the captivity of My people of Israel and they shall build the waste cities and inhabit them*" (14).

"*And I will plant them upon their land, and they shall no more be pulled up out of their land which I have given them, saith the Lord*" (15).

Micah.

"*I will surely assemble, O Jacob, all of thee; I will surely gather the remnant of Israel, I will put them together as the sheep . . . as the flock in the midst of their fold*" (ii, 12).

"*I will make her that halted a remnant, and her that was cast far off a strong nation*" (iv, 7)

"*Thou shalt beat in pieces many people*" (13).

"*And the remnant of Jacob shall be in the midst of many people as a dew from the Lord*" (v. 7).

"*And the remnant from Jacob shall be among the Gentiles in the midst of many people as a lion among the beasts of the forest, as a young lion among the flocks of sheep. Who if he go through, both treadeth down and teareth in pieces, and none can deliver*" (8).

Zephaniah.

"*I will save her that halteth, and gather her that was driven out. At that time I will bring you again, even in the time that I gather you . . . when I turn back your captivity before your eyes, saith the Lord*" (iii, 19–20).

Zechariah.

"*I will save My people from the east country and from the west country. And I will bring them and they shall dwell in the midst of Jerusalem, and they shall be My people and I will be their God in truth and in righteousness*" (viii, 7–8).

"*I will save the House of Joseph and I will bring them again to place them, for I have mercy upon them and they shall be as though I had not cast them off, for I am the Lord God, and will hear them*" (x, 6).

"*And they of Ephraim shall be like a mighty man*" (7).

"*I will gather them, for I have redeemed them: and they shall increase as they have increased*" (8).

"*And I will sow them among the people, and they shall remember Me **in far countries**, and they shall live with their children and turn again*" (9).

Malachi.

"*The Word of the Lord to Israel, I have loved you*" (Mal. i, 1).

"*For I am the Lord, I change not; therefore ye sons of Jacob are not consumed*" (iii, 6).

"*Remember ye the law of Moses My servant, which I commanded unto him for all Israel*" (iv. 4).

Habakkuk.

"*For the vision is yet for an appointed time, but at the end it shall speak and not lie*" (Hab. ii, 3).

"*O Lord, revive thy work in the midst of the years, in the midst of the years, make known*" (iii, 2).

¶ 5. Israel survives—what says history? [1]

On Assyrian tablets, the Israelites were called Beth-Khumri,[2] and represented on Persian monuments as the Sakai, and Saccaseni.[3] On an Assyrian tablet, Sennacherib

---

[1] Bibliography:

D. Davidson and H. Aldersmith, *The Great Pyramid: its Divine Message* (Williams & Norgate, Ltd.).

R. W. Morgan, *History of Britain: from the Flood to A.D. 700* (The Marshall Press, Ltd.).

And the following from Covenant Publishing Co., Ltd.:
Major B. de W. Weldon, *The Origin of the English*.
W. M. H. Milner, *The Royal House of Britain, an Enduring Dynasty*, with chart.
Professor E. Odlum, *God's Covenant Man*.
R. W. Morgan, *St Paul in Britain: the Origin of British as opposed to Papal Christianity*.
J. Dunham-Massey, *Tamar Tephi, the Maid of Destiny*.
L. G. A. Roberts, *British History Traced*.
M. H. Gayer, *The Heritage of the Anglo-Saxon Race*.
J. J. Morey, *The Parables of the Kingdom*.
Mrs S. Bristowe, *Sargon the Magnificent*.
Basil Stewart, *The Times of the Gentiles*.
Basil Stewart, *At Midnight a Cry!*
W. Pascoe Goard, *The Empire in Solution*.

[2] Beth-Khumri = children of Omri, the King of Israel who founded Samaria as his capital.

[3] Sak = Isaac; Saccaseni = the children of Isaac. The first mention of the word Sak in Genesis xvii, 19; xxi, 12, "*In Isaac shall thy seed be called*," and not used again for Israel until rejected by God: "*The high places of Isaac*" (Amos vii, 9); "*the House of Isaac*" (vii, 16).

ISRAEL IN HISTORY

records expeditions (*circa* 705 B.C.) against the Yasubi-Galileans in the mountainous regions of Media, called "'Ara," the revolt was led by Teuspa the Kimmerian, his people called Manda by Esarhaddon.

Professor Sayce states Cyrus was a Manda, and the Mandi were Gimirra[1]: "it would seem that the Mandix of Ecbatana were the Scythians of classical history—in the inscriptions of Darius, the Gimirra-Umurgah of the Babylonian text correspond with the Saka-Humuvarka of the Persian text, who are the Amyrgian Sakæ of Herodotus (vii, 64), who, he says, were the Scythians of the Greeks."

Rawlinson (reading of the black obelisk, found in ruins of Nimrod, now in British Museum) that "King Temember invaded Aria 670 B.C. to quell a rebellion, and captured their cities, Beth-Telabon, Beth-Everak, and their leader Esakska (=Isaac)." All these names are Hebrew; and this region called by the Persians Sakia, and their people Sacæ.

In the inscription on the Behistun Rock (found by Layard and Rawlinson, this rock is south of Sakiz, a town on west of the River Gozan) Darius tells that he fought against the Saka.

George Moore says that the word is spelt Tsak (Isaac without the initial yod), and the people living by the Chebar (probably the site of modern Sacho).

Thus:

From Hebrew Tsak (=Isaac) are derived Sac, Dak; Sacæ, Dacians; Saccaseni, Sak's sons, Saxons.
,, ,, Skth (=wanderer) are derived Skuth, Scyth: Scythians; so are mixed with eastern Scythians.
,, ,, Beth Khumri are derived Cumri; so mixed with Cymry, Cimbri, Kimmerians.
,, ,, Gutium, or land of Guta, is derived Getæ.
,, ,, Guta theod (people of Guta) is derived Goth.

Sailman's *Researches in the East* says Ortellius derives the origin of Goth from "Gauthei" because "the ten tribes were jealous for the glory of God."

[1] In 710 B.C. Sargon put down a revolt of the Medes: here Israelites from Samaria as Yasubi-Galileans seem to be aiding the Medes, and an indication of their later alliance with Cyrus.

When the threat to the Assyrian was made by the Medes (north), by Persia (east), and Babylon (south), with Egypt advancing on the south-west (625–608 B.C.), and with Scythians still ravaging, the Assyrians had their hands full and their armies engaged, and so must have withdrawn guards from such an extensive people as Israel; they may even have urged them to depart; the only route free of warring peoples was northwards over the River Araxes, at the time when the Scythians were expelled by Cyaxeres, King of Media (Herod., *Clio*, xv, 103–106).

Herodotus states that the defeated Scythians met on the borders of their country an army of escaped slaves; an indecisive battle on the River Araxes resulted, and it was agreed that these should be allowed to enter the country if they divided. This was done, one portion, migrating west, settled on the Danube as the Getæ, the other main portion, or Massagetæ, settled on the north and east of the Caspian Sea, round entry of the River Oxus.

Their names became mixed as Skyths with these Asiatic Scythians, as Cumri with the Cimbri and Cimmerians round the Euxine, so again it is made difficult to identify them among these races.

"*These are the ten tribes which were carried away prisoners out of their own land . . . and so came they into another land. And they took counsel among themselves, that they would leave the multitude of the heathen, and go forth into another country, where never mankind dwelt. That they might keep their statutes which they never kept in their own land. And they entered into the Euphrates by the narrow passages of the river. For the Most High then showed signs for them, and held still the flood, till they were passed over. For through that country was a great way to go, namely of a year and a half, and the same region is called Arsareth*"[1] (2 Esdras xiii, 40).

The name of one of the passes over the Caucasus is called "the gates of Israel."

When Armenia entered, they named the river Engl,

---

[1] (Hebrew) Ar = city. Thus the word = city on the Sereth, which river, rising in the Carpathians, flows into the Danube.

town Angl, mountain and town Sakh; at the head of the Euphrates are Mel-asgerd, Tav-asgerd, Penyal (Hebrew for "God shows his face"), Barachel ("God's blessing") and stream Israel-su.[1]

Rawlinson says that the European Scythians were not the same race as the Asiatic Scythians; Herodotus describes them as the Getæ, "former colonists of the Medes."

Herodotus states that the Scythians emerged from beyond the Euphrates, from the east across the Araxes (in fact the district of Gozan), flocked into Europe in the seventh century B.C., and later locates them in the north and west of the Black Sea (the region of the Getæ), that they became numerous, and that when Darius came against them they were 1000 years old (this would bring their origin to about 1490 B.C., the Exodus); their great hero was Zalmoxis.[2] He describes them as "the most righteous of nations," who said their God was the only true God. Here, east of the Araxes, they had multiplied and extended their territory for some centuries unknown to Europe, and were generally called Scoloti, but Greeks called them Scuthai or Scythians (Herodotus, *Melp.*, 5, 6, 7, 11).

The Sacæ sprung from a people in Media who obtained a vast and glorious Empire. On Scythians, "amongst others were two remarkable colonies that were drawn out of the conquered nations by those kings, the one they brought out of Assyria . . . the other out of Media"—formerly inconsiderable and few, they possessed a narrow region on the Araxes, but by degrees became powerful in numbers and courage. They extended their boundaries on all sides till at last they raised their nation to a great empire and glory . . . the mountainous regions about Caucasus, and all the plains to the ocean, the Palus Mœtus, with other regions near the Tanais. The Sakai, the Massagetæ . . . drew their origin from them (*Diodorus Siculus*, p. 127).

Sar-angai, Dar-angai—their origin from Beth Sak or Beth-Khumri (Rawlinson, App., Bk. vii, Essay 1).

---

[1] "*Set thee up waymarks.*"
[2] Zal = chief or leader. Moxis = Greek form of Hebrew Moses.

People called Aegli in Bactria close to the Sacæ, their real name Angai [1] (Herodotus, iii, 92).

The Aegli and Angai first met with on the northern route from Lake Baikal.

The Sachi lived near Lake Baikal with the Bretti from whom later separated (Alexander Del Mar).

Pliny: "Sakai who settled in Bactriana, in Armenia, were named Sacassani (Sakasina)."

Strabo calls part of Armenia they occupied Sacasena.

Stephanus: "a people called Saxoi, on the Euxine."

Ptolemy: "a branch of the Sakai was called Saxones."

Ancient work called *Varak* says: "The Sacæ possessed the Crimea."

Ancient maps show that the Tauric Sacæ, or Saxons, owned the Crimea, and also that the Sacæ had scattered out to the Sea of Japan.

An old marble tablet found in the Crimea: "Sargon, King of Assyria, came up against the City of Samaria, and against the tribes of the Beth-Khumri, and carried into captivity, into Assyria, 27,280 families."

An ancient cemetery found near Balaclava, on the tombstones:

(1) "I am Jehudi, son of Moses, son of Jehudah the mighty, a man of the tribe of Naphtali, of the family of Shimli, who was carried captive in the captivity of Hosea, King of Israel, with the tribe of Simeon, together with other tribes of Israel."

(2) "We must inscribe here the wonders which God has done for us; who can recount what has happened unto us all during 1500 years we have lived in this exile."

(3) "To one of the faithful in Israel, Abraham-ben-mar-Sinchah, of Kertch, in the year of our exile, 1682, when the envoys of the prince of Rosh Machech came from Kiow to our master, Chazar Prince David; Halaah, Habor and Gozan, to which places Tiglath-Pileser had exiled the sons of Reuben and Gad and the half tribe of Manasseh, and permitted to settle there and from which they have been scattered throughout the entire east, even as far as China."

---

[1] Englah = Hebrew for heifer or wild ox, unicorn.

(4) "This is the grave of Buki, the son of Isaac, the priest. May his rest be in Eden at the time of the deliverance of Israel."

Among the Scythian migrations into Europe then can be identified the Getæ and the Massagetæ, two branches of our race, as Israel. The Getæ were the tribe of Joseph with the two houses of Ephraim and Manasseh, who settled on the Danube and were later called the Dacians.

Dan had migrated before the Assyrian Captivity. A southerly branch, having coalesced with the Phœnicians and engaged in maritime trade, are found in Greece as the Danai, and in Ireland as the Tuatha de Danaan (1000 B.C.).[1] A northerly branch, having conquered Leshem and named it Dan,[2] had trekked north-west into Asia Minor, settled in Dardanum (Troy), and after the siege, where brother fought brother, separated. One branch crossed the Black Sea, settled near the River Danais or Tanais (=Don) and along the south of the Urals, and here called the Thyssa-Getæ; while the other main body followed up the rivers Danube, Danieper, Danester, Eridanus (Rhine), Rhodanus (Rhône), leaving their waymarks D N [3] over Europe, and settled in Denmark, whence later they entered Britain as the Danes.

The tribe of Levi was distributed among the other tribes; at Mount Sinai it was set apart,[4] consecrated to become the priests, and to them were entrusted the law and records, the charge of the Tabernacle and later the Temple. In Canaan they were given forty-eight cities among the other eleven tribes.

[1] "Connects the Danaos of Greece with Phœnicia, and the Tuatha de Danaans of Ireland came from the Danai of Greece" (W. E. Gladstone, *Juventus Mundi*, p. 136).

[2] "*And the coast of the children of Dan went out too little for them, therefore the children of Dan went up to fight against Leshem, and took it and smote it with the edge of the sword and possessed it and dwelt therein, and called Leshem Dan after the name of their father*" (Joshua xix, 47, also see Judges xviii, 1, 2, 25–29).

Verifying Moses' prophecy, "*Dan is a lion's whelp, he shall leap from Bashan*" (Deut. xxxiii, 22).

[3] "*Dan shall identify his people as one of the tribes of Israel, Dan shall be a serpent's trail*" (Gen. xlix, 16–17).

Eldad, a Jewish authority, says that Dan, rather than go to war with his brethren, Judah, went in a body to Javan (Greece) and Denmark.

[4] Numbers i, 47–53; Joshua xxi.

The Massagetæ, or main body, consisting of the other seven tribes, moved from the land of Guta (Gutium), south of the Caspian Sea, and settled on its north-east coasts round the River Oxus, and here are still in direct contact with their former neighbours, the Medes and Persians in the south.

¶ 6. In this region Hosea sees them:

"*For the children of Israel shall abide many days without a king* [of the Zarah-Pharez line] . . . *without a pillar* ["the guardian stone"] . . . *afterward shall the children of Israel return and seek the Lord their God, and David their King; and shall fear the Lord and His goodness in the latter days*" (Hosea iii, 4, 5).

"*My people are destroyed for lack of knowledge; because thou hast rejected knowledge, I will also reject thee, that thou shalt be no more priest to Me; seeing thou hast forgotten the law of thy God, I will also forget thy children*" (iv, 6).

"*My people ask counsel at their stocks, and their staff declareth unto them. . . . They sacrifice upon the tops of the mountains . . . under oaks*" (iv, 12, 13).

"*He is an unwise son, for he should not stay long in the place of the breaking forth of children. Though he be fruitful among his brethren, an east wind shall come, the wind of the Lord shall come up from the wilderness, and his spring shall become dry*" [1] (xiii, 13, 15).

And Isaiah:

"*The burden of the desert of the sea* [Caspian]. [The watchman's cry] *Babylon is fallen*" [2] (Isa. xxi, 1–10).

"*The burden of the valley of vision*" (xxii).

"*And behold, joy and gladness, slaying oxen and killing sheep, eating flesh and drinking wine: let us eat and drink, for to-morrow we shall die*" [3] (13).

"*For precept must be upon precept . . . line upon line*

---

[1] About 330 B.C. (time of Alexander) the River Oxus became diverted, the level of the Caspian fell, and the whole land became a desert. The Massagetæ migrated westwards, to the Crimea; north even as far as Lake Baikal, among the barbarian Asiatic Scythians.

[2] To the conquest of Babylon, allied with Cyrus.

[3] *Vide* Herodotus' story of Queen Tomyris, when Cyrus by stratagem evacuates his camp (p. 166).

... *here a little, and there a little ; for with stammering lips and another tongue* [1] *will he speak to this people*" (xxviii, 10, 11).

"*The tongue of the stammerers shall be ready to speak plainly*" (xxxii, 4).

Zechariah:

"*I scattered them with a whirlwind among all the nations whom they knew not. Thus the land was desolate after them, that no man passed through nor returned ; for they laid the pleasant land desolate*" (Zech. vii, 14).

And Ezekiel [2] is sent on a mission to them (*circa* 595 B.C.):

"*I have set thee for a sign unto the House of Israel*" (Ezek. iv, 3; xii, 6, 11).

"*I send thee to the children of Israel, to a rebellious nation* [ii, 3], . . . *be not afraid of them, neither be afraid of their words, though briers and thorns be with them and thou dost dwell among scorpions*" (6).

"*Go, get thee unto the House of Israel, and speak with My words unto them*" (iii, 4).

"*Go, get thee to them of the captivity, . . . and tell them, Thus saith the Lord God, whether they will hear, or whether they will forbear*" (11).

"*Then I came to them of the captivity at Tel-abib, that dwelt by the river of Chebar, . . . and remained there astonished among them seven days*" (15).

"*This shall be a sign to the House of Israel*" (iv, 3).

"*And say, ye mountains of Israel, hear the word of the Lord God, . . . I will bring a sword upon you, and I will destroy your high places*" (vi, 2).

"*Yet will I leave a remnant, that ye may have some that shall escape the sword among the nations, when ye shall be scattered through the countries*" (8).

[1] The Median dialect (Sanskrit)—"One primeval language": "The tribes of Simeon, Ephraim and Manasseh settled on the north-east of the Caspian Sea" (Forster).

[2] "The Book of Ezekiel bears the stamp of a single mind, is arranged methodically, and the form we have it is the same as from the prophet himself" (*Ency. Brit.*).

By this date Israel and Judah were separate houses, and so understood (*vide* Ezekiel iv, 4, 6): as Britain and America to-day.

"*And they that escape of you shall remember Me among the nations*" (9).

"*And they shall know that I am the Lord, and that I have not said in vain that I would do this evil unto them*" (10).

"*For they shall fall by the sword, by the famine, and by the pestilence*" (11).

"*Then shall ye know that I am the Lord when their slain men shall be among their idols . . . upon every high hill, in all the tops of the mountains, . . . under every thick oak, the place where they did offer sweet savour to their idols* [13]. *So will I make the land desolate, yea, more desolate than the wilderness*[1] . . . *in all their habitations; and they shall know that I am the Lord*" (14).

"*But they that escape of them shall escape . . . all of them mourning* [vii, 16]. *They shall cast their silver in the streets, and their gold*[2] *shall be removed* [19]. *And I will give it into the hands of strangers for a prey, and to the wicked of the earth for a spoil* [21] . . . *they shall pollute my secret place; for the robbers shall enter into it and defile it*" (22).

"*Wherefore I will bring the worst of the heathen, and they shall possess their houses*"[1] (24).

"*And the hands of the people of the land shall be troubled . . . and they shall know that I am the Lord*" (27).

"*All the House of Israel wholly are they unto whom the inhabitants of Jerusalem have said, Get you far from the Lord, unto us is this land given in possession*"[3] (xi, 15).

"*Thus saith the Lord, Although I have cast them afar off among the heathen, and although I have scattered them among the countries, yet will I be to them as a little sanctuary in the countries where they shall come* [16]. *I will even gather you out of the countries where ye have been scattered, and I will give you the land of Israel* [17]. *And they shall come thither, and they shall take away all the detestable things thereof and all the abominations thereof from thence*" (18).

---

[1] The land round the Oxus became a desolate wilderness, and occupied by heathen Turanian hordes, the ancestors of Attila's Huns.

[2] Where did the captive Israelites acquire wealth? As allies of Cyrus at the sack of Babylon.

[3] Will the conquering Massagetæ at Babylon unite with captive Jews and return to Jerusalem? No. The Jews thus reply, "*Get you far from the Lord, ours is this land.*"

EZEKIEL ON THE CASPIAN

In spite of the Jews' assertion, to Britain as Israel Palestine is restored,[1] and Divine command to them to cleanse Jerusalem of *"detestable things and abominations"* —that is, in the present time, 1936. Then follows His promise in vv. 19, 20, *which is to be fulfilled in 1936–1953*, when v. 17 above has come to pass:

"*And I will give them one heart, and I will put a new spirit within you, and I will take away the stony heart out of their flesh, and will give them a heart of flesh: that they may walk in My statutes and keep Mine ordinances, and do them, and they shall be My people, and I will be their God*" (19, 20).

"*They of the House of Israel say, The vision that he seeth is for many days to come, and he prophesieth of the times that are afar off. Therefore say unto them, Thus saith the Lord God, There shall none of My words be prolonged any more, but the word which I have spoken shall be done, saith the Lord God*" (xii, 27, 28).

"*Then came certain of the elders of Israel unto me and sat before me . . .*" (xiv, 1).

"*Nevertheless I will remember My covenant with thee in the days of thy youth, and I will establish unto thee an everlasting covenant. . . . And thou shalt know that I am the Lord*" (xvi, 60, 62).

"*Cast away from you all your transgressions whereby ye have transgressed; and make you a new heart and a new spirit: for why will ye die, O House of Israel?*" (xviii, 31).

"*Certain of the elders of Israel came to enquire of the Lord, and sat before me*" (xx, 1).

"*Thus saith the Lord God, In the day when I chose Israel, and lifted up Mine hand unto the seed of the House of Jacob . . .*" (5).

"*Yet ye say,*[2] *We will be as the heathen, as the families of the countries, to serve wood and stone*" (xx, 32).

"*As I live, saith the Lord God, surely with a mighty hand . . . will I rule over you* [33]. *And I will bring you out from the people, and will gather you out of the countries*

---

[1] "*And Jerusalem shall be trodden down of the Gentiles until the times of the Gentiles be fulfilled*" (Luke xxi, 24)—*i.e.* A.D. 1917.

[2] The elders of the Massagetæ: *vide* the parable of the prodigal son (p. 241).

wherein ye are scattered, with a mighty hand, and a stretched out arm, and with fury poured out" (34).

"*And I will bring you into the wilderness of the people,*[1] *and there will I plead with you face to face*" (35).

"*Like as I pleaded with your fathers in the wilderness of the land of Egypt, so will I plead with you* [36]. *And I will cause you to pass under the rod and I will bring you into the bond of the Covenant* [37]. *And I will purge out from among you the rebels, and them that transgress against Me, I will bring them forth out of the country where they sojourn, and they shall not enter into the land of Israel*[2]: *and ye shall know that I am the Lord*"[3] (38).

"*As for you, O House of Israel,*[4] . . . *Go ye, serve ye every one his idols, and hereafter also, if ye will not hearken unto me, but pollute ye My Holy Name no more with your gifts, and your idols*" (39).

"*For in Mine holy mountain, in the mountain of the height of Israel,*[5] *there shall all the House of Israel, all of them in the land, serve Me; there will I accept them* . . . [40]. *When I bring you out from the people, and gather you out of the countries wherein ye have been scattered*[6]; *and I will be sanctified in you before the heathen*"[7] (41).

"*And ye shall know that I am the Lord when*[8] *I shall bring you into the land of Israel, into the country for which I lifted up Mine hand to give it to your fathers* [42]. *And there shall ye remember*" (43).

"*And ye shall know that I am the Lord when*[9] *I have wrought with you for My Name's sake, not according to your wicked ways, nor according to your corrupt doings, O ye House of Israel, saith the Lord God*" (44).

[1] The wilderness of the steppes of Southern Russia.
[2] Into Britain—a remnant purged left behind on the route.
[3] Britain proclaimed Christian in A.D. 155 by King Lucius with consent of the Druids. *Triads of the Isle of Britain*: "King Lucius . . . first gave the privileges of the country and nation to all who professed the faith in Christ."
[4] As the Massagetæ.
[5] The British Isles.
[6] As Angles, Saxons, Danes, Normans.
[7] A Christian nation, before the miscellaneous religions of Asia.
[8] December 1917.
[9] Some further and imminent event which He will yet do—then Britain, as Israel, will realise that Jehovah is still the One God.

EZEKIEL

"*Then said I, Ah Lord God! they say of me,*[1] *Doth he not speak parables?*" (49).

"*I will lay my vengeance upon Edom by the hand of My people Israel,*[2] *and they shall do in Edom according to Mine anger and according to My fury*" (Ezek. xxv, 14).

xxvi–xxviii refers to the Phœnicians, and to Israel (especially Dan) who coalesced with them.

xxxiii, 1–7, to the *watchmen* of Israel: "*Turn ye, turn ye from your evil ways; for why will ye die, O House of Israel?*" (11).

"*Ye mountains of Israel,*[3] *hear the word of the Lord: Thus saith the Lord God, Because the enemy hath said against you, Aha, even the ancient high places are ours in possession*[4] . . . *because they have made of you desolate*" (xxxvi, 1).

"*But ye, O mountains of Israel, ye shall shoot forth your branches,*[5] . . . *for they are at hand to come* [8]. *For, behold, I am for you* . . . *ye shall be tilled and sown*[6] [9]. *And I will multiply upon you man and beast, and they shall increase and bring fruit* . . . *and will do better unto you than at your beginnings: and ye shall know that I am the Lord* [10]. *Yea, I will cause men to walk upon you, even My people Israel; and they shall possess thee, and thou shalt be their inheritance,*[7] *and thou shalt no more henceforth bereave them of men*[8] . . . [12]. *Neither will I cause men to hear in thee the shame of the heathen any more*"[9] (15).

"*And I scattered them among the heathen, and they were dispersed through the countries*" (19).

"*When they profaned My holy name, when the heathen said to them, These are the people of the Lord, and are gone forth out of his land*" (20).

[1] Ezekiel, "*a sign to the House of Israel*"—"they," to-day, can be taken as the Church and people of Britain.
[2] With Obadiah: referring to the British campaign against the Turks (as Edom) in Palestine in 1917.
[3] The British Isles; mountain, symbolic of a great nation; hill, of a small nation.
[4] During the Crusades, and in the days of the Moslems.
[5] The Colonies.   [6] Agriculture of Britain and the Empire.
[7] Since 1066, the land possessed by Israel.
[8] 1557, population 6 millions; 1800, 16 millions; 1933, 160 millions.
[9] No idolatry allowed in Britain.

"*Thus saith the Lord God, I do not this for your sakes, O House of Israel, but for Mine holy name's sake* . . . [22]. *And the heathen shall know that I am the Lord, when I shall be sanctified in you before their eyes* [23]. *And I will take you from among the heathen, and gather you out of all countries, and will bring you into your own land* [1] [24]. *Then will I sprinkle clean water upon you and ye shall be clean; from all your filthiness, and from all your idols, will I cleanse you* [25]. *A new heart also will I give you, and a new spirit will I put within you, and I will take away the stony heart out of your flesh, and I will give you an heart of flesh*" (26).

"*And I will put My spirit within you, and cause you to walk in My statutes, and ye shall keep My judgments, and do them* [27]. *And ye shall dwell in the land I gave your fathers,*[2] *and ye shall be My people, and I will be your God*" (28).

"*And I will multiply the fruit of the tree, and the increase of the field, that ye shall receive no more reproach of famine among the heathen*" [3] (30).

"*Not for your sakes do I this, saith the Lord God, be it known unto you*" (32).

"*And they shall say, This land that was desolate is become like the garden of Eden*" [4] (35).

"*Then the heathen* . . . *round about* [5] . . . *shall know that I the Lord build ruined places, and plant that that was desolate; I the Lord have spoken it, and I will do it*" (36).

"*Thus saith the Lord God, I will yet for this be enquired of by the House of Israel, to do it for them.*" [6] (37)

With such a preliminary in the preceding chapter, xxxvii opens with the marvellous vision of the valley of dry bones, to confirm that Israel is not lost, but survives; and the whole of Britain and the world will shortly see and understand it.

---

[1] Britain (completed in 1066): there follows in history the purification of these Isles with the movements of the Reformation, Protestantism, the separation of the British Church from Roman Catholicism.
[2] Palestine.   [3] None in Britain
[4] The accomplishing of it in the land of Canaan is to-day visible.
[5] Arabs, Moslems, Egyptians, Turks, Persians, Afghans, Indians.
[6] A challenge to the Church of England, of Scotland, and of all the Free Churches, a common purpose which should lead to unity.

"The valley was full of bones, and behold, there were many in the open valley, and they were very dry. . . . Man, can these bones live? O ye dry bones, Thus saith the Lord, Behold, I will cause breath to enter into you. There was a noise, and behold a shaking: and the bones came together, bone to his bone. And lo, the sinews and the flesh came up upon them, and the skin covered them from above. . . . Come from the four winds, O breath, and breathe upon these slain, that they may live. And the breath came into them, and they lived and stood upon their feet, an exceeding great army.

"These bones are the whole House of Israel" (11).

"Thus saith the Lord God, Behold, O My people, I will open your graves, and cause you to come up out of your graves, and bring you into the land of Israel [12]. *And I shall put My spirit in you, and ye shall live, and I shall place you in your own land: then shall ye know that I the Lord have spoken it, and performed it, saith the Lord"* (14).

The vision of the union of two sticks:

"Thus saith the Lord God, Behold, I will take the stick of Joseph, which is in the hand of Ephraim, and the tribes of Israel his fellows, and will put them with him, even the stick of Judah,[1] *and make them one stick, and they shall be one in Mine hand"* (19).

"And I will make them one nation in the land upon the mountains of Israel; and one king [2] *shall be king to them all: and they shall be no more two nations, neither shall they be divided into two kingdoms any more at all . . .* [22]. *I will cleanse them, so shall they be My people and I will be their God"* (23).

The wonderful verses 24–28 will be quoted later.[3]

---

[1] The Jews united to Britain (Jer. iii, 18), "*In those days the House of Judah shall walk to the house of Israel, and they shall come together out of the land of the North* [North = Europe] *to the land that I have given for an inheritance unto your fathers* [Palestine]."

"It is my conviction that Britain is the nation with whom God has from first to last identified Himself—I an Israelite of the House of Judah claim you as Israelites of the House of Ephraim. As believers in the faithfulness of our Covenant-keeping God, I call you to wake from your sleep" (Elieser Bassin, C.M., Ph.B., a Jew, born in Russia, of pious and wealthy parentage).

[2] Our Royal House, descended from the Zarah-Pharez line.

[3] See p. 414.

164 THE PILLAR IN THE WILDERNESS

Thus *these bones* are roughly tabulated:

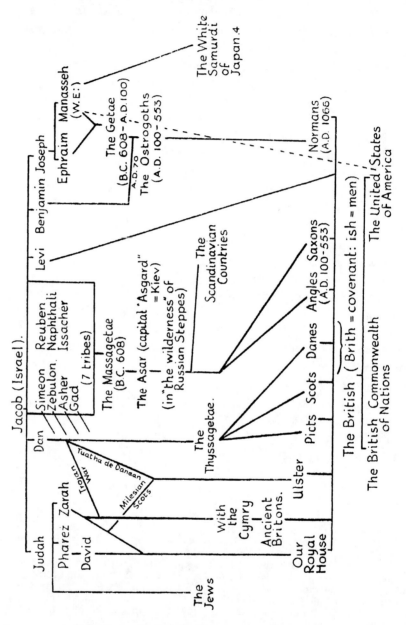

[4] Professor E. Odlum, *Who are the Japanese?*

THE MASSAGETÆ

¶ 7. 553 B.C. Cyrus, of the Persian tribe of the Pasargadæ, King of Anshan in Elam, revolted against the suzerainty of the Medes. In 550 B.C. he defeated Astyages, and plundered his capital, Ecbatana: and welded the tribes into the Persian Empire, with Susa, its capital.

546 B.C. Babylon, Egypt and Crœsus, King of Lydia, united against him; the last-named advancing alone was defeated and Sardis captured.

539 B.C. With the Massagetæ as his allies,[1] and Gobyras, Governor of Guta, an Israelite, Cyrus advanced against Babylon, that great city which fell in one night,[2] having previously drained off the water channels by damming up and diverting the reservoir north of the city. Why so easily? For the captive Jews opened the two-leaved gates of the city to their brethren, the Massagetæ.

Perhaps for this service, he released the Jews to return to Jerusalem, to rebuild the city.[3]

Cyrus, also for policy's sake, designates himself as the favourite of Marduk, and King of Babylon and King of the Countries (*i.e.* of the World). He offers marriage to the daughter of Queen Tomyris, of the Massagetæ, hoping thus to become King of Israel and the Jews and usurp the rights of David. He is refused, and goes to war with the Massagetæ, despite the warning of Crœsus not to aspire so high.

---

[1] Isa. xliv, 28. Of Cyrus: "*He is My shepherd, and shall perform all my pleasure; even saying to Jerusalem, Thou shalt be built, and to the Temple, Thy foundations shall be laid.*"

Isa. xlv, 1. "*Thus saith the Lord to his anointed, to Cyrus . . . to open before him the two-leaved gates . . . he shall build My City, and he shall let go my captives, not for price, nor reward.*"

[2] Dan. v. 1–31. "*The feast of Belshazzar, King of Babylon, and the writing on the wall, Mene, Mene, Tekel, Upharsin. . . .*"

Dan. v, 28. "*Thy Kingdom is divided and given to the Medes and Persians*" (the second of the beasts, a bear).

Dan. v, 30. "*In that night was Belshazzar, the King of the Chaldeans, slain.*"

Jer. li, 30. "*The mighty men of Babylon have forborne to fight, they have remained in their holds.*"

Jer. li, 59–64. "*Jeremiah wrote in a book all the evil that shall come upon Babylon . . .*" and to Seraiah [63], "*thou shalt bind a stone to it and cast it into the midst of the Euphrates . . . thus shall Babylon sink.*"

Possibly when the waters were drained, Cyrus re-found the book, and so learned the last of the Pharez line, Zedekiah.

[3] Ezra i.

The story, told by Herodotus, of Queen Tomyris is typically Israelitish.

Cyrus attempts to bridge the River Araxes: when she sends him a message: "O King, cease from thy labour, perchance this work of thine shall end in thine own harm. Nay, cease! Rule thine own people, and suffer me to rule over mine. But if thy desire be so great to make trial of the Massagetæ, then leave bridging the river, and we will go away three days' journey from this side of the river, that ye may cross over unhindered into our land.[1] Or if ye would rather that we cross into your land, then do ye likewise."[2]

Crœsus advises crossing, and the stratagem of leaving his camp, full of sheep and oxen, "a great feast with plenty of wine," guarded only by the weakest. A third part of the army of the Massagetæ came, slew the guard, and sat down to the banquet, *"let us eat and drink for to-morrow we shall die"* (Is. xxii, 13), and slept. Cyrus fell upon them, and took but few prisoners, which included Spargapises, the Queen's eldest son. The Queen sent a herald: "Thou bloodthirsty Cyrus . . . with such a vile drug thou hast mastered my son, and not in battle by the strong hand . . . give me back my son, and go away unharmed, content that thou hast done despite unto the third part of my army. But if thou wilt not do this, I swear by the Sun that, bloodthirsty as thou art, thou shalt drink thy fill of blood." Cyrus refused; Spargapises, when he came to, begged that his chains might be taken off and, when free, seized a sword and slew himself.

In the resulting battle, described as the fiercest and most savage that the world had yet seen, the Persians were defeated, and Cyrus slain (528 B.C.). Tomyris, with a goatskin full of blood, searched the battlefield, and finding the body of Cyrus thrust his head into the blood-bath,

---

[1] Suggestive of Cæsar's second landing in Britain, 54 B.C.—at the Gorsedd convened at Caer Troia, Avarwy's treacherous advice is accepted, and it is resolved "that it is beneath the dignity of the Britons to defend their country otherwise than by might of manhood, and that the landing of the Cesaridæ be unopposed."

[2] The sporting spirit of Briton in this age—we will play home or away, which you say!

and said: "Thou hast undone me, though I live thy conqueror, for thou hast slain my son by guile; but I, as I swore it, will give thee thy fill of blood."

Jer. li, 23. "*I will also break in pieces with thee the shepherd and his flock.*" (*Cf.* Isa. xliv, 28: Cyrus as the shepherd.)

¶ 8. Thus Israel, as the Stone Kingdom, has begun to "*break in pieces*" Nebuchadnezzar's image, first Babylon as gold, now against Persia as silver; and beginning to fulfil—

Jer. li, 20: "*Thou art My battle-axe and weapons of war, for with thee will I break in pieces the nations, and with thee will I destroy kingdoms.*"

Thus at last, though unconsciously, Israel begins to fulfil one part of its contract to the Covenant, to be His instrument to eradicate Cain-Bel influence among the nations who have drunken from the *golden cup* of their wine and therefore "*are mad.*"

Herodotus describes the dress and mode of living of the Massagetæ, their barbaric ostentatiousness, the trappings of their horses, ornamented with gold and brass,[1] their possession of abundance of fine gold, that they fought on horseback as well as they did on foot, they used bows and lances, but their favourite weapon was the battle-axe.[2]

Jer. l, 41–42, describes the "*people shall come from the north, and a great nation, . . . they shall hold the bow and the lance . . . against thee, O daughter of Babylon.*"

Jordanis states that the Massagetæ were the ancestors of the Goths (Ostrogoths) and were one people with the Getæ.

Alexander, having conquered Persia and Afghanistan, marches north-east from Samarcand to the River Jaxartes (thinking they had reached the Tanais = Volga), and

---

[1] Taken from the spoils of Babylon: horses not common in Palestine and Assyria (where chariots more than horseback), but they abounded on steppes of Russia.

[2] All characteristic of the Angles, Saxons and Normans.

were met on the opposite bank by the "Scythians of Europe" (the Massagetæ). The Massagetæ threaten to cross the river in rear while he is rapidly putting up defences, and after an indecisive battle his army retire, he and many having acquired disease by drinking foul water.

Lysimachus, when he succeeded Alexander, repeated the attack, but was defeated, taken prisoner, and released only after a heavy ransom.

¶ 9. Among the tribes of the Massagetæ were the Levites—not all had adopted the Babylonian god-system, the worship of the Sun, or of idols of wood and stone. For, on the restoration of the Temple at Jerusalem, Ezra sends for Levites to minister in the Sanctuary as of old.

Ezra viii, 17. *"And I sent . . . unto Iddo the chief at the place Casiphia [Caspian] . . . and to his brethren the Nethinims . . . that they should bring unto us ministers for the House of our God . . . and they brought us a man of understanding, of the sons of Mahli, the son of Levi . . . eighteen, . . . and twenty, . . . also two hundred and twenty Nethinims."*

During the Israel migrations across Europe, in contact with Druidism, many Levites would become Druidic priests. Or is it mere coincidence that at this time, and within this region, there arose three (or at any rate two) of the great religions of the East? Why? Israel in exile seeking their God.

(1) *Buddhism.*—The dates of Buddha given, 568–488 B.C. (*Ency. Brit.*).

"The Budii, said by Herodotus to be a tribe of the Medes," the word meaning "a separated people" (Moore, *Saxons, East and West*).

L. G. A. Roberts, *British History Traced*, p. 12, gives (i) Inscription in Hebrew graven in ancient Pali characters on wall of a rock temple in Kanari, twenty miles from Bombay, "Lo, the worship of Saka . . . as to Dan . . . yea Gotha" (in connection with the devastation by Cyrus); also George Moore, *Saxons, East and West*, (ii) At foot of Buddha sitting on marble throne at entrance of rock temple of Ajanta—a lion and unicorn (but with the two natural horns), crouching in peace.

These Budii, part of the Sacæ, were probably pushed south-east into India during or after Alexander's campaigns (330–300 B.C.).

(It has been seen that the Brahmins were descendants of Abraham by his wife, Keturah. They assumed the promises of the Covenant.[1])

(2) *Zoroastrianism.*—Re Zoroaster's dates (from the *Ency. Brit.*): Agathias, ii, 24, states uncertain, and adds:

"The Persians say that Zoroaster lived under Hystaspes . . . the father of Darius, or another."

"According to the Arda Viraf, 1, 2, Zoroaster taught some 300 years before the invasion of Alexander"—thus making his dates *circa* 630–550 B.C.

Some records make it earlier—the usual custom in the East to lengthen their ancestry and push back their mythology.

His doctrine from the Gāthās include: At the beginning of things, two spirits, representing Ormuzd and Ahriman, Good and Evil; the soul of man is the object of their war; the life of man in two parts—earthly and after death: a prophet sent to show man the way of salvation: eschatological concerning the last things, and ultimate triumph of good.

Zoroaster was the founder of the religion of the Persio-Iranian peoples, and of the wisdom of the Magi. His doctrines were the source of the Greek mystery-religion, Mithraism, which persisted until about A.D. 275.

(3) *Confucius*, 551–478 B.C., whose teaching was mainly by proverbs as a code of morals without stressing a deity —"What you do not like when done to yourself, do not do to others."

All religion is purer at the source than later when admixed with heathen people, whose "vulgar fancy requires sensuous, plastic deities which admit of visible representation"—thus the degeneration in application of Buddhism, Brahminism, Zoroastrianism—and even Christianity.

[1] Col. L. A. Waddell, *The Phœnician Origin of the Britons, Scots and Anglo-Saxons.*

CHAPTER VI

¶ 1. With the drying up of the River Oxus (*circa* 300 B.C.) the Massagetæ migrated, some far north to Lake Baikal, but the main body settled in the Crimea and South Russia, and founded their capital at Asgard (=Scythian Gerrhos; modern Kiev). Here in contact with their brethren, the Getæ, they possessed the land from which the Kimmerians had migrated, an empire extending from the Danube to the Urals, and gradually pushed northwards to the Baltic, where it coalesced with the migrations from Lake Baikal of the Bretti, Aegli, Angai, as Norsemen (north men).

The idol worship of Bel-Merodach has gone, and their religion is now Druidic, with admixture of Getic belief in Anses ("angel-men"), of Chaldean belief in sorcerers and magic,[1] and primitive nature worship of the north—as the Fatherhood of God, the immortality of the soul, with future reward and punishment, the consecration of valour —"seeking ever to die in battle rather than in peace" and so attain heaven (Valhalla, the Hall of Heroes).[2]

They, as the Asar,[3] and also called the Scythians, war with and defeat the Jotuns, who are pushed west over the Dneiper: with the Thursar, "people of the hoar frost," of the north (the Vandals): with Berg-Risar of the Urals (Sarmatians). When at war with the Vannir (Finns) Odin appears as their leader and gains the victory. (In the Saga, Odin appears as an elderly man, bald-headed, with one eye[4] and a sad expression.) Critias, writing of the

---
[1] "*My people ask counsel at their stocks, and their staff declareth unto them*" (Hosea iv, 12).

[2] Snorri Sturlason, in the Edda of the thirteenth century, collected the Odin traditions.

[3] "Asia-men," from Asia Minor and the Caspian.

[4] ? a reference to him in Zech. xi, 15-17, "*his right eye utterly darkened.*" See W. H. Milner, *The Royal House of Britain*, pp. 28-30.

THE ASAR

recording angels, says: "If the deeds of the Scythians are written in heaven there must be many scribes there."

The Romans went widely and conquered all nations. Odin skilled in witchcraft knew that his descendants would live in the northern part of the world. He set his brothers Vile and Vé to rule over Asgard. First, he went westwards to Gardariki (East Prussia), then southwards to Sakland (*Edda*).

¶ 2. Odin led the Asar from the Baltic across the Elbe into the country of the Cimbri, and into Scandinavia, expelling the Longobardi (the Lombards), leaving colonies in Jotun, the Jutes; in Saxon borders, the Saxons; thickest in the Holstein region of the Cimbri, the Angles: each ruled over by one of his numerous sons—from Wecta the house of Wettin: from Yngve, the Ynglings of Sweden; Baeldæg (or Balder) over the Angles, from whom descended Horsa and Hengist; Scaxnot, over the Saxons.

Here they lived apart from their neighbours, class prejudice against intermarriage with adjacent tribes, the Jotuns or Teutons.

The Asar "laid hearths, wrought wealth, raised on the plains altars, shaped tongs, made tools, played chess on the grass plot. They were cheerful, they did not lack gold."

The Jotuns "were helpless and fateless. They had no health, no mind, neither blood nor motion, nor proper complexion."

A.D. 450–550 saw the migrations of Angles and Saxons to Britain.

The Continental tribes had been broken between Israel as Goth-Angles the hammer, and Rome the anvil; "they moved along the line of least military resistance and poured into new homes through gaps in the Roman defences." Not until the Angles, the Saxons, the Jutes, the Danes had migrated to their "appointed place," [1] did these submerged peoples settle down to their habitations. Israel, "pushing with horns," [2] displaced—

---

[1] "*Moreover I will appoint a place for My people Israel, and will plant them, that they may dwell in a place of their own and move no more; neither shall the children of wickedness afflict them any more as beforetime*" (2 Sam. vii, 10).

[2] "*. . . his horns are like the horns of unicorns, with them he shall push the people together to the ends of the earth*" (Deut. xxxiii, 17).

1. The Gepids of N.W. Russia (? the Thursar), southwest into Hungary, where in the sixth century they were destroyed by the Lombards.

2. The Longobardi (the Lombards), pushed from Scandinavia, joined the Suevi of N. Germany, driven south into Bohemia and Hungary, and finally settled in Italy.

3. The Suevi driven from N. Germany into Swabia, from the Elbe to the Rhine "forced to flee" (says Strabo).

4. The Visigoths fled from north of the Danube before the Huns through Italy and France into Spain.

5. The Vandals from N. Hungary, pushed south, through Spain into N. Africa, where destroyed by the Romans in the sixth century.

6. The Burgundians from W. Russia into Gaul.

7. The mass of German tribes, Teutones, moved about until after the Anglo-Saxon migration—the Alemanni to Germany: and the Franks to France.

The etymological difference of the Anglo-Saxon race and the Prussian-Nazi is seen not only here but even earlier.

The Jotuns were also known as the Gutones, on the Vistula (mentioned by Pytheas of Marseilles in 300 B.C.). These were the Teutones (say Mullendorf and Grimm); who joined with the Cimbri of the Holstein region to attack Hungary in 115 B.C. and Rome in 103 B.C. (300,000 "barbarians" being slain at Aix and Milan).

The earlier origin of the Jotuns was probably from the Hittite-Assyrian co-mixture in Asia Minor, where they were invaded by the Cimmerians, and whence during the later Assyrian invasions (*circa* 750 B.C.) they migrated into Europe—their vulture or eagle emblem being the same.

Their even earlier origin as Aryan has been brought out in a lecture by Professor Wuest, Professor of Etymology at Munich (reported 31st August 1935), who fixed their beginning as "the people of Nordic race who settled in India 4000 years ago and subdued the inferior dark-skinned races there."

Now this was about the time that Abraham sent eastwards his sons by Keturah (Gen. xxv, 1–6, see p. 110), where they joined the east migrations of the Children of

THE GETÆ

the Sun. The Aryans are thus identified with the idolatrous house of Serug and Terah, who had adopted the Cain-Bel influence of Sumer. From them it was Abraham's purpose to separate his Covenant stock, Isaac (the ancestor of the Saxons).

The "remnant" of the Saxon race having chosen to be left behind on the Continent, and now become Christian, the split that Abraham made seems due for re-enactment shortly.

But follow Wuest's identification of his Aryan race with Cain (deified as the sun-god Merodach who ploughed the skies—hence their swastika emblem, see p. 336):

"The word is derived from the Aryan's great deeds as farmer and civiliser. Arya (in Indian language, upright and straightforward) described the manly and proud mien of the free peasant as he strode behind his plough, and, with gaze sharply fixed upon his goal, cut his furrow straight as a die. . . . In Latin it is the word 'arare,' which reminds us again of the noble farmer behind his plough."

The *Noelkischer Beobachter*, the official Nazi organ, paid tribute to his speech as a "deeply scientific lecture, its etymological facts were magnificently illustrated by texts from the age-old Indo-Nordic Books of Wisdom."

¶ 3. *The Getæ* (Joseph, comprising the Houses of Ephraim and Manasseh) settled along the Danube. They recognised God as a Spirit and are described as being nomads with flocks, growing little corn, a sad people and cunning, living in tents and huts surrounded by palisades, without attempt at building, so little of spoil to tempt an attacker. They made no aggressive wars, but if attacked became known as "the invisible Getæ." There are no Getic records, only those of their enemies—the Greeks and Romans.[1]

Herodotus tells that Darius with an army of 700,000 marched west into Thrace, and gave himself six days to conquer their land, but could never come to battle with them. On complaining of this he is told "we are but changing our pastures, as is our wont. Cities have we

[1] The Stone Kingdom—against the kingdoms of brass and of iron.

none, nor farms, for thee to spoil. But if thou art fain to see us fight, then go seek the tombs of our fathers, and lay hands on them—then shalt thou see whether we can fight or no. And for that thou callest thyself my master, I give thee leave to hang thyself."

Harassed by raids on his foraging parties—for small flocks would be purposely left unguarded—Darius is sent to clear his wits an indemnity of a bird, a mouse, a frog, and five arrows. He prides himself on this sign of submission, but Gobryas (still with the Persians) understands his people, and interprets: "Except ye become birds and fly up into the heavens, O ye Persians, or become mice and burrow into the ground, or become frogs and leap into the waters, ye shall not return from our land, but shall all be shot down by our arrows."

A sad and depleted army returns to the Danube, to find the bridge partially damaged, but is rescued by the boats of Histiæus of Miletus, amid Getic jeers and taunts.

The Getæ defied Xerxes, and then Philip of Macedon: they had crossed the Danube with an army of 150,000 and settled in parts of Thrace. Alexander marched against them, but having sacrificed to Zeus he thought better of it and retired to march on the Persians.

In 82–79 B.C. appeared Decinus, who was declared one of the Anses ("god-men"), as a leader and reformer. He taught them ethics, "nearly the whole of the philosophy," astronomy and botany, to the priesthood: in an outburst of zeal against national vice the people destroyed their vineyards.[1] On his advice, the Getic King ravaged the lands of the Teutones.

They were now an army 250,000 strong, possessing the country along the Danube from the Black Sea to Vienna known as Dacia; at the strategic centre for attack through Julian Alps to Rome, or through Thrace to Byzantium, threatening the Roman Empire to cut off east from west.

Against them Julius Cæsar was preparing a campaign

[1] Isa. xxviii, 1, "*the drunkards of Ephraim.*"

when he was assassinated; Augustus offered to marry the Getic King's daughter, and Rome, daily apprehensive, asked: "What news from Dacia?"

In the northern lands, Frodi, as king of the Asar, was the most powerful king when "peace all over the world, then Christ was born." His ships sailed the seas, his army was 300,000 strong. This peace was attributed to Frodi. "No man did harm to one another, even if he met the slayer of his father or his brother, bound or loose: no thieves nor robbers were then to be found, so that a gold ring lay for a long while and was not taken away."

In Britain, Cæsar had diplomatically ended his campaign, when his base in Thanet was about to fall, with a treaty at Verulam, and retired after a magnificent banquet given to him by Caswallon in London. The two sons, Cynvelin and Llyr, of Tenuantius were being educated by Augustus in his palace with his nephews, and the Briton walked the streets of Rome the only free man.

The Jews after one minor revolt against Rome were ready for another.

¶ 4. It was at such a moment that the Devil offered Him the third temptation: did he show Him "urbs Roma, orbs humana"? No, but four divisions of Israel—Getæ awaiting a successor to Decinus; the Asar, a religious leader; the kingdom of Siluria within ten years had accepted Christianity from Joseph of Arimathea, and fought a succession of Rome's best generals to a standstill; the Jews looking for their king.

No, the temptation was more subtle, as in the Garden of Eden: "Hath God said, to die on the Cross? Anticipate God's decrees 2000 years; here is Israel awaiting fusion; proclaim yourself their King"—the Dacian advance will hopelessly cut off the legions in Asia Minor; the Jews will rise: the seamen of the Cimbri from Brittia would swarm into the Mediterranean; the transport of Britons would gather the Cymric tribes of Gaul; and the Asar would march for the Rhine, gathering with them thousands of "barbarians,[1] who have their own bitter quarrels to settle with Rome."[2]

The Stone Kingdom will have broken in pieces the "iron and the clay," the last of the four beasts; and can now be set up, with their fore-ordained King.

But, the "seven times" of Israel's punishment had not lapsed. Israel had not *"enquired of Him,"* nor acknowledged their iniquities. Israel was not yet ripe to have the supreme power over other races placed in their hands.

Then Satan's object would have been achieved. For individual freewill, as Adam's, would still in operation deviate from Divine will, discord and death would still continue. The power of death over man (Gen. ii, 17) required His surrender of Spirit dominion over Matter,—that energy within the Ether *"created in the beginning,"*—while *"for three days in the heart of the earth,"* to create a new "magnetic" field or reservoir of Divine energy.

This the sacrifice of the Lamb fore-ordained, whereby His purpose is to be accomplished, to show to His spirit realm in heavens above and earth beneath the magnitude of Divine Love, whereby would be restored to man, whose freewill is attuned, His original gift of Spirit dominion over Matter, and the resurrection to life after death.

"Jesus said, Get thee hence, Satan, for it is written, Thou shalt worship the Lord thy God, and Him only shalt thou serve" (Matt. iv, 10).

His decision is given in monosyllabic words—not the arguable meaning of the polysyllabic—direct (accenting the disyllabic "worship only") and driving a concept into the heart of man, from which is wrung a murmur, "My Lord and My God."

The English Prayer Book, unique in Christendom, eliminates the intervening centuries and places the English peoples here, at the foot of the Cross; where "our forefathers" through their transgressions had failed their Messiah-King when He appeared on earth. At His

---

[1] This was the time of Hermann, a great German national leader, twenty years after the Roman defeat at Teutoberg.

[2] Weldon, *The Origin of the English*, first to bring out this interpretation.

Second Coming, which is imminent, will His Israel, this Britain as a people and nation, have still left unfulfilled His Covenant, "acknowledge thine iniquities, make thee a new heart and a new spirit, for I will yet be enquired of by the House of Israel to do it for them"?

¶ 5. In A.D. 69, Josephus, a Jewish leader and governor of Galilee, was writing to and conspiring with the Getæ on the Danube. He wrote (in a Roman prison, A.D. 93): "There are only two tribes subject to the Romans in Europe and Asia, the remainder are beyond the Euphrates to this day, an immense multitude and not to be numbered."

When the Jews revolted, the Getæ attacked Mœsia (now Bulgaria) led by King Decebalus (of whom, and not the Emperor Nero, Peter was thinking when he wrote to the strangers: "*Love the brotherhood. Fear God. Honour the King.*"—1 Peter ii, 17).

Decebalus crossed the Danube and drove the Romans through the Balkans: A.D. 86, the Emperor Domitian himself took the field, and an inconclusive peace was concluded by which an annual tribute was paid to the Getæ. Even the Romans mocked when Domitian returned as "Conqueror of the Getæ."

The Getæ were now holding the Danube merely as an outpost frontier, having retired to the Dneiper, and occupied Asgard as capital. While doing this they pushed the Marcommani from Bohemia into N.W. Hungary.

Yet Trajan's campaign in A.D. 98 advanced only sixty-five miles in eighteen months, and at the battle of Tapæ his losses were so great that he tore his garments for bandages.

After the last desperate battle, for which Trajan had built a stone bridge over the Danube, the Getæ set fire to their own city before retiring, Decebalus killed himself, and Dacia became a Roman colony, A.D. 106.

¶ 6. For a century the Getæ were quiescent, and became known as the Ostrogoths. But in A.D. 245, when Philip the Arabian refused to pay them indemnity, they advanced with an army of 300,000, under King Ostrogotha, and besieged Marcianople, but having no patience for long sieges were bought off and returned with much booty.

A.D. 251. A second raid through Mœsia to Philippolis under King Cniva nearly ended disastrously, but Decius overreached himself, he and his son were killed, and the Getæ returned with more booty and a further annuity.

A.D. 269. They sailed down the Dneiper and across the Black Sea, but not being skilful seamen, and after a disastrous battle at Naissus, a peace was concluded with Rome which lasted one hundred years, but they still retained the annuity.

A.D. 350. Misfortune came in the person of Ermaneric. By intrigue he had got himself made king, to the exclusion of Winithar, the elder line of the House of Amal. Allied by marriage with the Angles, jealous of his eldest son and led by Court intrigue, he killed his son and wife (the story told in Norse saga and the *Lay of the Nibelungen*), estranged the nation, and on his death-bed called in the Huns under King Balamber to his aid.

A.D. 406. The Ostrogoths, to get away from the Huns, crossed the Vistula and attacked the Gepids, then turning north of the Carpathians crossed Poland and attacked the Suevi between the Danube and Rhine. Thus they disturbed the Vandals and Marcomanni, who in turn attacked the Franks and lost 20,000 in the battle, and then turned south and flung themselves on Rome.

Into this vacated country came the Ostrogoths to settle, obliterating the Quadi and the Roman province of Pannonia.

Ermaneric's scheme for a great Pagan empire, to crush Rome, was later adopted by Attila, who collected his Huns, Ostrogoths, Gepids and some German tribes to advance westwards in A.D. 451.

Near Chalons, they were met by the Romans under Aetius, with Visigoths, German Saxons, Alans and Celts, and one of the decisive world battles was fought, with nearly 300,000 casualties. Attila was defeated and retired.

When he died, in A.D. 453, the Ostrogoths were again

THE OSTROGOTHS

independent, then being settled in Pannonia, along the middle Danube.

Two main campaigns developed — in the west, the Ostrogoths marched through the Black Forest, through Bohemia and Moravia, attacked the Suevi and Alemanni, who became their vassals, and crossed into Italy.

In the east, under Theodoric they captured Belgrade from the Gepids, and took Monastir. Bought off before Salonica, they fought through Thrace and Greece and occupied Constantinople. The Greek emperor, Justinian, offered Theodoric Italy if he would expel Odovacar, who had led a mutiny of mercenaries and deposed Augustulus, the last Roman emperor (A.D. 476).

So, in 488, Theodoric fought the Gepids again, collected his Ostrogoths, and with 250,000 men and 20,000 wagons crossed the Julian Alps and defeated Odovacar on the Isonzo and Adige.

Theodoric ruled Italy, Spain and the southern corridor of France until his death, in 526, with tolerance, the Italians had their religion and property safeguarded; to consolidate his empire he arranged marriages of the House of Amal with Vandals, Burgundians, Thuringians and Lombards.

It was not the Divine Plan that Joseph should remain in Italy to rule, for his birthright was in the appointed place, Britain. So followed a series of blows to the Goths to direct them northwards.

On Theodoric's death, the treachery of Theudad, his son-in-law, led to war with the Ostrogoths under King Witigis. Belisaurus, returning from his victorious campaign against the Vandals in N. Africa, pretended to accept Witigis' aid against the invasion of the Franks but, treacherously again, sent him as prisoner to Constantinople (A.D. 540).

"Thus this famous nation in almost the 2030th [1] year of its existence was destroyed" (Jordanis).

Yet 1000 men held out at Pavia and, more joining, they revolted under Bædula (known as Totila) and retook Rome.

---

[1] 2030—A.D. 540 = 1490 B.C. (about the time of the Exodus).

The eunuch Narses (palace chamberlain and a devotee of the Virgin Mary, being "supernaturally directed" by her) with an army was sent against them, and in a desperate battle in the Apennines Totila was slain.

They still resisted under another king, but he also was slain, on Mount Vesuvius. Narses stormed Cumæ, on the Bay of Naples, where the Goths holding their treasure made their last stand, and failed.

Negotiations were opened, and the Goths, leaving behind their treasure, were allowed to depart with their movable property, free men and never to serve under Roman banners (A.D. 553).

Whither?—over the Brenner, through the Alemanni and Suevi, their late vassals, beyond the Elbe, through the land vacated by Angles and Saxons, to Scanzia. There was no historian to record the march of the 10,000 Ostrogoths.

¶ 7. About A.D. 750, ballads appear that, on whichever side the Goth fights, that side wins.

A.D. 813. Normans, as the Northmen, issuing from the island Scanzia, which is called Norway, where dwelt the Goths, and Huns[1] and Dacians, attacked the Flanders coast with thirteen large vessels.

Duchesne (who in the seventeenth century collected Norman chronicles) states that the Normans were Dacians.

Dudo (Dean of St Quentin's in the tenth century), who wrote the earliest history of the Normans, states that they were Dacians; their ancestors came from Scythia beyond the Danube; describing Scythia, Caucasus, Caspian and the River Araxes.

A.D. 850. The Norman influence spreading over Russia, they refounded the city of their ancestors, Kiev, on the Dneiper, and, sailing in 200 ships across the Black Sea, captured Constantinople. They attacked the Mediterranean coast and established a Norman kingdom in Sicily (abandoned by their fathers, the Ostrogoths, in A.D. 535 to Belisaurus). But their expeditions eastwards gradually

---

[1] So named after an early hero of the north, who had the same name as a Hun of the Ostrogoth period.

dried up, and in A.D. 912 the Normans secured Rouen and Normandy.

The Norman was dark, short, clean-shaven, a diplomatist and a soldier, with a passion for rule, regulation and law. The Angle was a tall, blond Viking, with long yellow moustache and hair, a drunken pirate, obeying no order but his own. And such are the types in Britain to-day.

Thus in 1066 all the tribes had come to their "appointed place" (2 Sam. vii, 10), fulfilling—

"*For lo, I will command, and I will sift the House of Israel among all nations, like as corn is sifted in a sieve, yet shall not the least grain fall upon the earth*" (Amos ix, 9).

And from then onwards—

"*Thou shalt break forth on the right hand and on the left; and thy seed shall inherit the Gentiles, and make the desolate cities to be inhabited.*

"*In righteousness shalt thou be established: thou shalt be far from oppression; for thou shalt not fear: and from terror, for it shall not come near thee. Behold, they shall surely gather together, but not by Me: whosoever shall gather together against thee shall fall for thy sake. No weapon that is formed against thee shall prosper*" (Isaiah liv).[1]

Neither the great Armada of Philip, nor the Zeppelins and submarines of the Prussians: for 900 years her soil preserved—it is time that Britain on her knees every Sabbath gave thanks.

Such was the first appearance in England of the Northern pirates, called Danes or Normans, according to as they came from the island of the Baltic Sea or from the coast of Norway. They descended from the same primitive race as the Anglo-Saxons (Thierry, *The History of the Norman Conquests*, Bk. II).

The Jutes are Jews of the tribe of Dan, and the Jutes, Angles and Saxons were kindred nations (Vetus, *Chronicon Holsatiæ*, p. 54).

Tribe after tribe, Angles, Saxons, Jutes, Frisians, poured across the sea to make new homes in the Isle of Britain. Thus

---

[1] The whole of the second Isaiah, xl–lxvi (compiled by Benjamin in the Babylon Captivity and later in Galilee), is on the fulfilment, and for present-day Britain. The full fulfilment of the visions of the prophets is postponed 2520 years, the times of Israel's punishment.

grew up the English nation—a nation formed by the union of various tribes of the same stock. The Dane hardly needed assimilation, he was another kindred tribe, coming later than the others. Even the Norman was a kinsman.[1]

Britons, Anglo-Saxons, Danes and Normans were all relations; however hostile, they were all kinsmen, shedding kindred blood.[2]

The whole of the Engle people forsook their earlier seats for the shores of Britain.[3]

Thus the ethnological comparisons prove that the Anglo-Saxon peoples must be of an entirely different stock from the present German race, they do not belong to the Teuton stock.[4]

The invaders of Britain, of the Gallic and Mediterranean coasts, could therefore not have been the German tribes referred to by the Roman writers . . . Germania were not a seafaring people nor possessed of any civilisation, whereas the invaders of Britain were civilised.[5]

The researches of modern historians unequivocally favour the opinion that under the names of Keltai, Galtai, Gauls, Gaels, Gwyddyls, Celts, Cimmerii, Cimbri, Cymry, Brython, Llægrians, Scots, Picts, only one race, under different tribe or clan divisions, political organisations, and periods of existence is spoken of . . . hence, one people [p. 33] . . . we embrace under one general designation, Ancient Britons [p. 58]. As the Danes were brethren, though not loving, to the Saxons, Jutes and Angles, so were the old pure Normans brethren, or rather sons to the Danes [p. 271]. The enquiry . . . discovers a strong link of relationship between the Cymry and the English [p. 506].[6]

The names of the tribes entering Britain, of the pre-Captivity era, are all Hebrew in origin. Their language, customs and religion as well as their codes of law were all Hebraic.[7]

¶ 8. Here consider the pre-Captivity migrations of Israel, of Ephraim, Dan, Darda and Calcol: Jeremiah and Brutus.

---

[1] Freeman, *The Origin of the English Nation.*
[2] Sir E. Palgrave, *English Commonwealth*, i, 35.
[3] J. R. Green, *The Making of England*, i, 56.
[4] Professor W. J. Ripley, *The Races of Europe.*
[5] M. Paul du Chaillu, *The Viking Age*, vol. i, p. 14.
[6] Thomas Nicholas, *The Pedigree of the English People*, 1878.
[7] L. G. A. Roberts, *British History Traced*, p. 54, quoting *contra* Apion (Bk. I, sec. 14). Also Lyson, *Our British Ancestors.*

PRE-CAPTIVITY MIGRATIONS

It has been estimated that two-thirds (about three millions) of Israel went through the Assyrian and Babylonian captivities, and that a third had previously migrated, as the whole tribe of Dan.

771 B.C.—the first invasion of Israel by Assyria—to last captivity date, 717 B.C. At this threat to their land many of the maritime portions of Israel escaped by the ships of Tarshish westwards to Iberia (Spain); while another large portion trekked with their herds by land northwards to reach the shores of the Black Sea, and there mingled with the Kimmerians. On the migration of this race as Kelts along the Danube to Gaul and parts of Spain (700–400 B.C.) come many of escaped Israel (*vide* pp. 107, 189, *Triads of the Cymry*).

But even earlier are some migrations while Israel was still in Egypt:

(1) *Ephraim*.—Irish antiquaries state that the Partholonians or first settlers in Ireland were destroyed by a plague. They were succeeded by the Nemedians, a Japhetic people.

During their time came a Semitic people, being a colony from Africa—Camden says before the time of the Exode (1540 B.C.); they settled in the north of Ireland and declared they left Africa because they desired to escape the curse uttered by Noah on Ham. After three campaigns, they overthrew the Nemedians and obtained the complete dominion of the island.

These Formorians, as the Ephraimites were styled in Ireland, built the celebrated Round Towers, having landed at Tor [1] or Torinis or Torconning. The Formorian leader, More, returned to Egypt with sixty ships for reinforcements, to emancipate and transport them to the west, but he found the Exodus had taken place.

The Welsh called the Irish "Iddew," and the country "Iddeson" or Jewsland.[2]

"Beyond the pillars of Hercules the ocean flows round

[1] The sign of Ephraim was the bull.
[2] Roberts, *British History Traced*, p. 163, and quoting from *Precursory Proofs that Israelites came from Egypt into Ireland, and that the Druids expected the Messiah*, by Joseph Ben Jacob, 1816.

the earth, and in it are two very large islands, called British (one of which is called Albion, and the other Ierne), lying beyond the Keltoi." [1]

(2) *Dan.*—The "*trail*" of Dan—"Danaus, son of Bela,[2] a sojourner in Egypt; his brother was Egyptus. Danaus was informed by an oracle that his brother would slay him; he fled, taking with him his daughters (colonists) and came to Greece three years after the death of Joseph." [3]

Æschylus, *Suppliants* (sixth century B.C.), gives for landing of Danaus in Greece as tenth generation from Peleg.[4]

Eratosthenes gives the date of landing as 1466 B.C.—"seed divine, flying from his brother Egyptus since they feared an unholy alliance." They passed through Syria and founded Sidon.[5]

"Danaus having arrived in Argos made a law that those who had borne the name of Pelasgiotæ (throughout Greece) should be called Danai." [6]

They established themselves in Thessaly; hence embarked from Iolchis for Colchis (now Iberia).[7]

The last of the Danai to come to Greece was about the time of the oppression by Jabin, King of Canaan, 1285 B.C. Beloe's edition of Herodotus (1833) shows the story of

---

[1] Aristotle, *De Mundo*, c. 3 (fourth century B.C.).

[2] Bilhah, Jacob's concubine, was mother of Dan. "*There arose a new [i.e. Anti-Semitic] dynasty over Egypt which showed not favour to [the descendants of] Joseph*" (Exod. i, 8). This dynasty and not Joseph is probably referred to as Egyptus.

1656–1625 B.C. reign of Amenhotep III (Joseph died 1630) and his successor, Amenhotep IV (called Akhenaten) of XVIIIth Dynasty; during their reigns great changes in religion and period of Semitic supremacy; worship of Amen-Ra overthrown and replaced by that of Aten (or the Solar Disc)—monotheistic. Followed the XIXth Dynasty, started by Pa-Ramessu (Ramessu I). He with co-regent Horemhab and priestly intrigue restored Amen-Ra worship. (Amen = Ham deified.)

Followed as co-regent Seti I, whose son, Ramessu II, was the Pharaoh of the Oppression.

His son, Menephtah, was the Pharaoh of the Exodus, and left a stele (discovered by Petrie in 1896) on which "Israel is lost; his seed is not." His eldest son, Seti II, never reigned.

[3] Roberts, p. 27, from Petavius, *History of the World*.

[4] Peleg, son of Eber (word = colonist), Gen. xi, 16—from whom Dan is the tenth generation.

[5] Suggested in Isaiah xxiii.

[6] Euripides, *Strabo*, v. ii, 4.

[7] *Dan the Pioneer of Israel*, Col. Gawler.

PRE-CAPTIVITY MIGRATIONS

Hercules of Tyre is none other than that of Samson of Dan.[1]

The poem of Orpheus describing the journey of Jason and the Argonauts "mentions the island of Ireland, its woody surface, its misty atmosphere . . . the Argo will be traced, after passing through the Bosphorus and Palus Mæotis, as making its way by the Riphæan mountains into the Northern Ocean; and then returning by way of the Atlantic to Ireland, from whence the good ship proceeds between the Pillars of Hercules into the Mediterranean" (1250 B.C.).

In this journey of the Danai of Argos is described the migration of the tribe of Dan.[2]

W. E. Gladstone, *Juventus Mundi*, says the siege of Troy was undertaken by Danai against Dardanai, and that these were originally one. He connects the Danaos with Phœnicia, and the Tuatha de Danaans [3] of Ireland came from the Danai of Greece (p. 136).

On their entry into Canaan, the tribes of Dan and Asher were given a strip of the coast, and by 1296 B.C. had conquered and coalesced with the Phœnicians of Tyre and Ascalon. They gained the ascendancy and became the moving spirit and adventurers in the Phœnician shipping trade in the Mediterranean, to the Iberian peninsula (Spain), to the isles of Tarshish, to Ireland, and the Cassiterides Islands.[4] They discovered the advantage of possessing islands, which were free from the numerous land invasions of those times—Cyprus, Crete, Sicily, Malta, Sardinia, Corsica, Balearic Isles.

---

[1] Roberts, pp. 29-31.
[2] W. H. Milner, *The House of Britain*, pp. 5, 6, quoting *Transactions of the Royal Irish Academy*, vol. xvi, and Dr Latham.
[3] Meaning "the tribeship of Dan," their entry to north-east Ireland, *circa* 1000 B.C. "The Tuatha de Danaans were known as the Divine Folk" (J. Borwick, *Who are the Irish ?*).
[4] Tin at this time was available only in Cornwall; it is mentioned in Num. xxxi, 22. It has been thought that some of the tribe of Simeon settled first in Spain with the Zarah migration, and later migrated to Cornwall (Merton Smith, p. 7).
Num. xiv, 1-4 mentions a controversy in the Wilderness as if a large part had broken away, and between the numbering of Num. i, 23 and xxvi, 14 Simeon was reduced from 59,300 to 22,200 men; their lot in Palestine was a restricted one surrounded by Judah.

Thus by disobeying the Divine command [1] they brought the Phœnicians, of the race of Ham, into the southern parts of Ireland. This stock became known as the Venetians, Fenians, and finally as the Sinn Feiners, fulfilling the prophecy:

"But if ye will not drive out the inhabitants of the land from before you; then it shall come to pass, that those which ye let remain of them shall be pricks in your eyes, and thorns in your sides, and shall vex you in the land wherein ye dwell" (Num. xxxiii, 55).

Ezekiel xxvi–xxviii—the might of Phœnicia, and its fall:

"Tarshish was thy merchant by reason of the multitude . . . they traded in thy fairs [xxvii, 12]. *Dan also, going to and fro in thy fairs"* (19).

"The ships of Tarshish did sing of thee in thy market . . made very glorious in the midst of the seas [25]. *The isles that are in the sea shall be troubled at thy departure* [xxvi, 18]. *Though thou be sought for, yet shalt thou never be found again, saith the Lord God"* (21).

"And there shall be no more a pricking brier unto the House of Israel . . . when I shall have gathered the House of Israel from the people among whom they are scattered" (xxviii, 24, 25).

¶ 9. (3) *Darda.*—Genesis xxxviii, 27–30, gives the birth of the twin sons of Judah, a scarlet thread being bound on the hand of Zarah, which first appeared, but Pharez was born first.

The sons of Zarah, Darda and Carcol (mentioned in the genealogy of Chron. ii, 6, but not in the later), possibly resenting the birthright being passed to Pharez,[2] left Egypt before the bondage of Exod. i, 7–14; taking with them other discontented Israelites.

Darda, called Dardanus by Josephus, founded Ilion, or Troy

Sanchoniatho, an old Phœnician historian, identifies Saturn with Jacob—"Kronus, whom the Phœnicians call Israel, . . . had a special son Jehûd (*i.e.* Judah)."

[1] Exodus xxxiv, 12; Judges iii, 1–7. [2] Massey, *Tamar Tephi*, p. 42.

The following genealogies coincide with Jacob, Judah, Darda, Ericthonius[1]:

George Grote, *History of Greece*: Cronus: Zeus—*i.e.* Jupiter—Dardanus: Ericthonius.

Icelandic Genealogy of Odin: Saturnus of Krit (Crete): Jupiter: Darius: Ericthonius.

MS. in Royal Library, Windsor Castle: Saturnus: Jupiter: Dardanus: Ericthonius.

To Troy come the Northern Danite migration after conquering Leshem.

(4) *Calcol* (or Cecrops) founded Athens, in Greece, *circa* 1556 B.C.

From here, he led his followers westward, probably leaving part as the Etruscans, a colony in Italy; and landed as the "Ibharim"[2] on west of Spain, settling at Segubtoi (Hebrew, "the fortress of the wanderers"; =Saguntum, by Romans; Sagunto now).

They worked their way up the River Iberus (Ebro), via Araun-Gozan ("the cursed pasture," now Aragon), Zara-u-el ("the brightness of God," now Teruel), Segub-ragan ("the neighbour's fortress," now Segara); and established a stronghold at Zara-Gaza ("the strength of Zarah," now Saragossa). They still pushed on to the rich mineral Basque mountains, to the Yum-Birska ("the evil sea," now the Bay of Biscay), and there built another stronghold, Baal-boaz ("might of Baal," now Bilboa).

In this region the Zarahites are known to their brethren in Palestine, for they pay tribute to the kings of Judah, from Solomon to Uzziah (812–760 B.C.).[3]

---

[1] Milner, *The Royal House of Britain*, p. 31.

[2] Massey, *Tamar Tephi*, Ibhar (Hebrew, chosen or elected); Eber (Hebrew, coloniser).

[3] Rabbi Eliezer Bassin, Ph.B., *British and Jewish Fraternity*: "I then began to trace the Keltic Britons, and found them to be the first Israelite emigrants, who migrated from Palestine to Spain (then called Tarshish) perhaps as early as 1286 B.C. There can be no doubt that there were Hebrew colonists in Spain during Solomon's reign, for we read in 1 Kings x, 22, that Solomon had at sea a navy of Tarshish. To the fact of this Hebrew colony there exists the testimony of Philo, Josephus, Seneca and Cicero. Bishop Titcomb mentions the monument, found at Saguntum, with Hebrew characters: 'This is the tomb of Adoniram, the servant of Solomon, who came to collect tribute and died there' (1 Kings vi, 6; v, 14). From the voyage of Jonah we learn that ships traded regularly

At Sagunto have been found three large monumental stones, with Hebrew inscriptions—called

(1) "King Solomon's Stone";
(2) to Adoniram, Solomon's collector of tribute (1 Kings iv, 6);
(3) to some Israelitish official of King Uzziah.

During the Assyrian invasions, these colonies received large augmentations from Palestine as refugees.[1]

MSS. of these Israelitish settlements in N.E. Spain were collected by Don Gil Mentez, and were destroyed by the Carlist general, Iturralde, 1838: and another collection by Enriquez was burned at Trevino, in Alava, by Leon, 1838.

During the persecution of the Jews in Spain under Ferdinand and Isabella, A.D. 1492, in the Inquisition of Pope Alexander VI (Roderigo Borgia), these Israelitish families were unmolested until the persecution became hotter—when the ancient family of the Arbanals was charged with being the descendants of the Jews at the Crucifixion. "The Arbanals proved conclusively from authentic documents in their possession and to the satisfaction of even the Inquisitorial Court that their ancestors had left Jerusalem prior to 586 B.C." [2]

Some of these colonies coalesced with the Keltic migration, and became known as the Celtiberians. Thus in peace and prosperity the Zarahites, who considered themselves the "chiefest men of Israel," not having been through the Egyptian bondage, existed until the invasion of Spain by Hamilcar, of Carthage (238 B.C.). Before this date, refusing to submit to Carthage, nearly the whole had migrated to Ireland under Miletus, known as the Milesian Scots.[3]

---

between Tarshish and the coasts of Palestine (Jon. i, 3), for he found at Joppa a vessel ready for the voyage (Jonah desired to flee to the Hebrew Colony at Tarshish). . . . I came to the conclusion that some Israelites escaped to Spain at the time of the Assyrian invasion (720 B.C.) to join this Hebrew colony, and the Spanish Kelts, who were Israelites, migrated to Cornwall and Ireland."

[1] J. Dunham Massey, *Tamar Tephi*, p. 166.   [2] *Ibid.*, p. 162.
[3] Accompanying the House of Zarah may also have come some sons of Shelah, another son of Judah, into Ulster—"*The sons of Shelah, the son of Judah, were Er . . . and the families of the house of them that wrought fine linen*" (1 Chron. iv, 21).

THE MILESIAN SCOTS

Hear the tale of the times of old [1] ; hear of our race the renowned of the earth!

Our great fathers dwelt . . . beyond the sources of the great waters (*cf.* Joshua xxiv, 2, 3). Then did they spread themselves from the flood of Sgeind (Scinde = Indus, Cashmere) to the banks of the Tethgris (Tigris) . . . and then after reaching to Affreidgeis (Euphrates) they became the lords of all the lands on this side (Canaan) . . . (centuries elapse) . . . a multitude from the sun's rising (east) beneath the land of the first abode of our great fathers (at Ur in Sumer) poured in upon the land of our fathers that then lived (Canaan), like unto a swarm of locusts . . . their name Eis Soir (*i.e.* the Assyrians, 771 B.C., adding that Moses and the writer were both Scythians).[2]

A number of the leaders escaped to Armenia, where they became organised under princes, and many of the people flocked to them. After seventy-two years there was migration westwards; thirty-one years later they colonised Colchis, which they christened Iber. This was in the time of Prince Dorca, who died 13 years later (*i.e.* in all 116 years from 771 B.C.).

In these days multitudes of the Gaal passed over the Gaba-Casan . . . there they did raise up their tents . . . calling the land Iath Sciot in memory of our Race. . . . The remaining Gaal spread over the bosom of Ailb-bin (the Trans-Caucasian Albania). They excelled all people in the use of the bow (*cf.* Ephraim = England); forty-six years after the great Scythian migration beyond the Caucasus, the Gaal sent another great colony westwards over the sea. They first crossed northern Syria, thence they "passed through the floodgates that divide the world of water from the world of land" (the Straits of Gibraltar).[3] After a sojourn in Spain, they eventually arrive in Ireland. These were the Milesians.

In 550–450 B.C. occurred another Keltic migration, of

---

[1] W. H. Milner, *The Royal House of Britain*, pp. 6–8, quoting from *The Chronicles of Eri : being the History of the Gaal Scot Iber, or the Irish People, translated from the original MSS. in the Phœnician Dialect of the Scythian Language*, by Roger O'Connor, 2 vols., London, 1822.

[2] P. 17, Dr O'Connor says, "These are the Scythians who assumed the names of Goths, Getæ, Daci."

[3] About 609 B.C., dates arrived at by adding the reigns together—"like the Hebrews the Celtic nation were most careful to preserve the lines of descent of their kings and heroes" (James Simpson of Edinburgh).

the Picts from Gaul, into Ireland. They settled in the southerly coast-line, but these Pictish kingdoms were small except in Galway in the west. These as well as those of the Oestmen (Norse and Danes) soon declined and were subject to the Dan-Asher or "Iberian" Ardath (head king)[1] of all Ireland. On the arrival of the Milesians, they were pushed out of S.W. Ireland and allowed conduit to the north-east, whence they migrated *en masse* to S.W. Scotland (Galloway)."[2]

¶ 10. (5) *Jeremiah.*—It has been seen that Jeremiah the prophet, Simon Baruch his scribe, and the two daughters of King Zedekiah were left as the "remnant that escaped from Egypt," *circa* 584 B.C. (p. 135).

To Jeremiah:
"*Before I formed thee . . . I sanctified thee and I ordained thee a prophet unto the nations . . . be not afraid of them, for I am with thee to deliver thee, saith the Lord . . . see, I have this day set thee over the nations, and over the kingdoms, to pluck up and root out, and to pull down and destroy . . . to build up and to plant*" (Jer. i, 1-10).

"*I will make thee to pass by thine enemies into a land which thou knowest not . . . and if thou take the precious from the vile, thou shalt be as My mouth*" (Jer. xv, 11, 14, 19-21).

About this date, Spanish records describe the arrival of two Eastern princesses, accompanied by an ancient Prophet, and Brug his scribe, bearing with them a Sacred Stone, a Box and a golden Banner, inscribed with the royal Lion of the tribe of Judah.[3]

The younger princess married the eldest son of the reigning prince of the Zarahites.[4]

Thus, Jeremiah with his charges, knowing of the Hii-i-Yum ("the Isles which are beyond the Sea"), took

---

[1] Of only a few of these warrior-kings is recorded, as if it was strange exception, "he died in his bed."
[2] J. Dunham-Massey, *Tamar Tephi*, pp. 58, 59.
[3] Gayer, *Heritage of the Anglo-Saxon Race*, p. 68.
[4] J. D. Massey, *Tamar Tephi*, p. 46.

ship for Tarshish.¹ From Spain, he voyaged yet westwards.²

Now from ancient Irish chronicles, about the year 582 B.C. there arrived or was wrecked at a port in N.E. Ireland (now Carrickfergus), in a ship belonging to the Iberian Danaan, an aged man, named Ollam Fodhla, an "Egyptian" princess of beauty, dignity and charm, bearing the name Tamar or Tea Tephi,³ and a secretary, Simon Brug. They brought with them a massive and mysterious chest, which they regarded with reverence and jealously guarded, a golden Banner with the device of a Red Lion, and a large, rough Stone.⁴

Milner, from Irish chronicles—"this would make Ollam flourish 586 B.C., by which date Jerusalem had been sacked and Jeremiah had left Palestine. The epoch of Ollam Fodhla is the epoch of Jeremiah." ⁵ (Ollam = in Hebrew "the possessor of hidden knowledge," Fola = in Hebrew "wonderful," or in Celtic "revealer.") By his disciples "he was soe much again beloved and revered, that ever after his house, stock and family were by them in their Rimes and Poems preferred before any others. . . . His right name was Collawyn (in Hebrew "the long-suffering," "the patient")" (*Annals of Clonmacnoise*).

They were brought to Clothair to Eochaidh, Ardagh of Ulster, who had just been elected Heremon (in Hebrew = most high) or head king of Ireland. And he married Tamar Tephi, seated on the stone Lia Fail,⁶ brought by the Ollam.

---

¹ As Jonah, trying to escape from becoming a prophet, took ship for Tarshish (Jonah i, 3).

² "The Jews being most wise sages and learned philosophers, and knowing that the empire of the world should be settled in the strongest angle, which lieth in the west, seized upon these parts at an early period, and Ireland the first" (Camden's *Britannia*, p. 963).

³ In Hebrew, Tamar = palm, Tea = wanderer, Tephi = diminutive of affection, suggesting fragrance.

⁴ J. Dunham-Massey, *Tamar Tephi*, p. 61.

⁵ W. H. Milner, *The Royal House of Britain*, pp. 9, 10.

⁶ Lia-Fail (in Hebrew = the stone wonderful, the letters reading the same both ways.)

"The iron rings, the battered surface, the crack which has all but rent its solid mass asunder, bear witness to its long migrations" (Rogers).

"It is the one primeval monument which binds together the whole Empire" (Dean Stanley).

As dying, St Columba caused himself to be carried to the chapel at Iona where the Stone was kept, and laying his head on the stone, died.

"Eochaid sent a car for Lia Fail and he was placed thereon. . . . And Erimionn was seated on Lia Fail" and the crown was placed on his head, and the mantle on his shoulders, and "all clapped and shouted," and the name of that place, from that day forward, was Tara.

So its history—
Jacob's pillow—Israel's guardian stone—Lia Fail—The Stone of Destiny:

> In what land this messenger shall stay,
> A chief of Iber still shall bear the sway.

Now under the Coronation Seat in Westminster Abbey, on which all the Kings from Eochaid down have been crowned.[1]

> Tea . . . desired of her husband and kinsmen that the place she should most like of in the kingdom should be for ever after the principal seat of her posterity to dwell in . . . she chose Leitrim, which is since that time called Tara. . . . And in it she was interred.[2]

Tamar died when quite young, loved and venerated by all Ireland, was buried in a tomb sixty feet square, and with her the Box with its mystic contents, in the hill of Tara.[3]

> The nine laws established at this time . . . were the Laws of Eri, set in order by Ollam Fola, by which (with the addition of three others) the nations were ruled for 1000 years—should anyone fancy, from their similitude to the Laws of the Hebrews, called the Ten Commandments, that they are of modern date, let the fancy vanish. The Hebrews were Scythians as well as the Iberians![4]

Ollam Fola "built a fair palace at Tarrach only for

---

[1] Except Queen Mary, of the Tudors, who had a chair blessed by the Pope sent from Rome.
[2] *Annals of Clonmacnoise* and *Annals of the Four Masters*.
[3] J. D. Massey, *Tamar Tephi*, pp. 65, 66.
Some years ago Earl Balfour as Prime Minister was asked in the House of Commons to grant permission for search to be made for the Ark in the Hill of Tara—the request was refused. The Time had not arrived.
[4] Words of Dr O'Connor, quoted in *The Royal House of Britain*, p. 11.

THE HOUSE OF DAVID

the learned sort of this realm." It was a Royal University to be maintained by the Government of the day. It was a school of the Prophets.[1]

¶ 11. By the marriage of Eochaid and Tamar Tephi the two lines of Zarah and Pharez, the sons of Judah, were united. Thus was Jeremiah's mission (in Jer. i, 1–10; xv, 11–21, *"to take the precious from the vile"*) accomplished and the prophecy in Ezekiel's parable (Ezek. xvii, 22–24) fulfilled:

"Thus said the Lord God: I will take of the highest branch [Pharez] *of the high cedar* [Judah] *and will set it: I will crop off from the top of his young twigs* [Zedekiah] *a tender one* [female], *and will plant it upon an high mountain and eminent* [Britain], *in the mountain of the height of Israel will I plant it: and it shall bring forth branches, and bear fruit, and be a goodly cedar* [our Royal House]; *and under it shall dwell all fowls of every wing; in the shadow of the branches thereof shall they dwell* [the British Commonwealth of nations, under one King]. *And all the trees of the field* [the other Royal houses in Europe] *shall know that I the Lord have brought down the high tree* [Pharez], *have exalted the low tree* [Zarah], *have dried up the green tree* [Zedekiah, the last of Pharez' line], *and have made the dry tree* [Zarah exiled] *to flourish: I the Lord have spoken and done it."*

And W. M. H. Milner in *The Royal House of Britain: An Enduring Dynasty* has produced that wonderful genealogical chart (attempt at its contraction overpage)—where the king is, there will the people be found.

To David: *"Thine house and thy kingdom shall be established for ever before thee: thy throne shall be established for ever"* (2 Sam. vii, 16, following v. 10: *"I will appoint a place for My people Israel, and will plant them, that they may dwell in a place of their own and move no more"*).

"For thus said the Lord: David shall never want a man to sit upon the throne of the House of Israel" (Jer. xxxiii, 17 —the word "ish" in Hebrew meaning man or woman).

[1] Milner, pp. 11, 12.

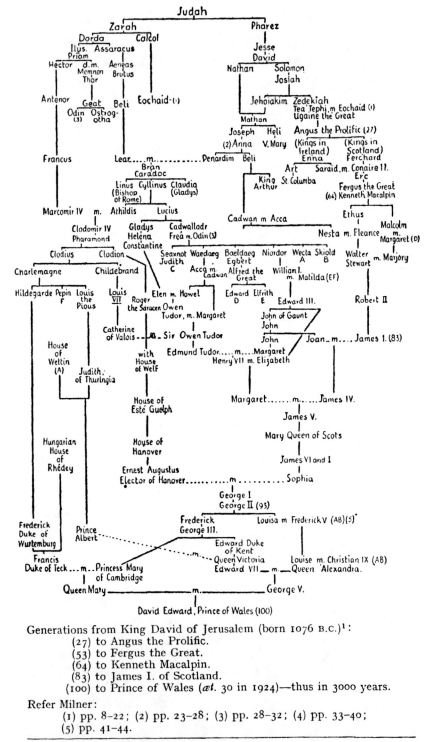

Generations from King David of Jerusalem (born 1076 B.C.)[1]:
    (27) to Angus the Prolific.
    (53) to Fergus the Great.
    (64) to Kenneth Macalpin.
    (83) to James I. of Scotland.
    (100) to Prince of Wales (æt. 30 in 1924)—thus in 3000 years.

Refer Milner:
    (1) pp. 8–22; (2) pp. 23–28; (3) pp. 28–32; (4) pp. 33–40;
    (5) pp. 41–44.

---

[1] Shows descendant, but Milner shows from son to son.

THE HOUSE OF DAVID

And the Prince of Wales, *æt.* thirty in 1924, becomes the 100th generation in 3000 years from King David, born 1076 B.C., the first king by Divine appointment. Thus, "*Judah, thou art he whom thy brethren shall praise . . . thy father's children shall bow down before thee. Judah is a lion's whelp . . . he couched as a lion. The sceptre shall not depart from Judah, nor a lawgiver from between his feet, until Shiloh come,*[1] *and unto him shall the gathering of the people be*" (Gen. xlix, 8–10, Israel's blessing).

Jacob here is speaking to his family: "*Jacob called unto his sons and said, Gather yourselves together, that I may tell you that which shall befall you in the last days . . . and hearken unto Israel your father*" (xlix, 1, 2).

And 3620 years later, "in the last days," the Zarah-Pharez King[2] of all Israel (God bless him) speaks on Christmas Day (1934) by broadcast to his family, all the House of Israel, throughout the earth:

> On this Christmas Day I send to all my people everywhere my Christmas Greeting. The day, with its hallowed memories, is the Festival of the Family. I would like to think that you who are listening to me now, in whatever part of the world you may be, and all the peoples of this Realm and Empire are bound to me and to one another by the spirit of one great family. . . . My desire and hope is that the same spirit may become ever stronger in its hold and wider in its range. . . . I send a special greeting to the peoples of my Dominions overseas. Through them the family has become a fellowship of free nations, and they have carried into their own homes the memories and traditions of the Mother Country. With them I bear in my heart to-day the peoples of my far-distant Colonies. The bond of the one spirit knows no barriers of space. . . . May I add very simply and sincerely that if I may be regarded as in some true sense the head of this great and widespread family, sharing its life and sustained by its

---

[1] "*Until Shiloh come*": Isa. xi, 1; Ezek. xxi, 27; Luke i, 32, 33.

[2] On whose Royal Coat-of-Arms is "*Dieu et mon Droit*" (God and my Birthright).

And of our King at this Jubilee it might have been written: "*Blessed be the Lord thy God, which delighted in thee to set thee on the throne of Israel; because thy God loved Israel to establish them for ever, therefore made he thee King over them, to do judgment and justice*" (2 Chron. ix, 8).

affection, this will be a full reward for the long and sometimes anxious labours of my Reign of well-nigh five-and-twenty years.

As I sit in my own home I am thinking of the great multitudes who are listening to my voice, whether they be in British homes or in far-off regions of the world. For you all, and especially for your children, I wish a happy Christmas. I commend you to "The Father of Whom every family in heaven and on earth is named." God bless you all.

Then all Israel, at the baton of Ephraim, first Britain, then Australia, then Canada, then South Africa, and finally in unison, in strains which circled the earth, sang the National Anthem, *God save the King*, giving to him an Empire's allegiance; just as Israel, in the early days of becoming a nation, gathered at Mizpeh and acclaimed its first king, "*and all the people shouted and said, God save the king*" (1 Sam. x, 24).

"*The Sceptre shall not depart from Judah . . . until Shiloh come*" (see Chapter XII).

"*And there shall come forth a rod out of the stem of Jesse, and a Branch shall grow out of his roots . . . with righteousness shall He judge*" (Isa. xi, 1–5).

"*Of the increase of His government and peace there shall be no end, upon the throne of David, and upon His Kingdom, to order it, and to establish it with judgment and with justice from henceforth even for ever*" (Isa. ix, 7).

"*Until He come, whose right it is, and I will give it Him*" (Ezek. xxi, 27).

"*The Lord God will give unto Him the throne of His father David and He shall reign over the House of Jacob for ever ; and of His Kingdom there shall be no end*" (Luke i, 32, 33).

His Second Coming is a definite expectation, and the indications from the Bible and from the Great Pyramid show its imminence.

¶ 12. (6) The line from Calcol has been traced to Ireland: that of Darda led to the kings in Britain (see genealogical table). *Circa* 1556 B.C., Darda founded Troy, to which came the Danites. Calcol had founded Athens, for the migration of Danaans in Argos.

THE ANCIENT BRITONS

Thus the Siege of Troy was fought between brothers.[1] Anchises and his sons fall, except Æneas, who sets fire to the city and cuts his way out to the Mount of Ida, and finally reaches Italy. He marries Lavinia, daughter of the reigning king, from whom are descended Romulus and Julius Cæsar. By his second wife, Edra, Brutus is born, who, when but fifteen years old, accidentally kills his father while hunting, and is banished from Italy.

Brutus joins Assaracus (Helenus), another Trojan prince, who had settled in Albania; they begin the second Græco-Trojan War, which ends victoriously; and Brutus marries Imogene, the daughter of the King of Greece.

With a navy of 332 ships he sets sail to find a kingdom, and at Legatta (Malta) he inquires of the oracle at the Temple of Diana.

The answer, written on the Temple of Diana in Caer Troia (London), has been translated by Pope:

> Brutus! There lies beyond the Gallic bounds,
> An Island, which the Western Sea surrounds,
> By ancient giants held, now few remain
> To bar thy entrance or obstruct thy reign.
> To reach that happy shore thy sails employ,
> There fate decrees to build a second Troy,
> And found an Empire, in thy Royal line,
> Which time shall ne'er destroy nor bounds confine.

Reinforced by colonies in Spain, and warring with the Gaels off the Loire for revictualling, he sets foot at Talnus (Totnes), 1100 B.C. He is welcomed by the descendants of Hu Gadarn, is made King, and founds Troynovantum (London); the whole island is regarded as one kingdom, and the established church Druidic: and the laws[2] promulgated are found in the *Triads of the Cymry*: they were later codified by Dyvnwal Moelmud:

Three things all Britons compelled to attend: Military Service, the Worship of God, the Courts of Justice.

---

[1] J. R. Morgan, *Kymry in Cambria*; Wm. James, *From Troy to Britain*; R. W. Morgan, *History of Britain from the Flood to A.D. 700*; R. W. Morgan, *St Paul in Britain*.

[2] Lord Chancellor Fortescue, *On the Laws of England*: "So the Kingdom of Britain had its original from Brutus and the Trojans who attended him from Italy and Greece, and were a mixed Government compounded of the regal and the democratic."

Three things the duties of man: worship God, do wrong to no man, fight for his country.

Three tests of Civil Liberty: equality of rights, equality of taxation, freedom to come and go.

Three persons who have a right to public maintenance: the babe, the old, the foreigner who cannot speak the British tongue.

Three orders exempt from bearing arms: the bard, the judge, the graduate-at-law—these represent God and His peace, and in their hand must no weapon be found.

Three ends of law: prevention of wrong, punishment of wrong inflicted, insurance of just retribution.

Three guarantees of society: security for life and limb, security for property, security of the rights of nature.

Three things that require the unanimous vote of the nation to effect: deposition of the sovereign, introduction of novelties of religion, suspension of law.

Dyvnwal also designed and partly made the royal British military roads through the island—pitched and paved; these being completed by Belinus, his son.[1]

Brennus, the younger brother, rebelled, but defeated by Belinus at Calater, fled to Gaul, where he married the daughter of Seguin, king of the Ligurians. On succeeding to the throne, he collected his forces, and landing near Battle was opposed by Belinus. Corwenna, their aged mother, intervened; throwing her arms round Brennus and kissing him, she besought him "to save her from the horrible spectacle of seeing the children of her womb engaged in impious hostilities against God, the laws of nature, their country and themselves."[2] Bareheaded and un-

---

[1] R. W. Morgan, *History of Britain*, pp. 43-45.
Nine roads are given:
1. Irish Road, or Watling Street, from Dover to Mona and Penvro.
2. Icknield Street from London, northward, through eastern districts.
3. Icknield Street from mouth of Tyne to St David's.
4. Pevensey to Edinburgh.
5. London to St David's.
6. Salt-mines of Cheshire to mouth of Humber.
7. Salt-mines to Portsmouth.
8. Torbay to Dunbreton on the Clyde.
9. Military road following the coast round the Island.

[2] *Ibid.*, p. 49.

armed, Brennus went with her to his brother, amidst the silence of the two armies drawn up for battle. Belinus dismounted from his chariot and, throwing away lance, met him half-way, and they embraced, to the thundering cheers of both armies; Britons and Gauls mingled to celebrate a day of joy and festivities.

The joined armies, 300,000 strong, landed at the mouth of the Seine, one battle decided any dissentient Gallic tribes. Within two years all Celtica was united under British rule; Belinus conquered the Teutonic tribes, and founded Aquileia, "where he was afterwards worshipped as a god." [1]

Brennus crossed the Alps 200 years before Hannibal, and defeated the Etrurians.

Plutarch writes: "His forces injured no man's property, they neither pillaged the fields, nor insulted the towns." [2] "The greater and more war-like Cimbri live in the Northern Ocean, in the very ends of the earth . . . as to their courage, spirit, force and vivacity, we can compare them only to a devouring flame. All that came before them were trodden down, or driven onwards like herds of cattle."

Brennus then advanced on Rome, and at the confluence of the rivers Allia and Tiber routed the Romans (6th June 490 B.C.) with such slaughter that it was noted on their calendar as "the black day." The capital held out for six months—during which, Fabius Dorso proceeded in his robes through the Cymric army to the Quirinal hill to offer sacrifice on the clan-day of his family, and returned in safety to the surprise of the Roman writers. It was ransomed for 1000 lbs. weight in gold, and when being weighed Brennus flung his sword in the opposite scale. When asked by the Roman Consul, "What means that act?" his reply was taken in silence, "It means, Woe to the vanquished!" [3]

He reigned for thirty years in Northern Italy, his Cymro-Keltic kingdom henceforth known as Cisalpine Gaul; he

---

[1] *Ibid.*, pp. 53, 57.
[2] The tradition of 490 B.C. is still that of the British Army; *cf.* the "Kultur" of the Prussians in August A.D. 1914.
[3] Morgan, p. 56.

founded Milan, Como, Brescia, Verona, Bergamo, Mantua, Trent, Vicentia ; proclaimed equality of laws, made roads, and encouraged literature.

¶ 13. In 57 B.C., Caswallon, with 50,000 men, "the second silver host," landed at Brest. Flûr, or Flora, daughter of Mygnach Gôr, to whom he was engaged, had been forcibly carried off by Morchan, a Regulus of Aquitania, at the instigation of Julius Cæsar, so affirm the Triads.[1] Caswallon stormed the castle of Morchan and brought Flora in safety to Caer Troia: one Roman Consul, Lucius Præconinus, was routed at Tolosa with the loss of 6000 men, and the proconsul Lucius Manilius fled, with the loss of his whole commissariat.

Cæsar states that the Britons were the first aggressors!

The Veneti of Armorica carried on a flourishing trade with Britain, and whenever a coastal city was besieged their navy ran in and evacuated the population for some other place. Cæsar first attacked their fleet (Cæsar, lib. iii), choosing a dead calm which nullified its manœuvring superiority, and after a battle from morning to night he gained the victory, which was stained by the massacre of the Venetine senate and every prisoner of war sold into slavery.[2]

The Druidic hierarchy imposed the Seal of Silence, and Cæsar obtained no information of Britain from Gaul. He sent Comius of Arras to demand tribute, the reply came [3]:

Caswallon, Pendragon of Britain, to Caius Julius Cæsar, Consul.

We have received your letters demanding tribute and submission on the part of this Island of Britain, to the Senate of Rome. The ambition of the Roman people we know to be insatiable; Europe is too little for them—they covet the riches of the nation whom the ocean itself divides from the rest of the world. But our possessions alone will not content them;—we must cease to be free, we must become their slaves. The Britons and the Romans derive their descent from the same Trojan origin [4]—such consanguinity should be the firmest

---

[1] Morgan, p. 69. [2] *Ibid.*, 70. [3] *Ibid.*, 76.
[4] See p. 197, and Milner's chart, p. 194.

THE ANCIENT BRITONS

guarantee of peace and equality between them. Our alliance we freely tend Rome; but as for submission, we have never hitherto known the thing, even by name. If the Gods themselves invaded our liberties, we would to the utmost of our power defend them—much more are we prepared to do so against the Romans, who are like ourselves but men.[1]

The first invasion lasted fifty-five days, and Cæsar was never able to advance more than seven miles from the spot where he landed (5th Aug. to 23rd Sept. 55 B.C.),[2] nearly losing his life when Nennius charged his own Tenth Legion, Cæsar covering its Eagle on the ground with his body and leaving his sword buried in the Briton's shield.

During the winter 600 additional transports and 28 ships-of-war were constructed, and secret negotiations made with Avarwy, a prince of the Coranidæ, who had applied for and received the protectorate of Rome.

At the threat of the second invasion, the Gorsedd was convened by Caswallon in London, who urged battle on the beach as before, and to prevent a single hostile camp on British soil. Avarwy urged that it was derogatory to the honour of the nation, and that Cæsar should be allowed to land and the battle fought on equal terms. The council resolved "that it was beneath the dignity of the nation of the Britons to defend their country otherwise than by the might of manhood, and that the landing of the Cesaridæ be unopposed."

Cæsar landed on 12th May 54 B.C., in the Isle of Thanet, where Avarwy commanded, made it his base, and in the ensuing battle for the passage of the Stour, as arranged, the Coranidæ under Avarwy deserted to the Romans. Caswallon lost heavily, but covered his retreat by the defence of Canterbury. A storm which wrecked his camp in Thanet delayed Cæsar. He re-formed his base and placed a guard of 10,000; then advanced with 38,000 Romans, and 20,000 Coranidæ under Avarwy.

A series of battles led to Chertsey, where Cæsar crossed the Thames, but five feet deep in a dry summer, behind

[1] Is it the language of the chief of blue-painted savages?
[2] Read R. W. Morgan, *History of Britain*, pp. 69–140.

elephants carrying a tower manned by archers. The chariot horses turned aside from the odour of the elephants advancing roaring and trumpeting. Caswallon fell back in guerrilla warfare towards Verulam, his depot, which he evacuated to lure Cæsar towards it. Then two of his lieutenants by a wide detour fell on Cæsar's base in Thanet, which Quintus Atius after an obstinate struggle just managed to save. If the camp and fleet at Thanet fell, as seemed likely at the second attack, the position of Cæsar's army with winter approaching was precarious.

Caswallon with treachery in Britain was ready to listen to overtures. So a treaty was concluded, and at the White Mount, Cæsar was entertained with chivalrous magnificence for seven days by Caswallon, a *fête* sung by all the bards. Avarwy had taken refuge with the Roman fleet. Cæsar embarked with all his forces, but with no hostages, with no tribute, in the presence of Caswallon and the British nobles (26th Sept. 54 B.C.).

His prestige had so suffered that the Arverni, Treviri, Eburones, Senones, Sicambri had risen to arms. Before he could go into winter quarters, the conquest of Gaul had to be re-enacted.

Aulus Gellius in account of Cæsar's invasion:

Horribilis ille Britannorum clamor, Tori pen i Caisar

("off with Cæsar's head").

Lucan records a common sarcasm of the Pompeian party against Cæsar:

Territa quæsitis ostendit terga Brittanis,

(when he lost his sword to Nennius).

Horace wrote that Britain was hardly touched by Cæsar's campaign,

Intactus aut Britannus ut descenderet sacra catenatus via,

and the Briton walked the streets of Rome the only freeman in Europe,

Invictus Romano Marte Britannus (Tibullus),

while for ninety-seven years no Roman ventured to set a

hostile foot in Britain (Morgan, *St Paul in Britain*, pp. 77–80).

King Tenuantius sent his two sons, Cynvelin and Llyr, to Rome, where they were educated by Augustus in his own palace with his nephews.

Cynvelin fought with the Romans under Germanicus; and later, when himself king, refused Augustus' demand for the restoration of estates to three Reguli of the Coritani, confiscated for treason in Cæsar's campaign. War threatened, but Augustus, on the advice of Horace (Ode 35, lib. v), abandoned his designs, reduced the heavy duty on British goods to light tariffs (*Strabo*, lib. iv, c. 5), and British nobles again resided amicably in Rome.

CHAPTER VII

¶ 1. *Encyclopædia Britannica*: "The Anglo-Israelite theory rests on premises which are deemed by scholars—both theological and anthropological—to be utterly unsound."

Quite; so it was anticipated; and the answer given—by One—"*The God of Abraham, of Isaac, of Jacob is not the God of the dead, but of the living*" (Matt. xxii, 32).

"*Shut up the words . . . even to the time of the end: many shall run to and fro, and knowledge shall be increased*" (Dan. xii, 4).

"*Blindness in part has happened to Israel, until the fulness of the Gentiles be come in*" (Rom. xi, 25).

"*For the vision is yet for an appointed time, but at the end it shall speak, and not lie*" (Habakkuk, ii, 3).

"*Thus saith the Lord, which giveth the sun for a light by day, and the ordinances of the moon and of the stars for a light by night, which divideth the sea when the waves thereof roar; The Lord of Hosts in His name. If those ordinances depart from before Me, saith the Lord, then the seed of Israel also shall cease from being a nation before Me for ever.*[1]

"*Thus saith the Lord, If heaven above can be measured, and the foundations of the earth searched out beneath, I will also cast off all the seed of Israel for all that they have done, saith the Lord*" (Jer. xxxi, 35–37).

"*Thus saith the Lord, If My Covenant be not with day and night, and if I have not appointed the ordinances of heaven and earth; then will I cast away the seed of Jacob, and David My servant . . . for I will cause their captivity to return, and have mercy on them*" (Jer. xxxiii, 25, 26).

[1] Are these the words of a local Jahwe who has failed to make his bedouin tribe, Israel, into a nation?

"*Woe be unto the pastors that destroy and scatter the sheep of My pasture! saith the Lord.*

"*I will gather the remnant of My flock out of all countries whither I have driven them, and will bring them again to their folds, and they shall be faithful and increase. And I will set up shepherds over them which shall feed them*" (Jer. xxiii, 1–4).

"*. . . in the latter days ye shall consider it perfectly*" (Jer. xxiii, 20).

"*Prophesy against the shepherds of Israel . . . that do feed themselves. . . . Neither have ye brought again that which was driven away, neither have ye sought that which was lost. . . . And they were scattered because there is no shepherd.*

"*My sheep have wandered through all the mountains and upon every high hill*[1]; *yea, My flock was scattered upon all the face of the earth, and none did search or seek after them.*

"*For thus saith the Lord God, Behold I, even I, will both search My sheep, and seek them out . . . I will seek out My sheep, and will deliver them out of all places where they have been scattered in the cloudy and dark day; and I will bring them out from the people, and gather them from the countries, and will bring them to their own lands, and feed them upon the mountains of Israel, by the rivers, and all the inhabited places of the country. I will feed them in a good pasture, and upon the high mountains of Israel shall their fold be. . . . I will seek that which was lost and bring again that which was driven away.*

"*Therefore will I save My flock, and they shall be no more a prey.*

"*And I will set up one shepherd over them, and he shall feed them, even My servant David. . . . And I the Lord will be their God, and My servant David, a prince among them; I the Lord have spoken it. And I will make with them a covenant of peace, and I will make them and the places round about My hill a blessing. There shall be showers*

---

[1] Mountains = large countries or nations, hill = a tribal area, high mountains of Israel = Britain.

of blessings . . . the earth shall yield her increase, and they shall be safe in their land, and shall know that I am the Lord, when I have broken the bands of their yoke, and delivered them out of the hand of those that served themselves of them. They shall dwell safely . . . they shall be no more consumed with hunger in the land.

"*Thus shall they know that I the Lord their God am with them, and that they, even the House of Israel, are My people, saith the Lord God; and ye, My flock, the flock of My pasture, are men, and I am your God, saith the Lord God*" (Ezekiel xxxiv).

"*Go and tell this people, Hear ye indeed, but understand not; and see ye indeed, but perceive not . . . lest they see with their eyes, and hear with their ears, and understand with their heart, and convert, and be healed*" (Isa. vi, 9, 10).

"*The Lord hath poured upon you the spirit of deep sleep, and hath closed your eyes; the prophets and your rulers, the seers hath He covered. And the vision of all is become unto you as the words of a book that is sealed. For the wisdom of their wise men shall perish, and the understanding of their prudent men shall be hid. . . . And in that day*[1] *shall the deaf hear the words of the book, and the eyes of the blind shall see out of obscurity, and out of darkness*" (Isa. xxix, 10, 11, 14, 18).

"*I will bring the blind by a way that they knew not; I will lead them in paths that they have not known; I will make darkness light before them, and crooked things straight. These things will I do unto them, and not forsake them. . . . Hear, ye deaf, and look, ye blind, that ye may see. Who is blind, but My servant . . . who so blind as the Lord's servant? Seeing many things, but thou observest not; opening the ears, he heareth not*" (Isa. xlii, 16–20).

"*Behold, the days come, saith the Lord God, that I will send a famine in the land, not a famine of bread, nor a thirst for water, but of hearing the words of the Lord. And they shall wander from sea to sea, and from the north even to the east, they shall run to and fro to seek the word of the*

---

[1] "That day" defined (pp. 394 *seq.*); the wise and prudent = the modern "shepherds."

THE ANGLO-ISRAELITE THEORY

Lord, and shall not find it. In that day, shall the fair virgins and young men faint for thirst" (Amos viii, 11-13).

"*O Lord, revive Thy works in the midst of the years, in the midst of the years make known*" (Habakkuk iii, 2).

¶ 2. The Matriculation English History, as taught, begins with the words of Julius Cæsar—Why? . . . "painted savages"!! [1]

And our University professors and dons still bow in admiration of Græco-Roman history and philosophy: until youth finds so little in Latin and Greek to interest.

Why not teach the real history of Britain, its own Druidic philosophy, from the purer source of Seth?

The Greek philosophy of Socrates, Plato, Aristotle was the reaction of the higher type of Japhitic-Semite mind to the crude and ancient Cain-Bel-Ham mythology of polytheistic gods—a striving for a better, but still a "run to and fro" way of life.[2]

Socrates demonstrates "the vagueness and hollowness of men's notions and beliefs about most things—that there were alternatives to many beliefs sincerely or vaguely held: nothing is to be taken for granted."[3]

"For the first time the reason and intellect of man began to grapple with natural science and an attempt by logical reasoning to grasp the meaning of life, of the world, of man's own being."

From Plato, "the Good, the True and the Beautiful exist in Forms (Ideas) . . . the dim imperfect shadows of Forms that are real."[4]

Bury writes: "Plato has left us the first systematic defence of Theism we know of and it is based entirely on his doctrine of the soul as the self-moved mover. And

---

[1] The ancient Britons stripped to the waist to fight—as now; not armour-protected, but with might of arm alone, not fearing cold steel; tattooed as now, or covered with some coloured antiseptic grease (*vide* methylene blue in modern surgery): the dye being obtained from the leaves of *Isatis tinctoria*, a plant with yellow flowers, growing to a height of three or four feet, and still found in Flanders and France.

[2] See pp. 330, 351.

[3] *Outline of Belief*, part iii, p. 196.

[4] *Ibid.*, part vi, p. 370.

Nietzsche spoke of Plato's ideals as 'Christianity before Christ.'" [1]

From Druidism all such can be derived more truly and less indefinitely.

Of Aristotle—"his whole life devoted to gathering facts, and classifying and ordering of knowledge." [1]

From the seventh century B.C. onwards, Greece was in contact with the knowledge of the Egyptian priests (and more directly so after the conquest of Egypt by Alexander the Great), and various schools had been founded in Alexandria—until in A.D.642 the famous library was burnt by order of the Caliph Omar.

> The Egyptian priests tried to keep such astronomical knowledge as they possessed to themselves. . . . They tried to impress Pythagoras with the vast extent of their own learning and at the same time seeking to obscure the real facts—and their ignorance of the derivation of the facts—by dogmatic and mystifying assertions . . . but they were ill-fitted for dealing in debate with an intellect so original, independent and penetrating as his . . . combined with his patience and readiness to comply with their regulations.[2]

From this source, and thus from the Pyramid, Pythagoras taught that the planets revolved around a common centre—which was not accepted by scientific circles in Europe until 2000 years later.

There also studied in Egypt—Thales, Plato, Euclid, Archimedes, Eratosthenes, and others.

Thus through the scientific knowledge that was of the Adamic race, embodied in the Pyramid by the House of Seth, though perverted by the Egyptian priesthood, there came to Europe through Greek philosophy this light to counteract the darkness that Cain-Bel influence had caused to cover knowledge.

Yet that influence took hold of such knowledge, and caused the mind of man, using Reason more than Spiritual Intuition, to soar to create on earth a man-made heaven —"*Your eyes are opened, ye are as gods, knowing good and*

[1] *Outline of Belief*, part iv, p. 245.
[2] See Davidson, *The Great Pyramid*, pp. 98-107.

evil"[1]; and from Socrates to this day man's mind is still thinking that it by Reason alone can found a Utopia.

Henry VIII of the Tudors (descendant of the Cymry and the old Druidism) did something to separate Britain, its Church and its Universities, from Continental religion and scholarship. Further separation was made by the Reformation to Protestantism.

But the divergence was not sufficient to redirect Britain to its own history, to the Druidism of Siluria and the Christianity of Glastonbury.

Thus has been produced the religion of the modern University undergraduate:

> Theology versus atheism is a dead issue, left to dons, misguided and academic, who devote themselves to the vain task of trying to express in words the unspeakable. Religion is simple, it is to be in love with life, to love the good things of the world; and morals simple too, they are the art of living happily together with other people. Man knows he will return, whence he came, to dust; it is the common fate of all men and animals too. His interest is in the real world, and in real living; and so he works for the building of heaven on earth, not in the clouds; on earth comrades can build it, if they will.

And his pattern is Karl Marx and Lenin; and he may add that, if he is good, he may reach a life after death, of which there is some evidence in Spiritualism.

Cain, with a cynical smile,[2] looks on, well content with the tares that he had sown in the Sumerian civilisation nearly 6000 years ago. He could not have wished it better put by the descendants of the Adamic race, knowing that they would reject the polytheism which was good enough for his aborigine subjects and their mixed and heathen descendants.

And the eyes of theologians are bent to the ground of Sumer, which is being dug by the Assyriologists, expecting to find in the grave of Cain the treasure of the Biblical source of Creation; instead of looking into the lidless coffer under the rejected Apex Stone of the Great Pyramid.

---
[1] Gen. iii, 5.
[2] See frontispiece of Mrs Bristowe's *Sargon the Magnificent*.

And we still see prelates and dons delving amidst Græco-Roman literature and in praise of its culture for an inspiration of God, getting more and more confounded by the metaphysical and the philosophical attributes of His nature, to the growing discontent of their congregations and undergraduates.

They have the idea of Christ, without all Nature's backing; so preach with lessening conviction to empty pews, and then only with the promises of punishment, of which they are uncertain, and of reward, which never satisfies the seeking mind of man—man who is not a passive creature, but who, "up and doing," demands "bread and water" and radiation *now*.

"*The kingdom of God is at hand.*"

And the ground of Europe is strewn with tares—in Britain, the vacant socialistic theory of the Varsity youth; in Russia, a generation who knows no god but a blood-stained State; in Italy, the revival of ancient Roman arms; in Germany, the revival of ancient Assyrian gods by other names; in France and the U.S.A., the worship of the god of gold, as did the Babylonian Nebuchadnezzar; at Geneva, a second Babel, who mention not the name of God.

England! O England! that feedeth thy youth on tares, that glorious youth who would yet die for a Cause, as it did on Flanders battlefields.

¶ 3. Thus has been traced the story of the ten tribes of Israel, after the Assyrian captivity, "scattered" and "gathered" into a land of their own.

Turning now to the tribes of Judah and Benjamin on their return from the Babylonian captivity on the decree of Cyrus "the shepherd," 534 B.C.

The remnant of the House of Judah, now for the first time called the Jews, settled in and around Jerusalem to rebuild the Temple and city. They were ruled by the priests, and later sent to the Massagetæ on the Caspian for Levites for the service of the Temple.

Samaria being occupied by the Assyrian colonists, the

THE RETURN FROM BABYLON

House of Benjamin settled on and around the shores of Galilee.

To the Hebrew of the eighth century B.C., immersed in idolatry, the idea of Spirit Dominion over Matter and immortality of the first Adam, forfeited by sin (deviation of freewill), given in the Prologue (Gen. i–xi) was dead. Relics of such conception had been carried by (1) the House of Cain, in their royal pit graves of Ur, and (2) by the House of Seth, in Druidism and the Pyramid symbology, which was retained by the Egyptian priesthood.

"*I am the Lord thy God, which brought thee out of the land of Egypt, from the house of bondage. Thou shalt have none other gods before Me*" (Deut. v, 6, 7).

Israel interpreted this that Yahweh was their own God, just as neighbouring nations had their own gods—the influence of Cain-Bel teaching.

"But this low nationalistic conception of God was overthrown by the monotheistic teaching of the great eighth-century prophets. Yahweh, they taught, was the God of all the earth, and there was no God beside Him, as such all nations were His." [1]

Then also "the individual in Judaism had no higher expectation than an unending existence in Sheol,[2] where social and national distinctions prevailed but not moral, for good or bad fared exactly alike." Sheol was "outside the jurisdiction of Yahweh" and "No individual retribution had been looked for . . . for Yahweh was concerned in the well-being of the nation as a whole, and not with its individual members . . . until monotheism had taught the worth of the individual soul, its immediate relation to Yahweh (first in the prophecies of Jeremiah and Ezekiel)." [3]

The concern of the Book of Job, "written about or before 400 B.C., from first to last is the current doctrine of retribution" (that good or ill fortune on earth reflects

---

[1] R. H. Charles, *Religious Development between the Old and New Testaments*, p. 65.
[2] *Ibid.*, p. 50. "Sheol (or Hades), an abode of misery and wretchedness, not Gehenna, the place of torment by fire . . . until after 100 B.C." (p. 121).
[3] *Ibid.*, p. 105.

man's righteousness or sin)—"*after my skin has been destroyed, without my body I shall see God*" (Job xix, 25-27)—appealing to the God of faith that this world was a moral chaos; so leading on, in Psalms xvi, xvii, xlix, lxxiii, to the idea of a future life.[1]

The seeming gap between the Old and New Testaments in the Book of His-Story is filled by the Apocalyptic eschatology.[2]

While the ordinary man saw only the outside of things in all their incoherence and isolation, the apocalyptist sought to get behind the surface and penetrate to the essence of events, the spiritual purposes and forces that underlie and give them their real significance.

Apocalyptic sketched in outline the history of the world and mankind, the origin of evil, its course, and inevitable overthrow, the ultimate triumph of righteousness, and the final consummation of all things. . . . It was thus a Semite philosophy of religion, ever asking, Whence? Wherefore? Whither? and it put these questions in connection with the world, the Gentiles, Israel, and the individuals. Apocalyptic . . . grasped the great idea that all history, alike human, cosmological, and spiritual, is a unity—a unity following naturally as a corollary of the unity of God preached by the prophets (Charles, p. 18).

From the Old Testament we pass to the Apocrypha and Pseudepigrapha of the Old Testament which were written between 180 B.C. and A.D. 100. . . . The two centuries preceding the Christian era . . . so far from being ages of spiritual stagnation and darkness . . . might with justice be described as the two most fruitful centuries in religious life and thought in the history of Israel.[3]

¶ 4. And this "light," the re-seeking of God as Creator, Redeemer, Restorer, of man's spiritual faculties, of individual immortality, and the expectation of the Messianic Kingdom, to "*make straight the path*" for the Coming of the Christ, emanated from the tribe of Benjamin.

---

[1] Charles, p. 110.

[2] Apocalyptic = revelation, in distinction to prophecy. Eschatology = the doctrine of the last things, the end of the present world order, and the hereafter.

[3] Charles, p. 115.

Moses' blessing of Benjamin—"*The beloved of the Lord shall dwell in safety by him; and the Lord shall cover him all the day long, and he shall dwell between his shoulders.*"[1]

And in the division of the Kingdom, Benjamin assigned to Judah—"*will give one tribe, that David, My servant, may have a light alway before Me in Jerusalem, the city which I have chosen to put My name there.*"[2]

This Apocalyptic literature was written probably for the most part in Galilee, the home of the religious seer and mystic. Not only was it the development of a religious but also of an ethical character. In both these respects the way was prepared by this literature for the advent of Christianity, while a study of the New Testament makes it clear that its writers had been brought up in the atmosphere created by these books . . . the hopeless outlook of the faithful individual in the Old Testament was transformed into one of joy . . . an everlasting existence in the unblessed abode of Hades . . . was transformed into the hope of a blessed immortality.[3]

It was from the apocalyptic side of Judaism that Christianity was born—and in that region of Palestine where apocalyptic and not legalism held its seat, even in Galilee.

Whence came Our Lord and eleven of His disciples.[4]

All the great Jewish apocalypses written before A.D. 10, which carried on the mystical and spiritual side of religion as opposed to the legalistic, Judaism dropped and banned after its breach with Christianity.[5]

Outside Israel there might be some true knowledge of God . . . but somehow it always failed prematurely or became corrupt.[6] In this respect Israel proved itself the chosen nation, for in it the succession of seers, prophets, wise men and apocalyptists was unbroken[7] and lasted till . . . the Greek and Roman empires had destroyed the old barriers that divided nation and nation, and so prepared the way whereby Christianity could . . . become the religion of the world.[8]

---

[1] Deut. xxxiii, 12. [2] 1 Kings xi, 36. [3] Charles, p. 9.
[4] Ibid., p. 33. [5] Ibid., p. 44.
[6] *Vide* the Sumerians and Egyptians, Brahminism, Buddhism, Zoroastrianism.
[7] By Benjamin, rest of Israel "were scattered" and Judah coming under the yoke of the Law.
[8] Ibid., p. 13.

The divergence in thought between the tribe of Benjamin and the Jews appears in this.

During the Babylonian Captivity a great Hebrew literature had been collected and re-edited. And now on the return of the Jews, from the time of Ezra and Nehemiah, "the Law was supreme, inspiration was officially held to be dead; God had spoken His last and final word through the Law, and the Canon was closed,[1] so that no new prophet could expect a hearing.

Thus the Apocalyptic books were the work of pseudonymous writers, whose assumed names are nearly always the names of ancient worthies in Israel—as Ezra, Baruch, Enoch, Moses, Isaiah.[2]

Expectation of God's intervention on behalf of His own people gradually developed into belief in the Messianic Kingdom. This with the belief in the immortality of the faithful led to the doctrine of the resurrection of the righteous (about 200 B.C.) to share together in the Kingdom. Under the savage persecution of Antiochus, to encourage the faithful . . . it was held that immediately after death men enter into a state of bliss or woe in Sheol which is but the prelude to their final destiny. . . . So a distinction was made between the righteous and the wicked and their judgment on the advent of the Messianic Kingdom . . . which is to be introduced suddenly or catastrophically by God Himself (Daniel, 1 Enoch 73–90), or by a gradual renewal of creation (Isaiah lxv, lxvi, Jubilees iv, 26; xxiii, 26–28).[3]

Later, with the non-appearance of the Messiah, the earth unchanged, untransformed, has now come to be considered as wholly unfit for the manifestation of this kingdom.

Thus about 64 B.C. the "Parables" of Enoch was written, in which developed a new and spiritual heaven and earth, into which flesh and blood could not find entrance, of which the righteous only partake, for everlasting, so that the resurrection and final judgment would take place before its advent.

---

[1] "The whole or substantially the whole Law was accepted by 444 B.C., the Prophets became part of the Jewish Scripture not improbably soon after 250 B.C., and 'the Writings' gradually obtained the same position within the next two or three centuries" (Professor George Gray).

[2] Charles, pp. 35–46.

[3] Ibid., pp. 51–54.

APOCALYPTIC ESCHATOLOGY

In Enoch is recorded the vision of *"One like unto the Son of Man coming with the clouds of heaven,"* who with his angels shall confound the kings of the earth, sit on the throne of God and judge the world.

After the fall of the Maccabees, of the House of Levi, from whom the Jews looked for the coming of the Messiah (Testament of XII Patriarchs), since the kingdom of the Messiah had not appeared from Judah—"the Pharisaic party was henceforth committed to political interests and movements, and looked for a militant Messiah, of the house and lineage of David (Psalms of the Pharisees, xvii, 23–36); a crucified Messiah was an impossible conception to the Judaism of that period."[1]

With them, God was the God of the Jews only, excluding the Gentiles; thus by refusing to part with the spiritual particularism of the past, and not accepting the universalism of the greater prophets, the Jews unfitted themselves for the reception of the teachings of Christ.[2]

Thus had developed in the land of Galilee—but not held by Rabbinic Judaism in the first century B.C.—the three chief notes of the coming kingdom of God. First, this kingdom was to be a kingdom within man—and so far to be a kingdom realised on earth. Secondly, it was to be world-wide, and to ignore every limitation of language and race. Thirdly, it was to find its true consummation in the world to come."[3]

The eschatological teaching of Benjamin has gradually been rejected by the legalistic element of the Jews, so that, in the Christian era, the former merges into Christianity,[4] while the latter having rejected it becomes Talmudic Judaism; and Legalism became absolute.

¶ 5. As an example of this eschatology, 2 Esdras is quoted, well described by Canon Charles: "This is the most profound and touching of the Jewish apocalypses."[5]

Esdras, alone and fasting, questions the purposes of the

---
[1] *Ibid.*, pp. 78–89.   [2] *Ibid.*, pp. 73–74.
[3] *Ibid.*, p. 71; quoting Christ's teaching on each: (1) Luke xvii, 20, 21; (2) Matt. xxi, 43; viii, 11, 12; xiii, 28; (3) Matt. xxii, 41; Mark ix, 1.
[4] As did the Druidism of Siluria.
[5] *Ibid.*, p. 249.

Most High in creating this world and in the scattering of His people.

And as the eschatology of Benjamin prepared for the Coming of the Messiah, so in these days, when the Times of the Gentiles are ending, many of the same queries of Esdras disturb the minds of men.

i. "*Thus saith the Most High, Have I not prayed you as a father his sons . . . that ye would be my people, and I should be your God; that ye would be my children; and I should be your Father? I gathered you together, as a hen gathereth her chickens under her wings; but now, what shall I do unto you?*

ii. "*Let them be scattered abroad among the heathen, let their names be blotted out of the earth, for they have despised my covenant. . . .*

"*Ask, and ye shall receive . . . the kingdom is already prepared for you, watch. . . . I will bring them out with gladness like a dove; stablish their feet, for I have chosen thee, saith the Lord.*

"*And those that be dead will I raise up again from their places, and bring them out of their tombs, for I have known my name in them . . . for I shall bring them out of the secret places of the earth, and show mercy unto them.*

"*I, Esdras, received a charge from the Lord upon the mount Horeb that I should go unto Israel; but when I came unto them, they would have none of me, and rejected the commandment of the Lord.*

"*And therefore I say unto you, O ye nations, that hear and understand, look for your shepherd, he shall give you everlasting rest; for he is nigh at hand, that shall come in the end of the world.*

"*Flee the shadow of this world . . . they that withdrew them from the shadow of the world have received glorious garments of the Lord.*

"*I, Esdras, saw upon the mount Zion a great multitude whom I could not number, and they all praised the Lord with songs. And in the midst of them there was a young man of high stature . . . and upon every one of their heads he set crowns.*

"*So I asked the angel and said, What are these, my Lord?*

"*He answered and said unto me, These be they that have put off the mortal clothing, and put on the immortal, and have confessed the name of God. . . .*

"*Then said I unto the angel, What young man is he that setteth crowns upon them?*

"*He answered, It is the Son of God, whom they have confessed in the world.*

iii. "*I, Esdras, was in Babylon, and lay troubled upon my bed, and my thoughts came up over my heart; for I saw the desolation of Zion, and the wealth of them that dwelt at Babylon. And my spirit was sore moved, so that I began to speak words full of fear to the Most High* [reviewing the story from the creation and Adam, "*unto him thou gavest thy one commandment, which he transgressed,*" Noah and the Flood; Abraham, Jacob, David].

"*Are the deeds of Babylon better than those of Zion? I have gone hither and thither through the nations, and I see that they abound in wealth and think not upon thy commandments. . . .*

"*Thou shalt find that men who may be reckoned by name have kept thy precepts, but nations thou shalt not find.*

iv. "*And the angel, Uriel, said to me, Thy heart hath utterly failed thee in regarding this world, and thinkest thou to comprehend the way of the Most High? Go to, weigh me a weight of fire, or measure me a measure of wind, or call me again the day that is past . . . thine own things, that are grown up with thee, canst thou not know, how then can thy vessel comprehend the way of the Most High?*

"*I said, . . . Wherefore is the power of understanding given unto me? For it was not in my mind to be curious of the ways above, but of such things as pass by us daily; because Israel is given up as a reproach to the heathen . . . and the law of our forefathers is made of none effect, and we pass out of the world as grasshoppers, and our life is as a vapour, neither are we worthy to obtain mercy.*

"*Then he answered me, . . . the world hasteth fast to pass away. For it is not able to bear the things that are promised to the righteous in the times to come. For the evil whereof thou asketh me is sown, but the gathering thereof is not yet come.*

"*For a grain of evil seed was sown in the heart of Adam from the beginning, and how much wickedness hath it brought forth unto this time! and how much shall it yet bring forth until the time of threshing come!*

"*I answered, How long? . . . wherefore are our years few and evil?*

"*And he answered, Thou dost not hasten more than the Most High, . . . for he hath weighed the world in the balance, and by measure hath he measured the times, and he shall not move nor stir them, until the said measure be fulfilled.*

"*Go thy way to a woman with child and ask . . . nine months, if the womb may keep the birth any longer.*

v. "*Behold, the days shall come that they which dwell upon the earth shall be taken with great amazement and the way of truth shall be hidden, and the land shall be barren of faith.*

"*But iniquity shall be increased above that which thou now seest, that which is after the third kingdom to be troubled; and the sun shall suddenly shine forth in the night, and the moon in the day . . . and the people shall be troubled . . . and he shall rule whom they that dwell upon the earth look not for* [1] *. . . and the Sodomitish sea shall cast out fish.*

"*There shall be chaos in many places; and understanding withdraw itself, and it shall be sought of many, and shall not be found; and unrighteousness and incontinency shall be multiplied upon earth.*

"*One land shall ask another, Is righteousness, is a man that doeth righteousness, gone through thee? And it shall say, No. At that time, men shall hope, but shall not obtain; they shall labour, but their ways shall not prosper.*[2]

"*I fasted seven days, mourning and weeping. Then . . . the thoughts of my heart were very grievous, and my soul recovered the spirit of understanding, and I said, . . . And now, O Lord, why hast thou given this one people over unto many . . . and scattered thine only one among many? . . .*

---

[1] See pp. 377, 409, 417–421.
[2] The present times.

of very grief have I spoken, while I labour to comprehend the way of the Most High, and to seek out part of his judgment.

"*And he said, Thou art sore troubled in mind for Israel's sake ; lovest thou that people better than he that made them ? even so canst thou not find my judgment, or the end of the love that I have promised unto my people. . . . The creature may not haste above the creator, neither may the world hold them at once that shall be created therein.*

"*Ask the womb of a woman . . . if thou bringeth forth ten children . . . at several times . . . at once? . . . even so have I given the womb of the earth to those that be sown therein in their several times.*

vi. "*In the beginning when the earth was made, before the outgoings of the world were fixed, or ever the gatherings of the winds blew, before the voices of the thunder sounded, and before the flashes of lightning shone, or ever the foundations of paradise were laid, before the fair flowers were seen, or even the powers of the earthquake were established, before the innumerable hosts of angels were gathered together, or even the footstool of Zion was established ; ere the present years were sought out, or even the imaginations of them that now sin were estranged, before they were sealed that have gathered faith for a treasure ;—then did I consider these things, and they all were made through me alone, and through none other : as by me also they shall be ended, and by none other.*

"*Then answered I, What shall be the parting asunder of the times ? or when shall be the end of the first, and the beginning of it that followeth ?*

"*And he said unto me, From Abraham unto Isaac, inasmuch as Jacob and Esau were born of him, for Jacob's hand held the heel of Esau from the beginning. For Esau is the end of this world,*[1] *and Jacob is the beginning of it that followeth. The beginning of a man is his hand, and the end of a man is his heel, between the heel and the hand seek thou nought else, Esdras.*

"*Behold, the days come, and it shall be that when I draw nigh to visit them that dwell upon the earth, and when I shall*

---

[1] Esau as Edom (Turkey) fell in December 1917: Ottoman Empire ended in 1924.

make inquisition of them that have done hurt unjustly with their unrighteousness, and when the affliction of Zion shall be fulfilled, and when the seal shall be set upon the world, that is to pass away; then will I show these tokens, the books shall be opened before the firmament, and all shall see together . . . and the trumpet shall give a sound, which when every man heareth, they shall be suddenly afraid. At that time shall friends make war one against another like enemies, and the earth shall stand in fear with those that dwell therein.

"And it shall be that whosoever remaineth after all these things that I have told ye of, he shall be saved, and shall see my salvation, and the end of my world.

"And they shall see the men that have been taken up, who have not tasted death from their birth, and the heart of the inhabitants shall be changed, and turned into another meaning.

"For evil shall be blotted out, and deceit shall be quenched; and faith shall flourish, and corruption shall be overcome, and the truth which hath been so long without fruit, shall be declared.[1]

"I wept again and fasted seven days . . . my soul was in distress [reviewing the creation], . . . Adam whom thou ordainest lord over all the works that thou hast made, of him come we all, the people whom thou hast chosen . . . for our sakes thou madest this [the first-born] world. As for the other nations which also came of Adam . . . these be lords over us and devour us. But we, thy people, whom thou hast called thy firstborn . . . are given into their hands. Why do we not possess for an inheritance our world? How long shall this endure?

vii. *"Said he unto me, . . . There is a city builded and set . . . full of good things, but the entrance thereof is narrow. . . . If this city now be given unto a man for an inheritance, if the heir pass not the danger set before him, how shall he receive his inheritance?*

"Even so is Israel's portion; . . . when Adam transgressed my statutes, then was decreed that now is done. Then were the entrances of this world made narrow. . . .

"For behold, the time shall come when these tokens of which

---

[1] The Second Coming—"*The Kingdom of God is at hand.*"

2 ESDRAS

I told thee before shall come to pass, that the bride shall appear, even the city coming forth, and she shall be seen that now is withdrawn from the earth. . . .

"*For my son Jesus shall be revealed with those that be with him, and shall rejoice them that remain four hundred years.*

"*After these years the world shall be turned into the old silence . . . like as in the first beginning . . . and the earth shall restore those that are asleep in her and the Most High shall be revealed upon the seat of judgment . . . truth shall stand, and faith shall wax strong . . . and then shall the Most High say to the nations that are raised from the dead, See ye and understand whom ye have denied, whom ye have not served.*

"*This is my judgment and the ordinance thereof.*

"*I answered, Who is there of them that be alive that hath not sinned, and who of the sons of men that hath not transgressed thy covenants? And now I see that the world to come shall bring delight to few, but torments unto many.*

"*And he answered, For this cause the Most High hath not made one world, but two. He that hath what is hard to get rejoiceth over him that hath what is plentiful, for I shall rejoice over the few that are saved, inasmuch as these are they that have made my glory now to prevail, and of whom my name is now named.*

"*I said, It were better that the dust itself had been unborn, so that mind might not have been made therefrom. But now the mind groweth with us, and by reason of this we are tormented, because we perish and know it. Let the race of men lament, and the beasts of the field be glad, for it is far better with them than with us, for they look not for judgment, neither do they know of torments or of salvation promised after death. For all that are born are full of sins and laden with offences.*

"*He answered, Now understand from thine own words for thou hast said the mind groweth with us. They therefore that dwell upon the earth shall be tormented for this reason, that having understanding they have wrought iniquity, and receiving commandments they have not kept them.*[1]

"*For how great a time hath the Most High been long-*

---

[1] Freewill deviating.

suffering with them that inhabit the world; and not for their sakes, but because of the times which he hath foreordained.

"*I said, . . . Shew further unto me whether in the day of judgment the just will be able to intercede for the ungodly, or to intreat the Most High of them.*

"*He answered, The day of judgment is a day of decision, and displayeth unto all the seal of truth. This present world is not the end, the full glory abideth not therein; therefore have they who were able prayed for the weak.*

"*But the day of judgment shall be the end of this time, and the beginning of the immortality for to come, wherein corruption is passed away, intemperance is at an end, infidelity is cut off, but righteousness is grown and truth is sprung up.*

"*I said, This is my first and last saying, that it had been better that the earth had not given thee Adam; or else, when it had given him, to have restrained him from sinning.*

"*For what profit is it for all that are in this present time to live in heaviness and after death to look for punishment?*

"*O thou Adam, what hast thou done? for though it was thou that sinned, the evil is not fallen on thee alone, but upon all of us that come of thee.*

"*For what profit is it unto us, if there be promised us an immortal time, whereas we have done the works that bring death? And that there is promised us an everlasting hope, whereas ourselves most miserably are become vain? And that there are reserved habitations of health and safety, whereas we have lived wickedly? And that the glory of the Most High shall defend them which have led a pure life, whereas we have walked in the most wicked ways of all? And that there shall be shewed a paradise whose fruit endureth without decay, wherein is abundance and healing, but we shall not enter into it, for we have walked in unpleasant places? For while we lived and committed iniquity, we considered not what we should have to suffer after death.*[1]

---

[1] The tendency of Freewill in a Spirit Being (being granted to man, Gen. ii, 7) in operation being to deviate from Divine Principles, God, with eternal foreknowledge of the necessity of a Redeemer, had appointed His own Divine Essence, to suffer, to show to the whole of His Spirit realm the magnitude of Divine Love, to illustrate His First Law of harmony, Love. The answer to Esdras' questionings is: "*The Lamb slain from the foundation of the world*" (Rev. xiii, 8).

"*He answered, This is the condition of the battle, which man that is born upon the earth shall fight; that, if he be overcome, he shall suffer as thou hast said; but if he get the victory, he shall receive the thing I say. . . . Choose the life, that thou mayest live* [1] *. . . so that there shall not be such heaviness in their destruction as there shall be joy over them that are persuaded to salvation.*

viii. "*The Most High hath made this world for many, but the world to come for few. There be many created, but few shall be saved.*

"*I said, Swallow down understanding, O my soul, and let my heart devour wisdom. For thou art come hither without thy will and departest when thou wouldest not, for there is given thee no longer space than only to live a short time.*

"*Touching man in general, thou knowest best, but touching thy people, will I speak, for whose sake I am sorry; and for thine inheritance, for whose cause I mourn; and for Israel, for whom I am heavy, and for the seed of Jacob, for whose sake I am troubled.*

"*O, look not upon the sins of thy people, for we and our fathers have passed our lives in ways that bring death, for thou because of us sinners art called merciful.*

"*He answered, As the husbandman soweth much seed upon the ground, yet not all that is sown shall come in due season, even so they that are sown in the world shall not all be saved.*

"*Thou comest far short that thou shouldst be able to love my creature more than I . . . yet in that thou hast humbled thyself, as it becometh thee, and hast not judged thyself worthy to be among the righteous, so as to be glorified . . . understand thou for thyself and of such as be like thee seek out the glory. For unto you is paradise opened, the tree of life is planted, the time to come is prepared, a city is builded, and rest is established. . . .*

"*For the Most High willed not that men should come to nought, but they which be created have themselves defiled the name of him that made them, and were unthankful unto him which prepared life for them. And therefore is my judgment now at hand.*

[1] *Cf.* Deut. xxx, 19.

THE PILLAR IN THE WILDERNESS

"*I said, . . . O Lord, . . . but at what time?*

ix. "*He answered, When thou seest a certain part of the signs are past . . . then the Most High will visit the world which was made by him. And when there shall be seen in the world, earthquakes, disquietude of peoples, devices of nations, wavering of leaders, disquietude of princes* [1] *. . . the beginnings are manifest in wonders and mighty works, and the end in effects and signs.*

"*And every one that shall be saved, and shall be able to escape by his works or by faith, whereby he hath believed, shall be preserved from the said perils, and shall see my salvation in my land, and within my borders, which I have sanctified for me from the beginning. . . .*

"*For there was a time in the world, even then when I was preparing for them that now live, before the world was made for them to dwell in, . . . so I considered my world, and lo, it was destroyed, and my earth, and lo, it was in peril, because of the devices that were come into it.*

"*And I saw, and spared them, and saved me a grape out of a cluster, and a plant out of a great forest . . . let my grape be saved, and my plant, for with great labour have I made them perfect.*

xi. "*The vision of the eagle and lion.*

xii. "*This is the interpretation of the vision. The eagle whom thou sawest come up from the sea is the fourth kingdom which appeared in vision to thy brother Daniel. Behold the days come, that there shall rise up a kingdom upon earth and it shall be feared above all the kingdoms that were before it. In the same shall twelve kings reign . . . in the midst of the time of that kingdom there shall arise no small contentions, and it shall stand in peril of falling, nevertheless it shall not then fall, but shall be restored again to its first estate.*[2]

"*And whereas thou sawest three heads resting . . . In the last days thereof shall the Most High raise up three kingdoms and renew many things therein; and they shall bear rule over the earth and those that dwell therein with much oppression above all those that were before them; for these are they that*

---

[1] Such are the present days.
[2] Rome, and Rome revived under Mussolini.

shall accomplish her wickedness, and that shall finish her last end.[1]

"*And whereas thou sawest that the great head appeared no more; it signifieth that one of them shall die upon his bed, and yet with pain. But for the two that remained, the sword shall devour them. For the sword of one shall devour him that was with him; but he also shall fall by the sword in the last days.*

"*And whereas thou sawest two under wings passing over unto the head . . . These are they*[2] *whom the Most High hath kept unto his end, this is the small kingdom and full of trouble, whom thou sawest.*

"*And the lion whom thou sawest rising up and rebuking the eagle for her unrighteousness, . . . this is the Anointed One, whom the Most High hath kept unto the end of the days, who shall spring up out of the seed of David, and he shall . . . reprove them for their wickedness and their unrighteousness, for at the first he shall set them alive in his judgment and when he hath reproved them, he shall destroy them.*

"*For the rest of my people shall he deliver with mercy, those that have been preserved throughout my borders,*[3] *and he shall make them joyful until the coming of the end, even the day of judgment, whereof I have spoken unto thee from the beginning.*

xiii. "*I dreamed . . . and lo, there arose a wind from the sea, that it moved all the waves thereof. And I beheld this wind caused to come up from the midst of the sea as it were the likeness of a man, and the man flew with the clouds of heaven, and when he turned his countenance to look, all things trembled that were seen under him, all they burned that heard his voice like as the wax melteth when it feeleth the fire.*

"*And there was gathered together a multitude of men, out of number, from the four winds of heaven, to make war against the man.*

"*And he graved himself a great mountain and flew upon it . . . he sent out of his mouth as it had been a flood of fire*

---

[1] Babylon, and her last three kingdoms—Austria, Prussia, Russia—whose emperors fell in 1918; their emblems—the eagle.
[2] Nazi Germany and Soviet Russia.
[3] "*Appointed place.*"

". . . so that of a sudden of an innumerable multitude nothing was to be perceived but only dust of ashes and smell of smoke.[1]

"Afterward I beheld the man come down from the mountain and call unto him another multitude which was peaceable.[2] *And there came much people unto him, whereof some were glad, some were sorry.*

"Through great fear I waked and prayed unto the Most High, for as I conceive in my understanding woe unto them that shall be left in those days! and much more woe unto them which are not left in those days, for they shall be in heaviness, understanding the things that are laid up in the latter days, but not attaining unto them.

"Yet is it better for one to be in peril and to come into these things, than to pass away as a cloud out of the world, and not to see the things that shall happen in the last days.

"He answered, He that shall endure the peril in that time shall keep them that be fallen into danger, even such as have works, and faith unto the Almighty.

"Know therefore, that they which be left behind are more blessed than they that be dead.

"Whereas thou sawest a man coming up from the midst of the sea, the same is he whom the Most High hath kept a great season, which by his own self shall deliver his creature, and he shall order them that be left behind. And whereas thou sawest that out of his mouth come wind and fire and storm, and whereas he held neither spear, nor any instrument of war, but destroyed the assault of that multitude which came to fight against him.

"Behold, the days come, when the Most High will begin to deliver them that are upon the earth. And there shall be astonishment of mind upon them that dwell upon the earth. And one shall think to war against another, city against city, place against place, people against people, kingdom against kingdom.

"And it shall be when these things come to pass and the

---

[1] *Cf.* Ezekiel xxxix, 1–6.
[2] Britain?—such describes her to-day; read further the interpretation as to identity.

signs shall happen which I shewed thee before, then shall My Son be revealed, whom thou sawest as a man ascending:[1]

"*And it shall be, when all the nations hear his voice, every man shall leave his own land and the battle they have one against another. And an innumerable multitude shall be gathered together, as thou sawest, desiring to come and fight against him.*

"*But he shall stand upon the top of mount Zion.*

"*And Zion shall come and shall be shewed to all men, being prepared and builded, like as thou sawest the mountain graven without hands.*[2]

"*And this my Son shall rebuke the nations, which are come for their wickedness, with plagues that are like unto a tempest, he shall destroy them without labour by the law, which is likened unto fire.*[3]

"*And whereas thou sawest that he gathered unto him another multitude that was peaceable; these are the ten tribes, which were led away out of their own land in the time of Osea the king; whom Shalmanaser the king of the Assyrians led away captive, and he carried them beyond the River, and they were carried into another land.*

"*But they took counsel among themselves, that they would leave the multitude of the heathen, and go forth into a further country where never mankind dwelt, that they might there keep their statutes, which they had not kept in their own land. And they entered by the narrow passages of the River Euphrates. For the Most High then wrought signs for them, and stayed the springs of the River, till they were passed over.*

"*For through that country was a great way to go, namely, for a year and a half; and the same region is called Arzareth.*[4]

"*Then dwelt they there until the latter time; and now when they begin to come again, the Most High stayeth the springs of the River again,*[5] *that they may go through; therefore*

---

[1] *Cf.* Luke xxi, 25-28.   [2] Daniel ii, 34, 35.
[3] *Cf.* Matt. xxv, 41-46.
[4] Ar = city, on the Sereth, a tributary of the Danube, the land of the Getæ. It being written after the time, this is recorded history of Israel's survival from Assyria; to become "the multitude peaceable" (= Britain to-day).
[5] Rev. xvi, 12—referring to the collapse of Turkey.

sawest thou the multitude gathered together with peace. But those that be left behind of thy people are they that are found within my holy border.

"*It shall be therefore when he shall destroy the multitude of the nations that are gathered together, he shall defend the people*[1] *that remain. And then shall he shew them very many wonders.*

"*Then said I, shew me this, wherefore I have seen the man coming up from the midst of the sea.*

"*He said unto me, Like as one can neither seek out nor know what is in the deep of the sea, even so can no man upon earth see my Son or those that be with him, but in the time of his day.*

"*This is the interpretation of the dream which thou sawest, and for this thou only are enlightened herein. For thou hast forsaken thine own ways, and applied thy diligence unto mine, and hast sought out my law.*

xv. "*Woe to the world and them that dwell therein! for the sword and their destruction draweth nigh, and nation shall rise up against nation*[2] *to battle with weapons in their hands.*

"*For there shall be sedition among men; and waxing strong one against another, they shall not regard their king, nor the chief of their great ones, in their might. For a man shall desire to go into a city, and shall not be able, for because of their pride the cities shall be troubled, the houses shall be destroyed, and men shall be afraid. A man shall have no pity upon his neighbour, but shall make an assault on their houses with their sword, and spoil their goods, because of the lack of bread, and for great tribulation.*

"*Behold, saith God, I call together all the kings of the earth, to turn themselves one against another, like as they do yet this day unto my chosen, so will I do also, and recompense in their bosom, my sword shall not cease over them that shed innocent blood upon the earth.*

"*Woe to them that sin and keep not my commandments.*

---

[1] *Cf.* Rev. xvi, 16.

[2] *Cf.* Matthew xxiv. Then follows prophecy and description of revolutions such as the French and the Bolshevik, typified in the "class war" and the Proletariat.

I will not spare them; go your way, ye rebellious children, defile not my sanctuary.

"For now are the evils come upon the whole earth, and ye shall remain in them, for God shall not deliver you, because ye have sinned against him.

"And then shall the dragons have the upper hand, remembering their nature.

"And after this shall there be stirred up great storms, fire and hail, and many waters, that all the plains may be full and all rivers. And they shall break down the cities and walls, mountains and trees, and they shall go on steadfastly unto Babylon and destroy her . . . then shall the dust and smoke go up unto heaven, and all they that be about her shall bewail her.

"And thou, Asia, that art partaker in the beauty of Babylon, and in the glory of her person, woe unto thee, because thou hast made thyself like unto her, thou hast followed her that is hateful in all her works and inventions.[1]

xvi. *"The fire is kindled, and shall not be put out, till it consumes the foundations of the earth.*

"Behold, famine and plague, tribulation and anguish, they are sent as scourges for amendment. But for all these things they shall not turn them from their wickedness, nor be always mindful of the scourges.

"Behold, victuals shall be so good, cheap upon earth, that they shall think themselves in good case,[2] *and even then shall evils grow upon earth, sword, famine, and great confusion.*

"For many of them that dwell upon earth shall perish of famine, and the other that escape the famine, shall the sword destroy. . . . The virgins shall mourn, having no bridegrooms, the women shall mourn, having no husbands, their daughters shall mourn, having no helpers.

"O my people,[3] *hear my word; make you ready to battle, and in those evils be even as pilgrims upon the earth. . . . He that buyeth, as one that will lose; he that occupieth merchandise, as he that hath no profit by it.*

---

[1] The heathen of Asia followed the civilisation of Cain-Bel, brewed in Babylon, of the Sumerians.

[2] And so was in the spring of 1914—then followed "the tribulation" (1914–1936).

[3] This then applies to present time.

"*Inasmuch as they that labour, labour in vain, for strangers shall reap their fruits and spoil their goods . . . the more they deck* [1] *their cities, their houses, their possessions, their own persons, the more will I hate them for their sins, saith the Lord.*

"*Behold, the Lord knoweth all the works of men, their imaginations, their thoughts and their hearts. Who said, Let the earth be made, and it was made; let the heavens be made, and it was made. And at his word were the stars established, and he knoweth the number of the stars:*

"*Who searcheth the deep and the treasures thereof: he hath measured the sea, and what it containeth:*

"*Who hath shut the sea in the midst of the waters, and with his word hath he hanged the earth upon the waters:*

"*Who spreadeth out the heaven like a vault; upon the waters hath he founded it:*

"*Who hath made in the desert springs of waters, and pools upon the tops of the mountains, to send forth rivers from the height to water the earth:*

"*Who framed man, and put a heart in the midst of his body, and gave him breath, life and understanding, yea, the spirit of God Almighty.*[2]

"*He who made all things, and searched out hidden things in hidden places, surely he knoweth your imagination, and what ye think in your hearts. How will ye hide your sins before God and his angels? Behold, God is the judge, fear him; leave off from your sins . . . so shall God lead you forth and deliver you from all tribulation.*

"*For there shall be in divers places and in the next cities a great insurrection upon those that fear the Lord, for they shall waste and take away their goods, and cast them out of their houses.*

"*Then shall be manifest the trial of mine elect,*[3] *even as the gold that is tried in the fire.*

"*Hear, O ye mine elect, saith the Lord; behold, the days of tribulation are at hand, and I will deliver you from them. Be ye not afraid, neither doubt, for God is your guide.*" [4]

[1] And this is what is being done at the present time.
[2] Refer Gen. ii, 7; with pp. 3, 11, 17, 267.
[3] Referring to the Anglo-Saxon race, A.D. 1914–1936.
[4] So meant directly for the Britain of to-day.

CHAPTER VIII

This Apocalyptic eschatology to His Benjamin as the "*light*" to Jerusalem and Judah was His direct Voice to Spiritual Intuition to prepare the way for His earthly mission.

On earth He did not adapt His teaching to a local eschatology, He had already inspired it, and on earth He confirmed it.

His Life on earth needs the highest that man can give in expression—the sublimest song of the poet, the deathless music of the composer, a picture never-fading by the artist. Mere words are inadequate. Let man attempt to write down from his highest concept what He actually did say and do—whether he be saint, theologian or a Milton—and it falls far short of the words mere fishermen, their memory intensified by Divine Spirit (John xiv, 26), transmitted from the Aramaic into Greek writing.

His Death and Resurrection, the Pharisee and the Roman could not disprove.

He spoke not in long phrases, as if the thought behind was in doubt and liable to argument, but using words as symbols to originate a concept, leaving to man's mind "*born anew*" to vision a whole thesis; as symbols to call to Spiritual Intuition rather than to Reason, to the "*breath of life*" in man—with which when man was first endowed the Creator was able directly to speak—rather than to the matter of his brain which had evolved. Yet by His miracles He attempted to force Reason to see.

In the long-sustained phraseology of the philosopher we hear the cymbals tinkling; in His words we hear in symbols the rushing of a mighty wind, whence it cometh, or whither it goeth, we know not.

Thus the true communication of thought is not in long speeches, nor in wordy arguments wherewith to convince

Reason, but is in transfer by the flash of a concept from mind to mind in a mathematical symbol, in a strain of music, in a view from the hill-top, in the beauty of spring, in the thrill of love.

Only in the flashes of inspiration, born of Spirit Intuition, does man come to meet the level of His plane and to knowledge of Him: not in long treatises of explanatory creeds, not in deeply reasoned sermons, not in personal appeals from religious experience; such pass away, but not His words.

If intuition is instinct raised to its highest power, then intuition raised to the nth power is Spiritual intuition or perception; raise again to the nth degree and man sees God and hears his "*he shall be my son*," and learns from Him that beyond the limit of human intelligence and reason—"*God is a Spirit, and they that worship Him must worship Him in spirit and in truth*" (John iv, 24).

In such words, tense with concept, calling to that Spiritual intuition of man which displaced his animal instinct 6000 years ago (that instinct which causes a salmon to return to its own river, and the migrating bird to return in the spring to its old place), "striking to the depths within us and speaking of the profoundest realities of our nature,"[1] John tells of the Incarnation of the Logos.

"*In the beginning was the Logos, and the Supreme Creative Power was with God, and was God. All things were made by Him, and without Him was not anything made that was made. In Him was life*" (John i, 1–3).

The whole of Evolution in a symbol.

"*And the Supreme Creative Power was made flesh, and dwelt among us (and we beheld His glory, the glory as of the only begotten of the Father)*" (John i, 14).

"*He was in the world, and the world was made by Him, and the world knew Him not.*"

"*He came unto His own* [this Israel-Britain, wandering round Europe for transgressing His Commands, only Benjamin and part of Judah to meet Him], *and His own received Him not.*"

---

[1] A. Noyes, *The Unknown God*.

THE CHRIST

The Incarnation of Supreme Creative Power in Matter demands an epic from man's sublimest conceptions to match the song of angels—

"*Glory to God in the highest, and on earth peace, goodwill toward men. For unto you is born this day in the City of David a Saviour, which is Christ the Lord. Ye shall find the babe, wrapped in swaddling clothes, lying in a manger*"[1] (Luke ii, 11–14).

In the opening music of Handel's *Messiah* comes man's tribute—hear it thrilling in—

> *Comfort ye, My people.*
> *Behold, your God.*
> *Arise, shine, for thy light cometh.*
> *For unto us a Child is born, unto us a Son is given.*

(working up in that glorious crescendo)

> *And the government shall be upon His shoulder,*
> *And His name—Wonderful, Counsellor,*
> *The Mighty God, the Everlasting Father,*
> *The Prince of Peace.*

(and diminuendo to quiet certainty of rest to a travailing world)

> *He shall feed His flock like a shepherd,*
> *. . . And ye shall find rest to your souls.*

And at each anniversary of His birth, there comes over mankind that suffused feeling of well-being, when Freewill attunes with all Nature; and throughout His Israel there rise the Christmas carols—

> *Noel, born is the King of Israel.*

> *. . . Awake, salute the happy morn,*
> *Whereon the Saviour of the world was born,*
> *Rise to adore the mystery of love,*
> *Which hosts of Angels chanted from above,*
> *Peace upon earth, and unto men goodwill.*

> *Hark ! the herald angels sing,*
> *Glory to the new-born King.*
> *Joyful, all ye nations, rise,*
> *Join the triumph of the skies,*

---

[1] From the Highest . . . to a manger.

> *Christ, by highest Heaven adored,*
> *Christ, the Everlasting Lord.*
> *Hail! the heaven-born Prince of Peace,*
> *Hail! the Sun of righteousness.*
> *Born to raise the sons of earth,*
> *Born to give them second birth.*
>
> *O come ye, O come ye to Bethlehem,*
> *Come and behold Him,*
> *Born, the King of Angels,*
> *Word of the Father,*
> *Now in flesh appearing,*
> *O come, let us adore Him,*
> *Christ the Lord.*

And that exquisite, less-known, full of this theme—

> *Born for us on earth below,*
> *See! the Lamb of God appears,*
> *Promised from eternal years.*
>
> *Lo! within a manger lies*
> *He who built the starry skies,*
> *He who, throned in height sublime,*
> *Sits amid the cherubim.*
>
> *Sacred Infant, all divine,*
> *What a tender love was Thine,*
> *Thus to come from highest bliss*
> *Down to such a world as this!*

Thus, the Apex Stone of Creation, the Son of man, born in a lowly manger, reproduces man's evolution from the animal kingdom, and in His Incarnation baptizes man with His Spirit.

"*Et homo factus est*" . . . contains all the wonder of the first birth of the spirit. One seems to be aware of a Universe hushed in awe over the first helpless man-child in whom it could be said that the transition from the sub-human had been accomplished; a child laid in the wild manger of the brute creation, æons before the discovery that the inns of this world had no room . . . while from all the starry heights to which his innocent face was upturned there shone the quiet annunciation and prophecy of the new bond between heaven and earth—"*Unto us a Child is born, unto us a Son is given*" . . . "*Ecce homo!*"[1]

---

[1] Alfred Noyes, *The Unknown God*, p. 320.

Man was made in the image of God, that of Christ, *"Who is the image of the Invisible God, the first-born of every creature* [the perfect model of God's creation]. *For by Him were all things created, visible and invisible; and He is before all things, and by Him all things consist"* (Col. i, 15–17).

"In Whom are hid all the treasures of wisdom and knowledge" (Col. ii, 3).

Has Science any tribute to Him, anything that Reason can attune with Intuition?

The Ether of Space is a theme of unknown and apparently infinite magnitude, and of a reality beyond the present conception of man. It is that of which everyday material consists, a link between worlds, a consummate substance of overpowering grandeur.

By a kind of instinct one feels it to be the home of spiritual existence, the realm of the awe-inspiring and the supernal. It is co-extensive with the physical universe, and is absent from no part of space. Beyond the furthest star it extends, in the heart of the atom it has its being. It eludes the human senses, and can only be envisaged by the powers of the mind. Yet the Ether is a physical thing, it is not a psychical entity, it has definite physical properties. It is not matter . . . its vibrations can be analysed, they bring to us information, and without them we could not exist. They may have been instrumental in bringing matter into existence. Ether is the universal connecting link, the transmitter of every kind of force. Action at a distance is wholly dependent on Ether. It welds the planets into a solar system, it is the vehicle of gravitation and light . . . it holds the atoms together.

It is the seat of prodigious energies—energies beyond anything as yet accessible to man. All we know of energy is but the faint trace, or shadow, or overflow of its mighty being. Hidden away in its constitution is a fundamental and absolute speed, a speed not of locomotion, but of internal combustion.

We know that all the bodies we see and handle are but elaborate and beautifully organised congeries of positive and negative electrons, held together and connected by the medium of which they themselves consist. The world, the stars, the heavens are nothing else. The mystery of existence is close upon us here. We grope in a kind of helplessness with our

> few animal senses, and we live our short lives, encouraged by a faith and hope that we are something more than appears, and that in the deep roots of our being we belong to another order of things. Speculatively and intuitively we feel to be more in touch with the ether than with matter. We are apt to identify ourselves with our bodies. Matter is not part of our real being; it is but an instrument that we use for a time and then discard. Our will, our mind, our psychic life, probably act directly upon the Ether, and only through it indirectly on Matter.
>
> Ether is our real primary and permanent instrument. Hence even our apparently most material sense (touch, as well as eye and ear) is dependent on this omnipresent medium, on which alone we can directly act, and through which all our information comes.
>
> It is the primary instrument of Mind, the vehicle of the Soul, the habitation of Spirit.
>
> Truly it may be called the living garment of God.[1]

Combining all the above, there is but one conclusion wherein Intuition and Reason attune—that the Ether is the medium between Spirit and Matter, and is that part of the creative essence of Spirit formed for the emergence of life to demonstrate His Purpose to the Freewill of His whole Spirit realm, that action was to be on One Universal Law, Love.

Within the Ether was created a reservoir of energy, that which binds the atom and pervades all space, which can be weighed and measured (see 2 Esdras iv), and the use of such power was natural to the whole Spirit realm, but possessed in highest degree and controlled by that part of the Perfect and Divine Essence, Christ, "*by whom all these things consist.*" Whereas Science seeks to yoke to her use the "etheric energy" in Space from the Matter or resultant end, He has the direct and instant command through the creative or Spirit end.

In and by the Ether, God acts—"action at a distance is wholly dependent on the Ether, which is the transmitter of every force": it is His real primary instrument in the

---

[1] Sir Oliver Lodge, *Ether and Reality*, pp. 173-179 (The Epilogue: "The Ultimate Physical Reality").

A FIELD OF MIXED GROUND

To face p. 236.

THE CHRIST

carrying out of the Universe's Laws, and in communicating with Spirit inherent in Matter. As Ether is the common medium which pervades all Space, so in such unity in Space is the symbol of the Logos in Oneness in all Vital Evolution, that Harmony, the Universal Law, is expressed in Matter.[1]

This source of energy, which gives through the Ether Spirit dominion over Matter, expressed as *"the breath of life"* breathed into man that *"man became a living soul"* (Gen. ii, 7), was granted to the ancestors of the Adamic race, and by them lost by Freewill deviating from principles of Divine Will (see p. 30).[2] The parable of such is given in the Garden of Eden story (Gen. iii), words taken as symbols of what actually occurred.

Of the two factors in man, Intuition and Reason, knowing that appeal to Intuition would always lead man back to God, then to Reason ("*Hath God said?*") appealed Evil—that deviation of Spirit Freewill prior to or coincident with Earth's creation (Jude 6), which produced discord in a world of Harmony. Thus in all Matter and in all mankind such tendency to discord in the awe-full Oneness of God is meet for annihilation ("*eternal destruction from the presence of the Lord*") at His appointed time.

Cain, driven from "*the presence of the Lord*"—gradually by the natural law of disuse, and by deliberately beclouding Divine Voice in the civilisation he founded and perverting knowledge gained by Spiritual Perception—lost Spiritual Intuition and stressed Reason. And so it has been throughout his descendants and throughout the Babylonian Kingdoms, man's mind in earthly environment is befogged from contact with his Spirit origin, and gropes in a "*world of shadows.*"

More and more, ever since, man seeks to rule the earth by Reason; through such he thinks he is making

---

[1] Refer pp. 18, 265–268.

[2] Traces of such God-given knowledge are seen in the Cain civilisation in Sumer before 3500 B.C., and in the construction of the Pyramid by the House of Seth before 2600 B.C.; it is seen to-day in the above conception of the Ether as the "ultimate physical reality" of Sir Oliver Lodge—in which Spiritual Intuition is aiding Scientific Reason.

discoveries (*vide* pp. 303-308), and deviating Freewill misdirects the knowledge to the means of man's destruction of his fellow-man, and to the accumulation of individual wealth and power—symbolised by the Pyramid's descending passage.

Only by Spiritual Intuition can this Displacement be visualised; and throughout His Bible story, Divine revelation to Spiritual perception is stressed.

The operation of freewill in a spiritual being is an essential for receiving and emanating Divine Love. And with eternal foreknowledge that freewill in operation would tend to deviate, and even before the processes of active creation began, God had already appointed that He Himself should suffer to effect the reconciliation, to illustrate to all His Spirit beings in Heaven and Earth the highest form of the outpouring of Divine Love, so to live in perfect unity and implicit faith in the perfection of His Providence (Davidson).

The Messiah—the Pyramid's displacement factor—was to bridge the gap of the existing displacement of man from the harmony of God's creation, to *redeem* man—*i.e.* to redeem his Freewill to harmony with Divine Will—but by the deliberate choice of each man.[1]

This then is His Purpose, through Christ—"*the mystery, which from the beginning of the world hath been hid in God, who created all things by Jesus Christ: to the intent that now unto the principalities and powers in heavenly places* [to His Spiritual realm in the heavens which may deviate, as well as to the lowest Spiritual realms on earth] *might be known the manifold wisdom of God, according to the eternal purpose which He purposed in Christ Jesus our Lord*" (Eph. iii, 9-11).

This world and all that is in it, having deviated from the Divine Law of Love and thus being in some way under the power of Evil, is due in the Oneness of His Perfection for total destruction, body and soul.

By His Incarnation, the injection of Divine Essence into this earth condition, and by His deliberate act in surrendering Spirit dominion over matter when "*for three*

[1] Refer pp. 247-250, 264-270, 374, 385, 424.

days and nights in the heart of the earth," meant that He "risked" such annihilation: He was not man as well as God if He had not.

And more, only by such stupendous act, which involved changing the energy distribution in the Cosmos, could Divine Purpose to attune Freewill be carried out, only by this could a "displacement factor" scientifically operate.

This Spirit power over Matter has been seen to be Christ's, *"the Second Adam," "the image of the Invisible God,"* the perfect model of God's creation. The control of this "etheric or cosmic energy" was as natural to Him (though He may not have known in earthly limitations the physical explanations of its Laws) as the use to the ordinary man-in-the-street of instruments and engines based on the utilisation of mechanical, electrical and heat energy, without knowing their scientific explanation, but feeling that they are based on certain and universal laws. When roused from sleep in a storm, He calmed the winds and the rain, His *"Peace, be still"* was as natural as a father's "Come, come, children, be quiet, go to bed now."

When we curse the weather, it is because we are bound to earthly cares, the trudge of routine, the daily grind of trying to exist—as cattle we are driven into a field in the morning and in the evening brought back to a stall—bound and out of tune.

Free from Matter's care (*"for after all these things do the Gentiles seek,"* Matt. vi, 30–32), we stride the hills with wind and rain on our faces, loving it, physically and mentally fit, in tune with Nature and her needs, we bless her. Then we touch the Power which man was destined to control.

"The stone which the builders refused is become the head stone of the corner.

"This is the Lord's doing, it is marvellous in our eyes" (Ps. cxviii, 22, 23).

The scientific presentation of the Pyramid's prophecy is beyond the ingenuity of man. The Great Pyramid's displacement factor[1] witnesses to the Divinity of Jesus Christ, "*by whom all things were created*," and to the truth of salvation through Him (Davidson).

Did His childish eyes while in Egypt gaze up at the white, smooth-sloped Pyramid at Gizeh, glistening in the sun, and did His child mind feel some vague affinity that in some remote past and in some way He was associated with its grandeur and its construction?

Did He feel as with boyhood's steps He trod the vales of Somerset[2] and climbed its hills that here was "*the appointed place*," and coming to the isle of Avalon, "abounding in all the beauties of Nature and necessaries of life," that here was consecrated ground?

And did such memories come before Him, as He quoted from the Canon of the Law and the Prophets to answer the Jews' demand, "*By what authority doest Thou these things?*"—"*The Stone which the builders rejected, the same is become the head of the corner. The Kingdom of God shall be taken from you and given to a nation bringing forth the fruits thereof*" (Matt. xxi, 23, 41–44).

From early boyhood His constant thought, "*Wist ye not that I must be about My Father's business?*"[3] was driving Him to accomplish some Purpose.

And with baptism, "*a voice from heaven, This is My Beloved Son, in Whom I am well pleased,*"[4] confirmed His Messiahhood and drove Him into the wilderness[5] to think —how to use this power within Him? how to accomplish His Mission? how to fulfil the Times, spoken by the mouth of all His holy prophets?

And the solution came, when hunger gnawed—

(1) "*If thou be the Son of God*," test that power and use that energy which binds the electrons in an atom, and from stones make bread:

—to limit such power, only to be used for His Father's business, not for His own needs: as man, to accept the

---

[1] See pp. 90, 96–100.
[2] This tradition is referred to on p. 277.
[3] Luke ii, 49.   [4] Matt. iii, 13–17.   [5] Matt. iv, 1–11.

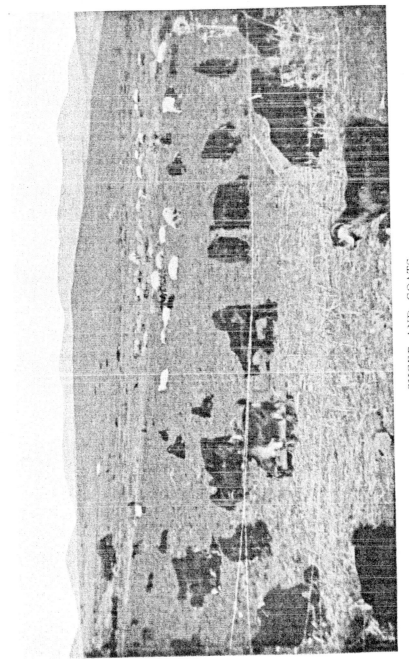

SHEEP AND GOATS

limitations of His fellow-men, who live "*not by bread alone, but by every word that proceedeth out of the mouth of God.*"

(2) How to accomplish His Mission as the Messiah?—the people looked for such to appear from heaven, then if He should appear from the top of the Temple into their midst, would they believe? Such would confirm His power within the Ether, "the vehicle of gravitation."

And the solution—Thou shalt not take the easy way: true belief in the Kingdom is from the heart and not through the eyes: "*Thou shalt not tempt the Lord thy God.*"

(3) From a *high mountain*, He visualises His Israel wandering Europe (see pp. 175, 176) awaiting a leader, and even the Jews desiring a hero-king. But the times of restitution are not yet for Him to found His earthly Kingdom; His Israel still worship earthly power and wealth, and His appearance would not change their heart "*to worship the Lord their God in spirit and in truth*"; they have first to acknowledge their iniquities.

His Israel as His Sheep are constantly in His mind; in every moment of His day, thoughts are always on His Father's business, the Kingdom of God. So that the daily ordinary things of life suggest the symbols of the Kingdom of Heaven in parables.

As one walks through Galilee to-day, there in the valley of Gennesaret is the field with stony ground, the wayside, the thorns, and the good ground; there, the mixed herd of white sheep and black goats, separating themselves naturally with a space between—there is even a black-and-white one half-way![1]

So in nearly all His parables there is a double meaning, one more or less visible, the other hid: (1) the one usually given in the churches; (2) the Israel idea.

Take the Prodigal Son (Luke xv, 11–32). The elder is Judah, the younger is Ephraim.

In 975 B.C. Ephraim-Israel demanded his portion of the Kingdom, and broke away from the Throne of David. Not

---

[1] H. V. Morton, in his book, *In the Steps of the Master*, gives several instances of local configuration and custom persisting to this day.

many days after Ephraim fell into idolatry and soon by the providence of God took his journey into the far country of Assyria. There he wasted his substance with riotous living. He forsook the law and true worship of God, the rules and customs of Israel; he spent all, and became spiritually a beggar, starving for want of moral and spiritual food. He had come to a land where there was a famine of true worship and common righteousness, and he began to be in want. He joined himself to the citizens of Guta, and tried to satisfy himself with their heathen customs, but remained unsatisfied and hungry, with an inherent desire to return to the true worship of God.

Then he found a home in England, and heard Christianity, and began to come to himself and repent. He compared his own lot with that of the hired servants (the nations of the four Gentile kingdoms carrying out the Divine purpose during Israel's punishment).

He saw their prosperity and compared it with his own lot, as a wanderer without a settled home. The repentance of the prodigal is the acceptance of Christianity, especially since the Reformation and 1800, with the Wesley revival, and founding of Bible and Missionary Societies.

The Father's welcome—as long ago as the first century He prepared for the son's return, the Christian Church was planted here by the apostles.

The feast is called after the lost son is found. He was thought to be dead and buried in a far country, as Ezekiel pictures him in the valley of dry bones; and as *"the lost sheep of the House of Israel."* The world will know that the prodigal son is found and restored, and then the feast is the same as the marriage feast after the cleansing of Israel and the return of her Lord to reign.[1]

Take the parable of the talents and pounds (Luke xix, 12–27).

"He called his ten servants, and delivered unto them ten pounds, and said unto them, Occupy till I come."

The Kingdom having been taken from the Jews and given to a nation bringing forth the fruits thereof, ten [2] signify the ten tribes of Israel—*"I am not sent but to the*

[1] J. J. Morey, *The Parables of the Kingdom*, pp. 174–179 (Covenant Publishing Co.).

[2] "Ten," symbolical of world completion without Christ.

lost sheep of the House of Israel." To them entrusted the ten pounds (in the same idea, the ten virgins, His Church without the bridegroom), His Kingdom without the King.

"The nobleman will soon return and He will expect great results from Israel-Britain as the custodians of His treasure." [1]

Even directly after the Sermon on the Mount, He adds, *"Think not that I am come to destroy the law or the prophets, I am not come to destroy, but to fulfil. For verily I say unto you, Till heaven and earth pass away, one jot or one tittle shall in no wise pass from the law, till all be fulfilled"* (Matt. v, 17, 18).

"The God of Abraham, of Isaac, of Jacob, is not the God of the dead, but of the living" (Matt. xxii, 32).

Of His miracles, the first *"did Jesus in Cana of Galilee"* (John ii, 1–11).

At the marriage feast *"the mother of Jesus saith unto Him, They have no wine."*

His reply may be given as: "Mother, the power within me to combine, arrange and direct energy in the Ether, you ask Me to use. Such can be used only for My Father's business, for the benefit of the spiritual in man. I will use it for others, on this first occasion, at your request."

"And when the ruler of the feast tasted . . . he saith, Thou hast kept the best wine until now."

And the second miracle was by the transmitting of etheric energy at a distance to heal, and can act only with faith attuned (John iv, 46–54).

Yet in what He performed, some energy radiated from Him; this is seen by His consciousness when the woman by act of faith touched but the hem of His garment.

"Except ye see signs and wonders, ye will not believe" (John iv, 48).

To the question of John (in prison, and perhaps hoping for a miraculous release): *"Art thou he that should come, or look we for another?"* His reply is—the miracles, *"Go and show John . . . the blind receive their sight, the lame*

[1] Morey, pp. 222–228.

walk, the lepers are cleansed, the deaf hear, the dead are raised up, the poor have the gospel preached to them," so that he might believe.

Of John the Baptist, prophet and apocalyptist, combined in one, He gave that wonderful testimony: *"What went ye out into the wilderness to see? A reed shaken by the wind? . . . Among them that are born of women there hath not risen a greater than John the Baptist: notwithstanding he that is least in the kingdom of heaven is greater than he"* (Matt. xi, 1-15).

John died before His work was accomplished with His victory over spirit annihilation, and did not see consummated "the Sacrifice" which God had fore-ordained.

"Except ye see signs and wonders, ye will not believe. . . . For He knew what was in man" (John ii, 25).

This explains why He would do no miracles in Nazareth, His home town.

And walking through Nazareth to-day, the spirit lacking is suggested in the raucous cries of mendicants, the whole business of buying and selling; the petty bullying of the weak by the strong; the front seat to the Jew, the back for the Arab, as the bus draws up.[1]

Looking from the north over Nazareth can be seen, to the right the Hill of Precipitation, where His townsfolk expressed their opinion of Him; and to the left, the top of Mount Tabor, said to be that of the Transfiguration, where expressed the Divine opinion, and whence He drew again power for His coming, stupendous act.

And so He came to Galilee—the miles He walked, never using a beast to aid progression except when on the colt of an ass at His triumphal entry to Jerusalem to fulfil prophecy. That lake and the surrounding beauty of Nature attracted Him. (*See Frontispiece.*)

Ye poets, ye artists, ye musicians, lack ye inspiration?

Then come ye to Galilee, and abide contemplative in the air He breathed, and bathe glamorously in the waters He

---

[1] There are two to-day in Palestine worthy of respect—the donkey and the British official (Ephraim): one balances on his back two impossible loads; the other, the Jew and the Arab.

trod. Look upon the constant changing and toning colours over sky and hill and sea with the storm coming up and as quickly passing, at dawn of day, and at setting sun. Did He with brush of cosmic power paint that picture on His lonely night vigils, never fading and still holding a charm and a peace that surpasseth earth's commotion? Here is the voice of Spirit calling to spirit that ye seek.

Man—His life forgotten—seeks rest and renewal of vitality in that acme of delight, a pleasure cruise—there to do something, play, dance or drink, and feel it is life. It is as well, dilapidated Tiberias, the site of Herod's palace, is not a trippers' paradise.

In Christ, Spiritual Intuition and Reason were perfectly attuned, hand-in-hand.

> The characteristic of greatness is the power of seizing upon the essential and eliminating the irrelevant. His spontaneous *obiter dicta* of penetrating brevity rather than in set discourse, illuminated with vivid illustration from the household or the farm, abounding in paradox, irony and humour. He aimed at leading men to understand and to originate, not merely to accept and obey.[1]

This is seen in His talk with Nicodemus, of the highest intellect and a master of Israel (John iii, 1–21), even when reported by a lesser intellect. Here He was trying to get through a concept on the lines of the highest scientific conceptions of to-day.

(Introductory) "*In Him was life, and the life was the light of men. And the light shineth in darkness, and the darkness comprehended it not.*

"*No man hath seen God; the only begotten Son, which is in the bosom of the Father, He hath declared him*" (John i, 4, 5, 18).

Life may exist in some imperceptible form, unknown to us, but without radiant or etheric energy it could not enter into relation with matter, it could not grow and develop and become

---

[1] Canon B. H. Streeter, *Foundations*, pp. 104–106.

conspicuous. Energy may not be necessary to abstract life, but it is essential to the display or manifestation of life in Matter.[1]

"Except a man be born again, he cannot see the Kingdom of God.

"For that which is born of Matter is Matter, but that which is born of Spirit is Spirit."

Through Einstein's theory of relativity, the four-dimensional world is conceived as a sort of blend of space and time, and that Matter is not an independent enduring substance. The theory of the electrical constitution of all Matter has abolished Matter, the old picture of a mechanistic world has gone, there is nothing now but energy. Man must try and forget his three-dimensional world, think of possibilities right outside actual human experience, think of the non-material shadowy four-dimensional continuum as a never-never land, a never-get-at-able place where the Great Operator works with entities a human being cannot see nor handle, nor as yet dimly understand. To do so perhaps we would need to be more than human, need to have other senses and more perfect eyes, a better brain and a different body. We have to acquire a new point of view, a new habit of thought, a new kind of consciousness . . . almost as if a new faculty of the mind were born and developed.[2]

A third and new faculty of knowledge? Along two streams in the evolution of man had developed two faculties of knowledge—(i) Instinct-Intuition (further emerged in the Adamic race, to Spiritual Intuition and Perception) and (ii) Intelligence-Reason (pp. 13-15).

Then to understand Spirit as the source of radiant or etheric energy, the electrical constituent of Matter, requires Spiritual Intuition raised again to the nth degree; it is beyond the human intelligence-reason faculty—we may "need to be more than human," may need an *incorruptible* body with *spirit* eyes; in fact, to be "*born again to everlasting life.*"

Christ then says, "*If I have told you earthly things*

[1] Oliver Lodge, *Ether and Reality*, p. 114.
[2] Sullivan and Grierson, *Outline of Modern Belief*, pp. 148-152.

[*i.e.* Energy-Matter], *and ye believe not, how shall ye believe if I tell you of heavenly things? This is the condemnation, that light has come into the world, and men loved darkness, rather than light*" (John iii, 12, 19).

Thus Christ stresses the necessity of faith, until with "*incorruptible body*" these things are understood.

"*For God so loved the world, that He gave His only begotten Son, that whosoever believeth in Him should not perish, but have everlasting life.*

"*For God sent not His Son into the world to condemn the world, but that the world through Him might be saved.*

"*He that believeth not the Son, shall not see life*" (iii, 36).

"*Search the Scriptures, they testify of Me.*

"*But if ye believe not its writings, how shall ye believe My words?*" (v, 39, 47).

"*This is the work of God, that ye believe on Him whom He hath sent.*

"*And this is the Father's will, that of all which He hath given Me, I should lose nothing, but should raise it up again at the last day*" (vi, 29, 39).

Christ goes on to speak of life in symbols which the Jews could appreciate, bread and water—the manna which came to them in the wilderness, and the water which spurted from the rock, both miraculously—which sustain the life of body.

"*For the bread of God is He which cometh down from heaven and giveth life unto the world, I am the bread of life*" (vi, 33, 35), leads to the next conception of that which will sustain eternal life in the Spirit realm, in symbols of the flesh and blood of His body:

"*And the bread that I will give is My flesh, which I will give for the life of the world.*

"*Verily, I say unto you, Except ye eat the flesh of the Son of man, and drink his blood, ye have no life in you.*

"*As the living Father hath sent Me, and I live by the Father, so he that eateth Me, even he shall live by Me.*

"*It is Spirit that quickeneth, Matter profiteth nothing; the words I speak unto you, they are spirit, and they are life*" (vi, 51, 53, 57, 63).

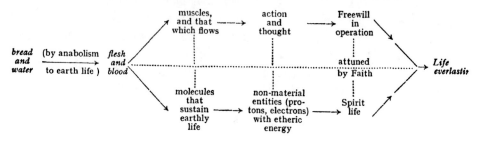

With some such but clearer concept, perhaps He was thinking, when "*My thoughts are higher than your thoughts, and My ways than your ways.*"

After the multiplication of the loaves and fishes in the feeding of the five thousand they flocked to Him in crowds; when He spoke to them in this way of His flesh and blood the Jews were shocked beyond measure; it was a conception beyond them, certainly not the mission of the Messiah they expected. So "*they sought to take Him, but no man laid hands on Him, for His hour was not yet come*" (vii, 30; viii, 20).

Christ carries on the conception:

"*I am the light of the world; he that followeth Me shall not walk in darkness, but shall have the light of life.*

"*Ye are of this world, I am from above.*

"*If ye believe not that I am He, ye shall die, in your sins.*

"*When ye have lifted up the Son of man, then shall ye know that I am He.*

"*He that sent Me is with Me, the Father hath not left Me alone.*

"*Now ye seek to kill Me, a man that hath told you the truth, which I have heard of God, this did not Abraham.*

"*Why do ye not understand My speech? even because ye cannot hear My word*" (viii, 12–43).

"*I know My sheep, and I lay down My life for the sheep.*

"*And other sheep I have which are not of this fold* [nations besides Israel], *them also must I bring, and they shall hear My voice; and there shall be one fold, and one shepherd.*

"*But ye believe not, because ye are not of My sheep.*

"*My sheep hear My voice, and I know them, and they follow Me*" (x, 14–27).

THE CHRIST

Now we have wandered from the scientific conception, but since the beginning is true, and a reality, then the significance of the words of Christ—the displacement factor of earth's creation—should be translated into a physical explanation. Such may be beyond human conception, and need for the present be represented by mathematical symbols, x, y, z ("for Nature lends itself much better to description in mathematical terms than in any other terms"[1]); where the Church has used as symbols words, as Redemption, Salvation, the Atonement.

The God who created this world and Universe is not "bluffing" man with a "fairy story" which his reason may or may not accept. Cain-Bel, stressing reason as is its wont, has interpreted Reality as a myth tale, that a Perfect Being descending to earth, dying and ascending, and what he actually said and promised, is not reasonable.

The God who created this world and Universe—man has to use mere words to express Divine Reality, and His Son used as symbols of Reality "*flesh and blood*" of His Body —injected something of Divine Essence into Matter, when for "*three days in the heart of the earth*" the Messiah was "*cut off*" from Spirit habitat, surrendered His Almighty Power over the Cosmos, sacrificed the Ether medium with its radiant energy, so that Evil as deviated Freewill in Matter was allowed to do as it wished, He was "risking" eternal destruction not only for Himself but for all mankind if He failed.

The substantiality of Matter is now explained as an association of non-material entities, protons and electrons: Matter and Radiation as constituent forms of etheric energy. Spirit being the source of such energy, Christ, by Whom these things consist, in His "Death" created a new, or rearranged, etheric energy, in some way altered the course of Nature—proved by His Resurrection, and so available to man whose spirit is attuned.

Thus His act to restore man's displacement created within Matter a new "spirit-magnetic" field, in which "filings" attuned may draw on a new "radiant" energy.

[1] *Outline of Modern Belief*, part xiv, p. 784.

With this new force available,[1] attuned Freewill after bodily death can combine from the protons and electrons of Space with its usual etheric energy an *incorruptible* body, which conforms to the environment of Space, and when in contact with Matter assumes the shape of the former *corruptible* body even to the reproduction of scars as "earth honours," but without disability, and which can pass through closed doors, ascend and descend.[2]

Such resurrection will not be until immediately before or at His Second Coming.

"*Behold, I come as a thief, Blessed is he that watcheth*" (Rev. xvi, 15).

"*Therefore be ye ready, for in such an hour as ye think not, the Son of Man cometh*" (Matt. xxiv, 44).

"*As in Adam all die, even so in Christ shall all be made alive . . . Christ, the firstfruits: afterward they that are Christ's at His coming*" (1 Cor. xv, 22, 23).

"*Behold, I show you a mystery* [secret]; *we shall not all sleep, but we shall be changed.*

"*In a moment, in the twinkling of an eye, at the last trump: for the trumpet shall sound, and the dead shall be arised incorruptible, and we shall be changed.*

"*For this corruptible must put on incorruption, and this mortal must put on immortality.*

"*Death is swallowed up in victory*" (1 Cor. xv, 51–54).

---

[1] "The nature of electrons is not known, but it is known that they are the source and absorbers of radiation. . . . Electrons and radiant energies are, so far as is known, the fundamental energies that constitute the Universe" (*Outline of Modern Belief*, part xiv, pp. 777, 780).

In a recent research, by bombarding the sodium atom with heavy hydrogen emanation, the sodium atom could be made radio-active for twelve hours, equivalent to one gramme of radium.

Also, see "the two tables" in Sir A. Eddington, *The Nature of the Physical World*, pp. 5–12.

[2] Apologies for crude interpretation (attempt at further elucidation in the Appendix); but in the concept a Reality, which must be left to a higher intelligence to clarify. Talking to a Dutch Roman Catholic priest, on the Jericho-Jerusalem road, as to a possible physical explanation of the Resurrection, I was met with a curt: "No, for such would destroy the very foundation of Christianity, Faith."

The Church's reply sounds similar to that given to Galileo and Darwin.

Such explanation to-day may be beyond human intelligence and perhaps will not be revealed until His Second Coming—which stresses Faith: but to the *seeking* mind that "*worships in spirit and in truth*" is promised "*all these things shall be added.*"

"WITHOUT A CITY WALL."—THE MOUNT OF OLIVES IN THE BACKGROUND

To face p. 250.

"We say this to you as a message from the Lord, that we, the living, the survivors until the appearance of the Lord, will not precede those who sleep. For the Lord Himself in command, with the voice of an archangel, and with a signal from God, will descend from heaven; and first the dead in Christ will rise again, then we the living remnant shall at the same time be carried up in clouds for an introduction by the Lord into the eternal condition, and then shall we always be with the Lord" (1 Thess. iv, 15–17). (Ferrar Fenton's translation.)

With His Second Coming, *"But the rest of the dead lived not again until the thousand years was finished. This is the first resurrection. Blessed and holy is he that hath part in the first resurrection; on such the second death has no power, but they shall be priests of God, and shall reign with Him a thousand years"* (Rev. xx, 5, 6).

To emphasise the above teaching, and His Power to accomplish so stupendous an act, which should change the composition of Ether-Matter and alter the course of Nature, so that death (result of deviation of Freewill) is swallowed up in victory (Freewill restored to Harmony in life eternal), He led up to the supreme miracle, the raising of Lazarus (John xi, 1–44). He deliberately waited for four days (one more than His own *"in the heart of the earth"*).

"He said, the sickness is not unto death, but for the glory of God, that the Son of God might be glorified thereby.

"Then said Jesus unto them plainly, Lazarus is dead. And I am glad for your sakes that I was not there, to the intent ye may believe; nevertheless let us go unto him.

"When Jesus came, He found that he had lain in the grave four days already.

"Jesus said, I am the resurrection and the life; he that believeth in Me, though he were dead, yet shall he live. And whosoever liveth and believeth in Me shall never die."

Martha to Mary, *"The Master is come and calleth for thee. As soon as she heard that, she arose quickly and came unto Him. Lord, if Thou hadst been here, my brother had not died.*

"Jesus said, Take ye away the stone. Martha saith unto Him, Lord, by this time he stinketh, for he hath been dead four days.

"Jesus lifted up His eyes and said, Father, I thank Thee that Thou hast heard Me, . . . because of the people that they may believe that Thou hast sent Me.

"He cried with a loud voice, Lazarus, come forth. And he that was dead came forth."

This act decided His death, for the chief priests and Pharisees with Caiaphas, *"One man should die for the people : from that day they took counsel together for to put Him to death."*

To this threat, His attitude has been attributed as meekness.

The word *meek* in the Sermon on the Mount (according to a sermon broadcast in December 1934) is better translated "debonnaire."

And it is so, when applied to Christ. In spite of this threat to life, He dared to continue to teach in the Temple before the chief priests and Pharisees, He rode right into their midst through the Golden Gate in triumph, before their eyes He whipped the moneychangers, who had been used to their booths by long Pharisaic custom, out of the Temple area.

He surrendered His supreme Spiritual Power over Matter to the material, to let the material do to Him what it wished, and it crucified Him: with strength sheathed, He stood before Pilate and took His scourging.

Such is not the "meekness" that appeals to the weak and sentimental, that holds up the pathetic and empties the churches; but the "debonnaire" of a man who, knowing the risk of annihilation, yet seeks not political avenues of evasion, but goes straight to what He believes is Righteous and True; and when the inevitable happens—as it will in the Cain-Bel government of this material world—moans not, but prays, *"Father, forgive them, for they know not what they do.*

"Blessed are the debonnaire, for they shall inherit the earth."

GOLGOTHA

THE CHRIST

The time was now come for Him to fulfil His great act, and He went up to Jerusalem. Mary, with premonition of the end, anoints His feet with costly spikenard. He had saved her brother from death after four days in the grave; if by any woman's power she could preserve His body to Resurrection, then she would do her utmost.

"*And Jesus said* [perceiving her intention, when none else did], *She hath done what she could, she is come aforehand to anoint My body to the burying. Wheresoever the gospel is preached, this shall be spoken for a memorial of her*" (Mark xiv, 3-9).

He feels the tremendous task in front of Him, which necessitated faith in His Father and in His Power to rescue:

"*Now is My soul troubled, and what shall I say? Father, save Me from this hour, but for this cause came I unto this hour. Father, glorify Thy name.*

"*Then came there a voice from heaven, saying, I have both glorified it, and will glorify it again.*

"*Jesus said, Now is the Crisis of this world: now shall the prince of this world be cast out.*

"*And I, if I be lifted up from the earth, will draw all men unto Me*" (John xii, 27-32).

Davidson works out the period of "*three days in the heart of the earth,*" verifying the exact anniversary, Thursday to Sunday, of the Passover (1486 B.C.)[1]—which was the symbol of Israel's deliverance by sacrifice of the Passover Lamb—now the deliverance of mankind by the sacrifice of "*the Lamb slain from the foundation of the world*" (Rev. xiii, 8). Only with His Divine Power surrendered could the betrayal compact of Judas[2] be allowed to take place.

And at the Last Supper, "*With desire, I have desired to eat this passover with you before I suffer*" (Luke xxii, 15). He emphasised for all time the symbolism of bread-wine, flesh-blood, in this His act of Cosmic significance.

"*I go to prepare a place for you, I will come again and receive you unto Myself, that where I am, there ye may be also.*

---

[1] Davidson, pp. 490-506. See p. 125.
[2] Luke xxii, 3-6.

"Believe Me that I am in the Father, and the Father in Me, or else believe Me for the very work's sake.

"If ye love me, keep My commandments.

"This is My commandment, that ye love one another, as I have loved you. Greater love hath no man than this, that a man lay down his life for his friends.

"And now I have told you before it come to pass, that when it is come to pass, ye might believe" (John xiv–xvi).

Then His prayer to His Father, in perfect faith (John xvii):

"Father, the hour is come; as Thou hast given Thy Son power over all flesh, that He should give eternal life to as many as Thou hast given Him.

"Now they have known that all things whatsoever Thou hast given Me are of Thee.

"Sanctify them through Thy truth, Thy word is truth . . . that they all may be one.

"Father, I will that they also, whom Thou hast given Me, be with Me, where I am.

"I have glorified Thee on the earth, I have finished the work which Thou gavest Me to do.

"And He went, as was His wont, to the Mount of Olives."

And in Gethsemane the crisis was not the dread of bodily death, with the ignominy of a Roman Crucifixion before a shrieking mob which, in spite of all His teaching and miracles to induce belief, yet howled for His extinction—that was expected in a Cain-Bel-influenced world; but the "cutting off," in some way beyond human conception, from Divine Spirit, which necessitated a Faith greater than Abraham's.[1]

His unswerving faith in God assured Him that His deliverance would be effected, and that the Cross and what followed it would lead triumphantly to the establishment of His Kingdom. The Resurrection manifests to us the omnipotence

---

[1] Abraham's act of the sacrifice of his son, to whom the Promise of the Covenant, symbolic of God's act in the sacrifice of His Son. So to all men at some time comes this test of Faith—in which way Freewill to operate: it is the natural consequence of Spirit resident in Matter, to pierce "the World of Shadows" which seems to lie as a thick pall between Spirit and Matter.

THE GARDEN TOMB

THE CHRIST

of the Father. Death is the point in human experience at which Nature invades our personality most inexorably; it is the final collision and dispute in which Nature appears to have the last word. By what happened to Christ it is shown that God in His supremacy can use even death for His transcendent purpose. Christ held triumphantly to His faith in God amid the torture, bodily and spiritual, of the Cross, where in some awful sense He was bearing the sins of the world.[1]

Because the Father promised the Son that if He would have faith in Him that He would raise Him to life again, and He could by His faith win for all His redeemed the priceless gift of eternal life, not only of the soul but also of the body.[2]

"Who in the days of His flesh, when He had offered up prayers and supplications with strong crying and tears unto Him that was able to save Him from death, and was heard in that He feared; though He were a Son, yet learned He obedience by the things which He suffered. And being made perfect, He became the author of eternal salvation unto all them that obey Him" (Heb. v, 7–9).

When God promised to raise our Lord from that death of the soul, He was pledging Himself to an act of such stupendous power as had never before been exercised in this Universe. The creation of this and all it contains was as nothing in comparison to it.[3]

"What is the exceeding greatness of His power to usward who believe; according to the working of His mighty power, which He wrought in Christ, when He raised Him from the dead, and set Him at His own right hand in the heavenly places" (Eph. i, 19, 20).

When the vision of the suffering Messiah was Divinely revealed to them, David and the prophets sank back aghast: (in the idiom of their day)—

"I am counted with them that go down into the pit. Lord, why castest thou off my soul? Why hidest thou thy face from me" (Ps. lxxxviii).

---

[1] H. R. Mackintosh, *Christ and Faith in God*, in "God and the World through Christian Eyes" Series, pp. 146, 147, 149 (Student Christian Movement Press).

[2] Hamilton-Hunter, *The Origin of Evil*, pp. 93–96 (Covenant Publishing Co.).

[3] *Ibid.*

"*Save me, O God, for the waters are come in unto my soul.*

"*Let not them that wait on thee, O Lord God of Hosts, be ashamed for my sake, let not those that seek thee be confounded for my sake, O God of Israel* [pathetically appealing that He might not fail man].

"*Let not the waterflood overflow me, neither let the deep swallow me up, and let not the pit shut her mouth upon me*" (Ps. lxix).

"*In my distress I called upon the Lord, and cried unto my God. Then the earth shook and trembled,*[1] *the foundations of the hills moved and were shaken. He bowed the heavens also and came down and darkness was under his feet. He did fly upon the wings of the wind. At the brightness that was before him his thick clouds passed. Then the channels of waters were seen, and the foundations of the world were discovered at thy rebuke, O Lord.*

"*He sent from above, he took me, he drew me out of the great waters, he delivered me from my strong enemy*" (Ps. xviii).

And the cup that He drank in Gethsemane was to do more than to neutralise the golden cup of Babylon: it was to restore the displacement of the Spirit in man to Divine Harmony.

"*Father, all things are possible unto Thee, take away this cup from Me; nevertheless not what I will, but what Thou wilt*" (Mark xiv, 36).

And when betrayed, He said, "*But this is your hour, and the power of darkness*" (Luke xxii, 53).

And Caiaphas says to Him, "*I adjure Thee by the living God that Thou tell us whether Thou be the Christ, the Son of God. Jesus saith unto him, Thou hast said; nevertheless I say unto you, Hereafter shall ye see the Son of Man sitting on the right hand of Power and coming in the clouds of heaven.*

"*Then the high priest rent his clothes, saying, He hath spoken blasphemy; what think ye?*

---

[1] *Cf.* Matt. xxviii, 2–4: "Behold, there was a great earthquake, for the angel of the Lord descended from heaven, and came and rolled back the stone from the door."

"IN THE HEART OF THE EARTH"

"*They answered and said, He is guilty of death*" (Matt. xxvi, 63–66).

And the world has judged between them.[1]

Then on the Cross, in utter, forlorn loneliness, "*Eloi, Eloi, lama, sabachthani (My God, My God, why hast Thou forsaken Me?)*," but with Faith unshaken, "*It is finished; Father, into Thy hands I commend My spirit,*" and having said thus He died, then descended into Hades to consummate His work. What actually was achieved when Spirit free of body is impossible for human conception to imagine.

And His body was taken by Joseph of Arimathea and Nicodemus, anointed and laid in a "*new tomb hewn out in the rock*" (Matt. xxvii, 60), aboveground and with pure air.[2] And in His Resurrection, by Divine Power and the "radiant" energy resulting from the creation of the new spirit-magnetic field within Matter, in His case the actual protons and electrons of His earthly body were transformed into His *incorruptible* body.

"*For thou wilt not leave my soul in hell; neither wilt thou suffer thine Holy One to see corruption*" (Ps. xvi, 10).

Thus His last message to Caiaphas and the Jews, even

---

[1] Let a one-time conqueror of Europe, his glory in ashes, exiled to a lonely Atlantic isle, judge. Napoleon, after surveying the deeds of Alexander and Cæsar "with the hope of rivalling their exploits," says: "There is just one Name in the whole world that lives; it is the name of One who passed His years in obscurity, and who died a malefactor's death. . . . Here is One who is not a mere name, who is not a mere fiction, who is a reality. He is dead and gone, but still He lives—lives as a living energetic thought of successive generations, as the awful motive-power of a thousand great events. He has done without effort what others with lifelong struggles have not done. Can He be less than Divine? Who is He but the Creator Himself; who is sovereign over His own works, towards whom our eyes and hearts turn instinctively, because He is our Father and our God" (Newman's *Grammar of Assent*, pp. 490, 491).

[2] They can have their sect-divided Church, with its priestly vestments and tribute of kings, the gold and the gilt; but outside the tense turmoil of contending factions that is Jerusalem, the resting-place of His Body finds a quiet sojourn in the site of the Garden Tomb, on the Nablus road, found by General Gordon and essentially British.

Here it is now tended by a gardener, in well-worn khaki, who has served Israel in several wars; keen on the verity of the site and his job, he plants as tribute the natural Palestine flowers that He once looked on. If you send for these five post-cards, double or treble it (for he refuses the change as backsheesh), as tribute for the Garden, that it may neither toil nor spin, but be arrayed in glory greater than Solomon's.

when they had killed Him, appealing to their reason, if their spiritual intuition was darkened, to believe in Him—

If Lazarus, dead four days in a dark, airless underground tomb, with body decayed, was raised to life, then why not Myself, for two nights in a new clean tomb? Now will you see and believe? [1]

To the Jews who had asked for a sign, "*Jesus answered, Destroy this temple, and in three days I will raise it up; but He spake of the temple of His body*" (John ii, 19–21).

And the place on earth the Logos chose [2] wherein He should act for the redemption of man's Freewill was that very spot where as Eden the Adamic race emerged and deviated from Divine Will.

On the site where was "*breathed into his nostrils the breath of life, and man became a living soul*" (Gen. ii, 7), exactly there at Bethlehem "*the Logos was made flesh . . . to give his flesh for the life of the world*" (John i, 14; vi, 51).

Where stood "*the tree of the knowledge of good and evil . . . pleasant to the eyes, to be desired to make one wise*" (Gen. iii, 6), there was Calvary.

Where in the garden the woman said, "*The serpent beguiled me and I did eat,*" there the Magdalene said, "*Rabboni.*"

And exactly where it was said to the first Adam, "*Dust thou art, and unto dust shalt thou return,*" therefore there it was said of the Second Adam, "*Why seek ye the living among the dead? He is not here, for He is risen.*"

The seeking yet doubting mind of Thomas says, "*Unless I see the marks and put my hand on the prints, I will not believe*" (John xx, 25).

It is certain that at His Second Coming, which is near at hand, He will show the marks and the prints, and demonstrate the physical reality of His Death and Resurrection.

---

[1] "*And they shall look upon Me whom they have pierced, and they shall mourn for Him as one mourneth for his only son*" (Zech. xii, 10).
"*Behold, He cometh with clouds, and every eye shall see Him, and they also which pierced Him*" (Rev. i, 7)—yet to be fulfilled by the Jews.
[2] See pp. 19–21.

"IN THE HEART OF THE EARTH"

"Seek ye the Kingdom of God and His righteousness, and all these things shall be added unto you" (Matt. vi, 33).

The reason for the acceleration in scientific knowledge among mankind since 1844, and especially since 1918,[1] is that human intelligence may be sufficiently advanced to receive such concepts at His hands.

The Son of man holds the keys to man's existence and the nature of the Universe.

To them *"that worship Him in spirit and in truth,"* He will *"add all these things"*—the origin of evil, the emergence of living from non-living matter, the step by which man came from the animal world, the explanation of the Ether and its inherent energies, for *"by Him were all things created, visible and invisible, and by Him all things consist, and in Him are hid all the treasures of wisdom and knowledge."*

The philosophic Bridge between Spirit and Science vanishes (pp. 7, 8).

Science of the twentieth century is on the verge of Light, is looking directly into the screen of a shadow world steadily enlightening. It will see directly the coming *Light*, and murmur with conviction from the deepest *"thoughts of his heart—My Lord and My God."*

His answer will be the same, *"Because thou hast seen, thou hast believed; blessed are they that have not seen, and yet have believed."*

---

[1] For the significance of the datings see Davidson, *The Great Pyramid*, as to the Grand Step epoch and antechamber entrance: the Era of Inductive Science—"intellectual effort with mind attuned to synchronise with a stream of energy from the Godhead, whether realised or not" (quoted p. 304).

APPENDIX TO CHAPTER VIII

(I) Conceptions suggested by the Relativity theory:
From Sullivan and Grierson, *Outline of Belief* Series, part xv, summing up:

> Time itself is universal, time which is not the past, the present, the future, is everlastingly co-existent with Space, and the two together, time and space, form the animating principle of the universe; together the creator of "events" (in technical phraseology).
>
> These two things, blended together, form as it were a sort of universal matrix (as Eddington calls it) that determines the forms of other entities that come into being. It is from Space-Time that things emerge (p. 832).
>
> Since Time is an intrinsic factor in the real nature of things, it is called the Fourth Dimension.
>
> According to the Relativity theory, we live in a four-dimensional world, which our minds split up into three dimensions of space (length, breadth and thickness) and one dimension of time.
>
> Future and past are distinguished merely by a difference of sign; there is no essential difference between them. The equations which enable us to determine the future (as when we predict eclipses) can equally well be made to work backwards (p. 833).
>
> In this Space-Time continuum, as *one* reality, things happen . . . it is the source of creative activity . . . from it "world stuff" comes into being; it is the source of the stuff out of which all existents are made; possibly to this source may be traced back the electric particles we call electrons and protons . . . it is the "primordial form" of the universe, it is associated with the distribution of matter (p. 834).

(The origin of all this can hardly be better expressed in words as symbols than in Genesis i, 1, 2—see p. 3.)

THEOCENTRIC REALITY

Suppose in this year of grace, 1935, you were transported to a distant star, the light from which took 120 years to reach the earth, and that you had an enormous telescope to look through; then, looking down to-day on, say, the plains of Waterloo you would witness the battle taking place, just as, had you been a spectator present, you would have been looking on in 1815 (p. 835).

Jeans states that the remotest objects visible in the heavens are so distant that light, travelling at 186,000 miles per second, takes 140 million years to come from them to us. Light from our sun, 93 million miles away, takes about eight and a third minutes to reach the earth.

Then suppose you die in 1936 at age 60, and are transported instantaneously to a distant star, the light from which takes 60 light-years to reach the earth, and imagine looking through an enormous telescope focussed on the spots on earth where you lived. Then you would see your childhood and life with all that you did passing in review.

Then if you are taken to a star 1939 light-years away, you would see the life of Christ on earth unfolding; if to a star 5935 light-years away, then you would see the emergence of the Adamic race, and what actually happened in the Garden of Eden.

Such then is a scientific phantasy based on light.

Is such possible in terms of consciousness?—an earth "event" created by Freewill, so changing the course of etheric energy in the space-time continuum (or must we imagine such to be limited by the stratosphere, as the wireless waves)? For example, take such a maximum disturbance as the bombardment in the Great War, with its colossal energy changes, destruction of material, and the mass death of men.

Thus on the vault of the Universe might be recorded as on a scroll the events that have happened on the earth—these "events" in Divine Consciousness (i.e. *Light*): where at some particular point the course of each life can be searched.

And thus (assuming the velocity of Spirit is instantaneous), after death, the spirit of a man can catch up his light-years and see his life in retrospect, and learn at each particular point where Freewill in action has deviated.

An "event" on the earth goes away at the velocity of light, like a ripple begun in the middle of a lake.

In terms of consciousness, and beyond the stratosphere, is there fading (as when the lake ripple reaches towards the margins) in the recording?

Earth's history for man from the Adamic race may have been fixed at some 6000 years, so none may be lost—the time set with His Second Coming, when is "*the first resurrection.*"

Thus the Judgment Day is not an arbitrary verdict by some Divine "judge," but "a day of decision" when the deviation of each Freewill is seen, to be compared to Christ's, to be weighed in values of Reality, or as Esdras (2 Esdras vii) puts it:

"*The day of judgment is a day of decision, and displayeth unto all the seal of truth; even as now a father sendeth not his son, or a son his father, or a friend that is most dear, that in his stead he may be sick, or sleep or eat, or be healed; so never shall one pray for another in that day, neither shall one lay a burden on another, for then shall all bear every one his own righteousness or unrighteousness. This present world is not the end; the full glory abideth not therein; therefore have they who were able prayed for the weak. But the day of judgment shall be the end of this time, and the beginning of the immortality for to come, wherein corruption is passed away, intemperance is at an end, infidelity is cut off, but righteousness is grown, and truth is sprung up.*"

And thus also—the parables of the pounds and talents.

What is the speed of thought? Is the velocity of Divine Spirit instantaneous or limited to that of light?

Being both transcendent and immanent—perhaps either at will. Conceived as at the velocity of light (as limited to the four-dimensional world), the Most High when He created the Space-Time continuum, and knowing the course that once started such would run—in which Freewill in operation would deviate from Divine Principle that has

THEOCENTRIC REALITY

to be taken by Faith (Freewill in Spirit being essential to receive and emanate Divine Love), and from which "world stuff" and "events" would emerge—then *acted* (transcendentally) to restore Freewill to harmony, that He Himself should suffer, by sacrifice as the highest form of service, to show His created beings in heaven and on earth the highest outpouring of Divine Love ; *acted* millions of years ago, even before man appeared on this earth, as to reach the earth at a particular moment in its history, 4 B.C. = "Divine foreknowledge."

To Christ as the Logos in this Space-Time continuum has been entrusted the salvation of the race of Adam, and "*all the families of the earth*"—"*Before Abraham was, I am*"—in it, before and after His earthly Incarnation; who "purchased" the right to exercise His influence by His "cutting-off" (see pp. 253–257, 303, 323–325, 377, 415–417).

(II) Conceptions suggested by the Quantum theory: From the *Outline of Belief* Series:

> Physicists have no knowledge of the existence of an atom except when it is radiating energy. . . . An atom only reveals its existence by interacting with the physical universe, and these interactions are always accompanied by energy changes, that is, with the movements of electrons (part v, p. 287).
>
> A quantum is a tiny finite packet or pulse of energy (that is, of radiation of definite wave-length), the minimum amount sufficient to disturb an electron . . . occurring not as a continuous process but in little jerks.
>
> Every time an electron drops nearer to its neutron a pulse of radiation is emitted. Every time a pulse of radiation is received and absorbed, the electron jumps up to an orbit further from the neutron (part iv, p. 240).
>
> Any method of observing the electron affects the circumstances of that electron, and renders its future uncertain, we alter its behaviour in an unpredictable way. By "observing," the more accurately we know the position of the electron, the less accurately we know its velocity . . . its movement is determined by no known laws, as if having a kind of freewill of its own. Thus it happens the atom is regarded as a collection of events rather than as a substantial entity" (part v, pp. 287–289).

Is it not possible that Spirit Freewill (not acting on certain Divine Principle but deviating), to which was entrusted the creation of this earth, so combined the proton and electron in the atom that when it was strung in the 666 carbon chain, to emerge as protoplasm, then life has to die?

In the Perfect Oneness of Harmony, all deviation by its own act leads to destruction—that is, death.

So, in the behaviour of the electron, is this indeterminancy of orbit, or wobble as to which path to take.

To restore life, it was necessary for Christ, "*by Whom all things consist*" (*i.e.* so that universal laws have not become confounded), to be "*in the heart of the earth*" (words as symbols of a Reality)—therein some spirit action within matter to alter the course of etheric energy—to cause proton and electron to combine in "the other way," so that life when it emerged and attuned to harmony would be life eternal.

His action entailed Faith as necessary—as all action by Spirit Freewill in the Space-Time continuum, and by Freewill on earth, this "world of shadows" which screens between Spirit and Matter, exampled by His own—to show the whole Spirit realm how Freewill is to act in the Divine Universe of Harmony, where God is all in all.

(III) Present-day thought, trained by novel-reading and argument, runs along a space-time line. The author of a novel has a beginning and an ending, the argument has a thread running from point to ensuing point.

Such was first emphasised by Socrates, who starts from a definition to deduce to a logical concept.

If the deduction be but slightly wrong (as is probable, for each Freewill is liable to deviate from harmony), the ensuing point is slightly shifted off the true line (unknown to the Freewill); so that the ultimate view is logically displaced, yet persisted in as the necessary considered truth.

Whereas thought (especially scientific thought) is formed as within concentric orbits—or, rather, spheres—like unto the formation of an atom; and includes in each space-time orbit as subject-matter the relevant past, present and future of the implication in science, philosophy, religion and history as four-dimensional; with a definite central nucleus, stabilised and attuned.[1]

So that each orbit is trained to be visualised at any particular moment as a whole; the thought-matter being jerked up or dropped back, as consciousness receives impressions from external objects and other minds (being jerked up to receive and absorb), or produces in expression and experiment (after being attuned to the whole, being dropped to nucleus as energy is emitted in action).

Scientific thought thus receives, absorbs and learns from previous work, and then experiments; more new knowledge is assimilated, then attuned or rejected, then action is emitted in further research. The orbits are expanding and retracting continuously, until truth is perfected—never proceeding in straight lines to a "dud" issue, as is the way of argument, which way of thinking breeds bigotry and persecution.

(IV) What is modern psychology but the study of conflict in the mind of Freewill which deviates yet seeks harmony?

Using the above simile, Freewill may thus be likened in its mental make-up to the atom.

An external impression (as radioactivity of absorbable wave-length) is received and absorbed by an "association of ideas" or mind (the electron), is allotted (or jerked up) to its appropriate orbit, until assimilated and put in true perspective by the whole.

The "concept" is "dropped back" towards the Ego (the proton or nucleus) when thought, wish, or impulse is expressed in action (as radiation emitted by the electron), and thus mind is ready to receive again.

If the impression is received but not attuned to the whole Freewill the impulse to expression is "repressed." The wish cannot be expressed in action (the electron is unable

---

[1] With such idea, the index at the end has been attempted.

to emit radiation) until Spirit (as that within the ether which pervades the whole structure) provides the energy, and the Ego or Body (the proton or central nucleus) has declared its will or decision in which way to act.

Spirit—as the origin of the Space-Time continuum or "primordial form" within the Ether which binds the atom and pervades all the universe and space, "the something to which truth and beauty and goodness matter"—is the source of radiant energy and sustains life.

Without that Spirit, the ego "dies"; with it, the ego "acts" (Gen. iii, 19; ii, 7).

(a) If several such unattuned repressions ("wobble" electrons) and an ego (proton) unattuned to harmony in its environment, then the Freewill (the whole atomic structure) tends to be unstable.

(b) If the external mental impression is not received or assimilated (*i.e.* does not cause the electron, as association of ideas or mind, to jerk up) then the Freewill make-up is inadequate in consciousness, and loses the chance of expression on its environment (as radioactivity), and thus is deficient in action.[1]

(c) Freewill (as the whole atom) shows its existence only when expressing itself (radiating energy) and interacting with the environment: in action it tends to deviate from harmony (unless attuned to something definite in the Cosmos,—that something is the Divine Law of Harmony), and so leads to its own discord and chaos.

In (a), (b) and (c), Freewill misuses cosmic energy, the gift of Divine Spirit.

Now this Spirit has been defined in Genesis ii, 7.

"*And the Lord God formed man of the dust of the ground* [the evolution from matter [2]] *and breathed into his nostrils the breath of life*" (the injection of Spirit, Divine Essence,[3] which thus enables mind, as the electron, to associate ideas, to think to the past and to the future, to appreciate "forms" —beauty and goodness and truth—and to combine with

---

[1] With this idea, reread the incident in Bethune Baker's *The Way of Modernism*, pp. 43-45.
[2] See p. 17.
[3] See p. 1—Christ as the Logos.

THE "ATOMIC" MAKE-UP OF FREEWILL 267

ego or body as the proton); *"and man became a living soul"* (so the whole unity as Spirit Freewill or Personality, capable of receiving and emanating Love, the first Law of Divine Harmony, and capable of immortality).

On this concept "body-mind"[1] as proton-electron may be considered in the simplest form of life as a single electron with one orbit round a proton; man as evolved from the ape ancestry as the carbon atom with its six electrons; finally the white race after the action of Genesis ii, 7, as the complicated uranium atom[2]; or put diagrammatically:

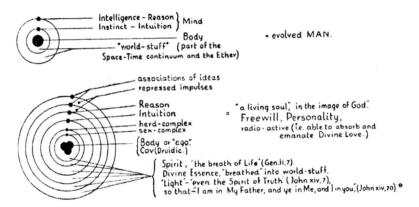

Thus, the atom, the universe of stars, and man are uniform in composition—leading to a final conception, when deviated Freewill is finally destroyed, to God as All in All.

Just as the atom combined with others forms a molecule, and molecules combined form protoplasm; just as planets together form a solar system, and stars with stars make a spiral nebula, an island universe; so man with man forms a community, a tribe, a race.

---

[1] See pp. 13, 14.

[2] In uranium: "92 electrons each playing its appropriate rôle in a symmetrically co-ordinated atomic system . . . inside its nucleus counted 238 positives and 146 negatives . . . and the atom changes to something else if any one of these positives or negatives drops out" (Professor Millikan).

* As proof of Intuition—with the concept (1) of the Ether as medium between Spirit and Matter, or, as put scientifically, that etheric energy is required for the manifestation of Life in Matter, and (2) of the "atomic" make-up of Freewill as above—read *The Ultimate Physical Reality*, quoted p. 235; then with the concept thus enlarged read John xiv-xvi.

They are all created under a Universal Law, and if man's Freewill deviates from such (as "indeterminancy" of electron within atom causes union of such atom with another to be loose, and the molecule unstable) he causes discord in the community; and if a king, a dictator, or any form of leader, such leads to discord, chaos, war and race decadence.

So do civilisations rise and fall.

Admit as central—the One Universal Law of Love, in the Perfect Oneness of Divine Harmony.[1]

Freewill (only limited [2]—if such word is here applicable —by this One Law) in action which deviates from such Harmony is evil, and brings its own discord, and ultimate inevitable annihilation from the Perfect World ("*eternal destruction from the presence of the Lord*").

So, since destruction of it is not immediate, are wars produced.

Suppress war, if other Freewill can; that is, ensue peace.

Once within war, a Freewill has to attune itself to harmony with the impressions received from the local environment—shells, trenches, mud, discomfort, the prospect of pain and death.

The Freewill which isolates himself from such impressions (as the "Conchie") remains deficient in consciousness and in action.

If such external impressions are unattuned by Freewill, as fear of death (from thoughts of family at home, the sight of wounds and dead around, the uncertainty of survival) contending with preservation of life, so that impulses arise to satisfy nature and shirk duty, then the consequent repressions while still seeking to attune lead to neurasthenia and shell-shock.

Similar conflict of impulses, unattuned to harmony within the local environment, occur in the ordinary civilian life, resulting in "nerves," nervous breakdown, etc.: while all deviation is sin.[3]

---

[1] See pp. 23–25, 247, 248.
[2] "*Our wills are ours to make them thine.*"
"*Whose service is perfect freedom.*"
[3] See p. 358.

Happiness is thus the action of Freewill in harmony with its environment. And "real health demands that the individual shall be at peace within himself, and not constantly impelled to escape into some objective activity or distraction." [1]

The injunction is laid on each "ego" to attune Freewill to Divine Harmony, thus to absorb and emanate Divine Love [2]; and given to man was One example.

Christ attuned His own Freewill to the local environment—of earthly life, limitation of material means, opposition both spiritual and physical—and lived His life in Love and perfect Harmony with His Father.

His teaching to Spirit Freewill in earthly environment is given in the Sermon on the Mount (Matt. v), with—

"*What shall it profit a man if he gain the whole world, and lose his own soul?*"

"*For where your treasure is, there will your heart be also.*" (Where your Freewill directs, there will be the measure of your soul.)

"*If thine eye be single, thy whole body shall be full of light.*" (If thine ego be attuned, thy whole Freewill will be "radioactive"—capable of absorbing and emanating Divine Love.)

"*Whosoever would save his life, shall lose it.*"

"*This is My commandment, That ye love one another, as I have loved you.*

"*Greater love hath no man than this, that a man lay down his life for his friends.*"

Revealing Divine Purpose to attune Freewill by Faith:

"*For God so loved the world, that He gave His only begotten Son, that whosoever believeth in Him should not perish, but have everlasting life.*"

When Adam deviated (Gen. ii, 16, 17; iii), this Spirit source of energy injected into world-stuff becomes in some way lessened (such is seen in every man) or withdrawn,

---

[1] Dr Crichton-Miller (at Oxford, 12th August 1935), who continued: "Modern life is replete with opportunities of escape—from chewing gum to morphine; from dancing to ocean cruises . . . civilisation will resolve itself into an elaborate and increasing escape from reality."

[2] See pp. 247, 374, 385.

so that the body or ego dies—"*for dust thou art and unto dust shalt thou return.*"

Christ's action "*in the heart of the earth three days*" was in some way directed to restore this "*breath of life*," Divine Essence, back into the atomic make-up of Freewill, so that such power is again available to attuned Freewill to fashion proton-electron as non-material entities into "*an incorruptible body*" and immortality: and in the act He risked annihilation for Himself.

Without such power "*life everlasting*" is impossible, though "ego" may live in some way (which spiritualists seek) until the final destruction of "*the first world,*" [1] but not coming to "*the second world.*"

And His Resurrection proved the efficacy of His action.

[1] See pp. 251, 424.

CHAPTER IX

¶ 1. The historical figure had disappeared: the footprints on the shore of Galilee had vanished (Noyes, p. 345).

The Apostles expected that the message of Pentecost and based on His Resurrection could not but be accepted by all Judea, Galilee and beyond to Israel, and that, with repentance, "the restitution of all things" would take place immediately, and so was their expectation of Christ's return.[1]

"Therefore let all the House of Israel know assuredly, that God hath made that same Jesus, whom ye have crucified, both Lord and Christ. Repent . . . for the promise is unto you and to your children and to all that are afar off" (Acts ii, 36–39).

"Repent thee therefore, and be converted, that your sin may be blotted out, that so might come the times of refreshing from the presence of the Lord; that He might send Jesus Christ, whom the heavens must receive until the times of restitution of all things, which God hath spoken by the mouth of all His holy prophets since the world began" (Acts iii, 19–21).

"And He shall confirm the Covenant with many for one week; and in the midst of the week He shall cause the sacrifice and the oblation to cease; and for the overspreading of abominations, He shall make it desolate even until the consummation" [2] (Dan. ix, 27).

During the week the Gospel was preached to the Jews (Acts ii), then extended to the Samaritans (Acts viii), then to the Gentiles (Acts x, 45): finally, when the Jews were

---

[1] *Cf.* the expectation of the angels that with His birth there would be *"peace on earth, goodwill toward men"* (Luke ii, 14).

[2] See Davidson, *The Great Pyramid*, pp. 307–309, 358; and fulfilling Christ's prophecy of the fall of Jerusalem, A.D. 70, when it was made desolate with the overspreading of abominations until the consummation, 1917–1936.

filled with envy, contradicting and blaspheming, then Paul and Barnabas said:

"It was necessary that the word of God should first have been spoken to you: but seeing ye put it from you, and judge yourselves unworthy of everlasting life, lo, we turn to the Gentiles, for so hath the Lord commanded" (Acts xiii, 45-47).

Throughout Asia Minor, in the great cosmopolitan Greek towns, and in Rome itself, what are known as "Mystery Religions," derived from Egypt and the East, were widely popular. They seem to have had their roots far back in the dim past, primitive nature worship.

The Gentiles accepted Christianity as a superior kind of Mystery Religion.[1]

By the fourth century B.C. the crude mythology of Cain-Bel had given way to allegories: there were no longer gods in whom anyone believed as a hard fact—religion to be true must be universal, not the privilege of a particular people. Thus, as a mental reaction to polytheism, developed Greek philosophy, but with no direct revelation of the One God—the good is useful, virtue is happiness, the only possible virtue is true knowledge.

Through Socrates, Plato, Aristotle, it defines the indefiniteness of human aspirations, the alternatives to many beliefs in each of which are fallacies, delusions and inconsistencies; thus it asserts the aim of man is to think, to "desire to know" some "way of life."

This atmosphere early Christianity absorbed, and out of the One direct Christ has made of Him numerous creeds, which some uphold as necessary of belief, and others deny—out of which confusion is born persecution. This philosophic reaction on early Christianity in Paulinism, Christology, is the theme of later theological writers.

As Cain perverted the story of the Creation and gave to the Sumerians a galaxy of gods and goddesses, acceptable enough to pervert pre-Adamite man and his succeeding mixed races, so one can almost read his hand in the subsequent philosophy, based on Greek tradition, "Hath

[1] *Foundations*, by B. H. Streeter and others, p. 181.

God said? . . . ye are as gods; enshrine knowledge and reason," in recent Agnosticism and Rationalism.

The intellectual and æsthetic tradition of Europe looks back to Athens, not to Galilee: and no amount of special pleading will make it plausible to maintain that Science, Philosophy or Art owes as much to Jesus as to Hippocrates, Plato, or Praxiteles.[1]

For these days, there is less inspiration value in the intellectual tradition of Athens than in the eschatology of Galilee, the fringe of which has not yet been touched.

Burdened with Plato's philosophic attributes of Deity, modern theology is intellectually bogged in busily answering Dualism, Pantheism, Agnosticism, and the crude groping uncertainty of after-life in Spiritualism. From the pure monotheism of Seth, in Druidism, it could have derived a more certain faith in God and immortality; it could have seen in Darwin's theory the Creator's plan; and in the Fourth Gospel, unadulterated by Platonism, the creative and redemptive power of Christ.

There is more of scientific inspiration in the chapters of John than in the writings of Hippocrates. In the "flesh and blood" of John vi and viii are the non-material entities and etheric energy that make up the substantiality of Matter, and which ("*except ye eat . . . ye have no life in you*") sustain immortality.

The "flesh" of Hippocrates is all body-tissue (skin, muscles and vessels) without discrimination, and when moribund its appearance is even yet known as the "*facies hippocratica.*"

Praxiteles' statue of Hermes (carrying the child Dionysus is his arms) "is a masterpiece. But the young child can hardly be regarded as a success; he is not really childlike" (*Ency. Brit.*, 13th ed., vol. 22, p. 256).

Art had never been forced to contemplate the child until He came; then looking on a naked infant for the first time, for centuries it continued to draw him as a subthyroidic adult.

The traditional Christian view of salvation and redemp-

[1] Canon B. H. Streeter, *Reality*, p. 201 (Macmillan & Co.).

tion has remained without that "born again" to a new concept; for the scientific mind demands "to see the marks."

Just as official religion before the coming of Christ could not contemplate a crucified Messiah; so official religion to-day has not considered the Universe's Creator, Divine Spirit, "suffering" within matter, to effect the restitution of Freewill to Divine Harmony, nor such implication in physical terms.

As then, eschatology was the forerunner of His First Coming, so to-day His Church cannot see "the era of inductive science" as the preparation of man's mind to accommodate itself to the effects of His Second Coming.

The Modernist holds that eschatology was not a Divine revelation, but wrung from Benjamin and Judah by their sufferings under Antiochus, in hope that a just God would ultimately vindicate His people.

> From the persecution of Antiochus dates the rise and prevalence of apocalyptic with its characteristic eschatology. The vivid directness of the ancient prophet is replaced by a complicated symbolism, to our modern taste fantastic and bizarre, influenced somewhat it is thought by Babylonian and Persian models.[1]

This Schweitzer developed—"man feared that to admit the claims of eschatology would abolish the significance of His words for our time"; then the question was raised whether His teaching (on the Kingdom of God and the end of the present world order) was not in fact an "interim-sethik"—the morality for a short period of transition [2]; and also read into the words of His apostles and of Paul their expectation of an immediate Second Coming.

Then:

> I am coming more and more to feel that to water down and explain away the apocalyptic element is to miss something

---

[1] *Foundations*, p. 88.
[2] Yet, "*because they thought that the Kingdom of God should immediately appear*" (Luke xix, 11) He gave them the parable of the pounds (Luke xix, 12-27).
Also the parable of the talents (Matt. xxv, 14-30 "*. . . after a long time*") and "*the gospel must first be published among all nations*" (Mark xiii, 10).

ESCHATOLOGY AND MODERNISM

which is essential. Much of the unique moral grasp of the New Testament is in one way directly a result of the eschatological background of the period.[1]

The Modernist has but to look at the Great Pyramid, built more than 2000 years before Antiochus, in which was embodied in stone, and in scientific laws—

(1) The Birth of the Messiah, "the Son of Man (Enoch xlvi and xlviii), would be accomplished on Saturday, 6th October 4 B.C. (15th day of the month Tisri), and that the Passion of the Messiah would be accomplished on Friday, 7th April A.D. 30 (15th day of the month Nisan).

(2) Its Displacement Factor witnesses to the Divinity of Jesus Christ, "*by whom all things were created.*"

(3) The whole of that eschatology from Daniel to Revelation, as Apocalyptic, and confirmed by Christ, was dated in the Pyramid's internal structure, and is being fulfilled to the day: "the times of Restitution of all things," "in the latter days," "in that day," was to be preceded by Chaos, the beginning of "the tribulation" being 4th/5th August 1914—10th/11th November 1918.

(4) And warns him to consider the meaning of the King's Chamber, the entrance date of which is September 1936.

Yet the theologian still goes on, relying on Science's dictum that the earth will last millions of years, and expecting that man will still be evolving his progress towards the Sermon on the Mount.[2]

He has accepted its dictum that there is no evidence of any historical "Fall." The Bible relates His-Story of the action of Spirit on Matter—"the born again" concept to understand.

Science has not explained—indeed it is not within its province to—the emergence of Spiritual Intuition, of Freewill with choice to operate in accord with or in deviation from Divine Harmony (though such "in embryo" within

[1] *Foundations*, p. 119.
[2] "*Knowing this first, that there shall come in the last days scoffers . . . saying, Where is the promise of His coming? for since the fathers fell asleep, all things continue as they were from the beginning of creation*" (2 Peter iii, 3, 4).

creation, as Psalm cxxxix, 16-19). Neither has science explained the sudden emergence of the high intellectual standard in the land of Sumer of 3500 B.C., which founded the world civilisations at Ur in Chaldea and at Thinis in Egypt, nor that of the knowledge of the Universe's Laws embodied in the Great Pyramid before 2600 B.C.

And so the theologian debates the Work of Jesus Christ:

(1) The Liberal view. The work of Christ was to teach men the true nature of God, and set an example of the highest kind of human life, which it behoves all men to follow.

(2) The Conservative view. It is the *death* of Christ that is fundamental, it saves men from the burden of guilt and from the consequent wrath of God.

(3) The "inclusive view." "Neither . . . by itself is satisfactory; but we agree with the Liberal that it was the moral character of Jesus, as shown during His whole life, which is of the first importance; and with the Conservative that death was necessary and in no sense accidental to the fulfilment of His work." [1]

Why bisect Him? The whole includes the part. Let them argue; for argument was begotten out of Greek philosophy, and trained by modern education.

¶ 2. In the nineteenth year of the reign of Tiberius Cæsar, A.D 35,[2] occurred the first persecution of the Church by Saul of Tarsus, which caused it to be "scattered abroad," except the Apostles, from Jerusalem (Acts viii, 1).

A Vatican MS.[3] dated A.D. 35 states that in that year "Joseph of Arimathea, Lazarus, Mary, Martha, Marcella, their maid, and Maximin, a disciple, were put by the Jews

---

[1] *Foundations*, p. 267.

[2] For British history as it might be taught see R. W. Morgan, *History of Britain from the Flood to A.D. 700* (Marshall Press, Ltd.); W. M. H. Milner, *The Royal House of Britain*; L. S. W. Lewis, *St Joseph of Arimathea at Glastonbury*; R. W. Morgan, *St Paul in Britain, or the Origin of British as opposed to Papal Christianity* (the above, Covenant Publishing Co.).

[3] Quoted by Cardinal Baronius (historian and librarian to the Vatican) in *Ecclesiastical Annals*, on which he spent thirty years, and who wrote: "Better silence than a lie mixed with truth."

in a boat without sails or oars and floated down the Mediterranean; the vessel drifted finally to Marseilles; from thence Joseph with his company crossed into Britain, preached the Gospel, and finally died there."

Perhaps there is some truth in the strange tradition that still lingers, not only among the hill folk of Somerset, but of Gloucestershire and in the West of Ireland, that St Joseph of Arimathea came to Britain first as a metal merchant seeking tin from the Scillys and Cornwall, and lead, copper, and other metals from the hills of Somerset, and that Our Lord Himself came with him as a boy.[1]

> *And did those Feet in ancient time*
> *Walk upon England's mountains green?*
> *And was the Holy Lamb of God*
> *On England's pleasant pastures seen?*
> *And did the Countenance Divine*
> *Shine forth upon our clouded hills?*
> *And was Jerusalem builded here*
> *Among those dark Satanic mills?"*
>
> (William Blake, 1757–1827.)

"Joseph of Arimathea is by Eastern tradition said to have been the younger brother of the father of the Virgin Mary." [2]

"Anna, the cousin of the Virgin Mary, assigned as the ancestor of the Tudor princes, was the daughter of Joseph of Arimathea, reputed to be the founder of a British dynasty." (Her daughter, Penardim, married Llyr, King of Siluria, from whom Bran, Caradoc, Gladys (Claudia), Linus, Lucius, Helen (who married Constantius) and Constantine the Great were descended; and from her son, Beli, King Arthur of Avalon is eighth in descent.)

It was because she was there already that Joseph went to Britain. Anna, married to a prince of our land, was here ready to welcome that earliest band of missionaries who brought the Gospel of her cousin's Son to Britain within five

---

[1] Rev. L. S. Lewis, *St Joesph of Arimathea at Glastonbury*, p. 17, and giving Blake's verses as "The Glastonbury Hymn."
[2] R. W. Morgan, *St Paul in Britain*, p. 119.

278 THE PILLAR IN THE WILDERNESS

years of His Ascension, and who by the courtesy of a British King established in Britain the first of all the national churches of the world.[1]

William of Malmesbury says that St Philip sent twelve missionaries from France to Britain (such is confirmed by Freculphus, Bishop of Lisieux, A.D. 825–851), of whom the leader was St Joseph of Arimathea, and that the King gave them Ynys-vitrin (Avalon, or Glastonbury) and twelves hides of land.

The Domesday Survey (of A.D. 1088, folio, p. 249b, 449): "The Domus Dei, in the great monastery of Glastingbury, called the Secret of the Lord. This Glastingbury Church possesses, in its own villa, xii hides of land which have never paid tax."

Thus, about A.D. 38, Joseph of Arimathea, at the request of the Druids of high rank whom he knew,[2] and to see his daughter, Anna, came, and with his eleven companions landed and rested at Wyrrall (Weary all) Hill at Glastonbury (then an island), later made over to them in free gift by King Arviragus. He is said to have planted his staff in the ground, and from it there sprang the famous Holy Thorn. A branch of this tree, in flower, was always sent to the King on Old Christmas Day up to the reign of Charles I, "a custom happily revived in the last few years." An irate Puritan tried to cut it down, and mortally wounded it; a splinter pierced his eye, causing his death. Thirty years later the tree died, but not before three thorns had been budded from it, and still in May and again on Old Christmas Day these flower.[3]

This, the first aboveground church in the world, was 60 feet by 26 feet, built "Gallico more" of timber pillars and framework doubly wattled inside and out, and thatched with straw.[4] And thus it stood until reconstructed by

---

[1] Rev. W. M. H. Milner, *The Royal House of Britain*, pp. 23–28, quoting from a very ancient MS. in the Heralds' College, London, Pedigrees in Jesus College, MS. 20, thirteenth century, MS. tenth century in British Museum, and other sources.
[2] See *Freculphus apud God*, p. 10.
[3] Quoting from L. S. Lewis, *St Joseph of Arimathea*, pp. 11, 12.
[4] Morgan, *St Paul in Britain*, p. 124.

St David, whose uncle, Maelgwyn of Llandaff, *circa* A.D. 450, writes:

> Joseph of Arimathea, the noble decurion, received his everlasting rest, with his eleven associates, in the Isle of Avalon. He lies in the southern angle of the bifurcated line of the Oratorium of the Adorable Virgin. Moreover he has with him two silver white vessels filled with the blood and sweat of the great Prophet, Jesus.

So the lays of the Holy Grail have Glastonbury's two stories of St Joseph and King Arthur inextricably mixed.

St Joseph died A.D. 76, and on his tomb:

"*Ad Britannos veni post Christum sepelivi. Docui. Quievi.*" [1]

And from the Royal charters of the church and monastery from King Arthur, and the nephew of its second founder, to Edward III:

> The first church in the kingdom, built by the disciples of Christ (charter of Edgar).

> This was the city which was the fountain and origin of Christ's religion in Britain, built by Christ's disciples (charter of Ina, or Ivor).

> We know that Christ, the true Son, afforded His light to our island in the last year of Tiberius Cæsar, A.D. 37 (Gildas, British historian, A.D. 520–560).

St Gregory of Tours, A.D. 544–595, in *History of the Franks*, p. 133; Haleca (Archbishop of Saragossa) in his *Fragmenta*, and the *Chronicon of Pseudo Dexter* all state that Joseph of Arimathea was the first to preach the Gospel in Britain.

> The Church of Avalon in Britain no other hands than those of the disciples of the Lord themselves built (Publius Discipulus).

> The Mother Church of the British Isles is the Church in Insula Avallonia, called by the Saxons Glaston (Ussher).

> If credit be given to ancient authors, this Church of Glastonbury is the senior church of the world (Fuller).

> It is certain that Britain received the faith in the first age from the first sowers of the Word . . . from the disciples of Christ Himself soon after the Crucifixion . . . in Britain the Church of Glastonbury is the most ancient (Sir Henry Spellman, *Concilia*).

---

[1] "When I had buried the Christ I came to Britain. I taught. I slept."

¶ 3. In A.D. 36, Bran had resigned the crown of Siluria to his son, Caradoc, to become Archdruid.

In A.D. 37, the British fleet under Arviragus watched the ludicrous spectacle of a Roman Emperor, Caligula, who had succeeded Tiberius, addressing his army of invasion, bidding them follow his example of wreathing "this garland of green seaweed around my immortal brow," and filling his helm "with these smooth and brilliant shells"—"To war beyond the bounds of nature is not courage but impiety . . . to each of you my fellow-soldiers I promise a year's extra stipend," and then retiring.

In A.D. 42, King Guiderius sent to Rome complaining of the Emperor Claudius' encouragement of the intrigues of Beric and Adminius (two reguli of the Brigantes and Coritani). The Roman fleet transports for the army under Aulus Plautius, noted for discipline and rapid marches, were ready for invasion, when the army refused to move, "we will march anywhere in the world, but not out of it," and embarked only when it was addressed by a eunuch, Narcissus, who offered to lead them into Britain.

Guiderius fell in the first battle, and was succeeded by his brother, Arviragus, who was the first to give his vote for Pendragon to his cousin, Caradoc.

Caradoc's campaign for ten years against the might of the Roman Empire, extending from the Euphrates to Spain and with no other war in hand, against her best generals, Plautius, Geta, Vespasian, Titus, Ostorius, is a British epic.[1]

In the Triads: "Three have been our hero-kings—Cynvelin, Caradoc, Arthur. Except by treachery, they could not be overthrown." "Caradoc, son of Bran, whom every Briton, from the king to the peasant, followed when he lifted his spear to battle."

> The Silures reposed unbounded confidence in Caractacus, enumerating the many drawn battles he had fought with the Romans, the many victories he had obtained over them (Tacitus, *Annal.*, lib. ii, c. 24, A.D. 80).

During a truce of six months, when the first four generals were recalled, Caradoc visited Rome for the marriage of his

[1] Morgan, *St Paul in Britain*, pp. 87-102.

sister Gladys (Pomponia Græcina) to Plautius. In the Iolo MSS. it is recorded that he appeared before the Senate, and stated that he had ordered "every tree in Siluria to be felled, that the Romans might no longer allege it was the British forests, and not British valour, which baffled them." Dion narrates: "When Caractacus was shown the public buildings of Rome he observed: 'It is singular a people possessed of such magnificence at home should envy me my soldier's tent in Britain.'"

The campaign was continued against Ostorius, and after one battle, in Shropshire, Caradoc at her repeated solicitations took refuge at Caer Evroc (York) with Aregwedd (or Aricia), queen of the Brigantes, and grand-niece of that traitor Avarwy. Here, by her orders, he was seized while asleep, loaded with chains, and delivered to Ostorius.

He and his family were taken to Rome, A.D. 52, and Bran, the Archdruid, his father, voluntarily surrendered himself as hostage.

Three million people lined the streets of Rome to view Caradoc's arrival.

"Rome trembled when she saw the Briton, though fast in chains,"[1] and his speech before the convened Senate in Latin noted by Tacitus—such a speech never before by a Roman enemy, its spirit like unto the old Republican times:

> Had my government in Britain been directed solely with a view to the preservation of my hereditary domains or the aggrandisement of my own family, I might long since have entered this city an ally, not a prisoner; nor would you have disdained for a friend a King descended from illustrious ancestors[2] and the dictator of many nations. My present condition, stript of its former majesty, is as adverse to myself as it is a cause of triumph to you. What then?
>
> I was lord of men, horses, arms, wealth; what wonder if at your dictation I refused to resign them? Does it follow, that

---

[1] "*Roma catenatum tremuit spectare Britannum.*"

[2] The Druidic custom was that when the child was fifteen years old, at a public reception, his ancestry was read. That of Caradoc, from the Pantliwydd MSS. of Llandsannor, went back thirty-six generations (to beyond Brwt) to 1080 years—that is, 330 years before Rome was founded in 753 B.C.

because the Romans aspire to universal dominion, every nation is to accept the vassalage they would impose? I am now in your power, betrayed—not conquered. Had I, like others, yielded without resistance, where would have been the name of Caradoc? Where your glory? Oblivion would have buried both in the same tomb. Bid me live, I shall survive for ever in history one example at least of Roman clemency.

The Roman custom at such triumphs was for the captive king or general to pass "under the yoke" on the Sacra Via, then to be flung into the Tarpeian dungeons, there to die starved, strangled, or beheaded, and the body dragged by hooks into the Tiber.

More by fear of the consequences in Britain than by clemency, Caradoc was the sole exception—to reside in free custody in Rome for seven years, never to bear arms again against her. This he faithfully kept after his return in A.D. 59 to reside at Aber Gweryd, now St Donat's Major, in Glamorganshire, even when, in the following year, all Britain, including the treacherous Iceni-Coritani, rose in religious war after the Menai massacres.

But his capture infuriated Britain, and the war was intensified under Arviragus, whom the Silures elected Pendragon. In A.D. 53, Ostorius was defeated near Caerleon and asked to be recalled, wearied in mind and body. Didius Gallus and Veranius also retired with nervous exhaustion; then Suetonius Paulinus followed.

"In Britain after the capture of Caractacus the Romans were repeatedly conquered, and put to the rout by the single state of the Silures alone" (Tacitus, *Annal.*, lib. v, c. 28). And Juvenal records the question asked in the Roman Forum after every battle in Britain:

"Hath our great enemy, Arviragus, the car-borne British King, dropped from his battle-throne?"

This firm resistance to Rome was mainly due to the national religion, Druidism (R. W. Morgan).

When the Romans effected a footing in Britain, they found in Druidism a constant and implacable enemy to their usurpation. They would have been glad to introduce their religion, but to that point there was an invincible obstacle in the

BRITISH CHRISTIANITY

horror and contempt of the natives for a religion formed by a corruption of their own allegories.[1]

¶ 4. The story of Caradoc's family in Rome is told by Morgan.[2] Gladys, his sister; Gladys, adopted by Claudius, and then named Claudia, and Eurgain, his daughters, and Linus, his son, had been converted to Christianity at Avalon. Claudia, aged seventeen years, in A.D. 53 married Rufus Pudens, whom she had met when he was stationed as praetor castrorum at Regnum (now Chichester). Their palace, with 200 male and 200 female servants, all born on the hereditary estates of Pudens in Umbria, on the Mons Sacer, became the home of Caradoc's family, the Palatium Britannicum.

Here Claudia wrote several volumes of odes and hymns, and the house became the rendezvous of the poets and authors of Rome, Martial the epigrammist writing several poems celebrating her beauty, grace, wit and fascination. Here Rufus, Bran, Caradoc, and the rest of the family, became Christians; the house was then called the Titulus, or Hospitium Apostolorum, and Hermas[3] became the pastor. Here resided the Apostles Peter and Paul when they came to Rome—Paul being sent to Rome, for his appeal to Caesar, in the second year of Nero—*i.e.* A.D. 56.

> Claudia was the first hostess or harbourer of St Peter and St Paul at the time of their coming to Rome.[4]

> It is delivered to us by the firm tradition of our forefathers that the house Pudens was the first that entertained St Peter at Rome, and that there the Christians assembling formed the Church, and that of all our Churches the oldest is that which is called after the name of Pudens.[5]

The children of Claudia and Pudens were brought up at St Paul's knees, the eldest was baptized by him Timotheus, after his friend, Timothy. The family[6] ministered to him

[1] Cleland, *Ancient Celtica*, p. 13.
[2] R. W. Morgan, *St Paul in Britain*, pp. 102–110.
[3] Rom. xvi, 14.
[4] Robert Parsons, the Jesuit, *Three Conversions of England*, vol. i, p. 16.
[5] Baronius, *Annales Ecclesias*.
[6] In letter to Timothy, 2 Tim. iv, 21: "*Eubulus* [cousin of Claudia] *greeteth thee, and Pudens and Linus and Claudia.*"

up to the last, and Rufus Pudens [1] and all the children who suffered martyrdom [2] were buried by his side in the common family cemetery in the Via Ostiensis.

The apostles having founded and built up the Church at Rome committed the ministry of its supervision to Linus; this is the Linus mentioned by Paul in his Epistle to Timothy.[3]

Clemens (who succeeded Cletus within twelve years of the death of Linus, as third bishop [4] of Rome) writes: "Sanctissimus Linus, frater Claudiæ."

(*Re* the Titulus)—In this sacred and most ancient of churches, known as that of Pastor, dedicated Sanctus Pia Papa, formerly the house of Sanctus Pudens, the senator, and home of the apostles, repose the remains of 3000 blessed martyrs, which Pudentiana and Praxedes, virgins of Christ, with their own hands interned.[5]

The Titulus became known as the Church of St Pudentiana, the baths adjacent, later known as Thermæ Timothinæ and Thermæ Novatianæ, with all the palace grounds, were bequeathed by Timotheus to the Church in Rome, and were its only possessions till the reign of Constantine.[6]

¶ 5. It appears that Bran left Rome with Aristobulus, his son Manaw, Ilid (a Hebrew), and Cyndaw, before Caradoc. He was accompanied by Eurgain, and her husband Salog, lord in her right of Caer Salog (Salisbury), a Roman patrician.

Ilid established his mission under the protection of Bran, his grandson Cyllinus (eldest son of Caradoc), Salog and Eurgain in the centre of Siluria on the spot in Glamorganshire known from then till now as Llan-Ilid. At this Llan, or "consecrated enclosure," the Princess Eurgain founded and endowed the first Christian cor, or choir, in Britain. From

---

[1] Rom. xvi, 13, "*Salute Rufus chosen in the Lord, and his mother and mine*" (? Paul the elder half-brother by Hebrew marriage, Rufus, younger half-brother by later Roman marriage).

[2] St Timotheus suffered martyrdom, at 90 years, in A.D. 144; St Praxedes, A.D. 144; St Novatus, A.D. 139 (5th persecution); St Pudentiana, A.D. 107 (3rd persecution); Pudens in A.D. 96; Linus in A.D. 78 or 90. Claudia died a natural death in A.D. 97.

[3] *Opera* of Irenæus (A.D. 180), lib. iii, c. i.

[4] Baronius, *Ad Maii*, 19.

[5] Morgan, *St Paul in Britain*, pp. 171, 172, quoting authorities.

[6] *Ibid.*, pp. 102-104.

BRITISH CHRISTIANITY

this Cor-Eurgain issued many of the most eminent teachers and missionaries down to the tenth century.[1]

Aristobulus, whom Paul saluted writing to the Romans, was Bishop of Britain.[2]

It is perfectly certain that before St Paul had come to Rome Aristobulus was absent in Britain.[3]

Aristobulus was the father-in-law of St Peter, and his wife Christ healed (Matt. viii, 14, 15) (Morgan, p. 131).

Aristobulus was one of the 70 disciples and a follower of St Paul the apostle, along with whom he preached the Gospel to the whole world. He was chosen by St Paul to be the missionary bishop to the land of Britain, inhabited by a very warlike and fierce race. By them he was often scourged and repeatedly dragged as a criminal through their towns, yet he converted many of them to Christianity. He was there martyred, after he had built churches, ordained deacons and priests for the island.[4]

Beyond Siluria, among the Ordovices, the protection of Bran did not avail Aristobulus, who coming from Rome, the national enemy, fell perhaps a victim to this fact rather than a martyr to religion.[5]

Simon Zelotes, the apostle, was also a missionary to Britain.

Simon Zelotes traversed all Mauretania, and the regions of the Africans, preaching Christ. He was at last crucified, slain and buried in Britain.[6]

Crucifixion was a Roman penalty . . . not known to the British laws. Simon Zelotes suffered, as tradition affirms, in the vicinity of Caistor, under the prefecture of Caius Decius, whose atrocities caused the Boadicean War.[7]

And St Paul: "*a chosen vessel unto Me, to bear My name before* . . . *the children of Israel*" (Acts ix, 15).

"*Whensoever I take my journey into Spain, I will come*

---

[1] *Ibid.*, p. 135.
[2] Dorotheus (Bishop of Tyre, A.D. 303), also Hippolytus (A.D. 160) mentions in his list Aristobulus as Bishop of the Britons.
[3] Alford (a Jesuit), *Regia Fides*, vol. i, p. 19.
[4] Martyrologies of the Greek Church for March 15th. Confirmed by the Adonis Martyrologia for March 17th: "Aristobulus, Bishop of Britain, brother of St Barnabus the Apostle, by whom he was ordained Bishop."
[5] Morgan, p. 137.
[6] Dorotheus, *Synod de Apostol.*   [7] Morgan, p. 130.

to you, for I trust to see you on my journey, and to be brought on my way thitherward by you" (Rom. xv, 24).

During the six years between his liberation from his first imprisonment in Rome to his martyrdom at Aquæ Salviæ in the Ostian road, in A.D. 67, Paul fulfilled his intention to journey to West Europe; then to visit Aristobulus and Caradoc in Britain—in Siluria, beyond the bounds of the Roman Empire, hence the silence of the Greek and Latin writers upon it.[1]

> Paul, liberated from his first captivity at Rome, preached the Gospel to the Britons and others in the West.

> When Paul was sent by Festus on his appeal to Rome, he travelled after being acquitted into Spain, and thence excursions into other countries and to the islands surrounded by the sea (Theodoretus, Bishop of Cyropolis, A.D. 435).

The Divine hand preventing Paul's journeys into Asia and Bithynia directs him to Britain (Acts xvi, 6, 7).

> That we possess as substantial evidence as any historical fact can require of St Paul's journey to Britain.[2]

> The same view is substantially maintained by Baronius, Alford, Archbishops Parker and Ussher, Stillingfleet, Camden, Gibson, Cove, Nelson, Allix, etc.[3]

From *The Triads of Paul the Apostle*, among ancient British Triads:

> In three places will be found most of God: where He is mostly sought; where He is mostly loved; where there is least of self.

> The three chief considerations of a Christian: lest he should displease God; lest he should be a stumbling-block to man; lest his love to all that is good should wax cold.

Tradition assigns to St Paul the foundation of the great abbey of Bangor.[4] Its discipline and doctrine were known

---

[1] Morgan, *St Paul in Britain*, p. 175.
[2] Bishop Burgess, *Independence of the British Church*.
[3] Morgan, summing up from the sources of Eusebius, Theodoret, Clemens, ancient MS. in Merton College and William of Malmesbury (pp. 174, 175, 158, 162).
[4] "Preceded that of any other monastery in Europe or Asia by over a century," destroyed by the Saxons, A.D. 607.
"It was the national University for Agriculture, Theology, Science, and Literature; at one time 10,000 teachers and students connected with it" (Morgan, *History of Britain*, p. 134).

as "Pauli Regula" (the Rule of Paul), and over each of its four gates was engraved his precept: "If a man will not work, neither let him eat."

And Peter: "How came St Paul not to salute St Peter in his Epistle to the Romans? Peter, banished with the rest of the Jews from Rome by the edict of Claudius, was absent in Britain."[1]

Bede, a canonised saint and earnest adherent of the Roman Church, states (*History*, lib. iii, c. 29) that the remains of the bodies of the Apostles Peter and Paul were, at the solicitation of King Oswy to Pope Vitalian (his letter to Oswy being extant), removed from Rome to England, and deposited at Canterbury, A.D. 656. Morgan adds: "Their remains then, if any, repose in British soil."[2]

Thus the end of Caradoc's military career, and his family's captivity in Rome, "a catastrophe which at the moment appeared not only irretrievable but to militate against the justice of Heaven in the government of nations," led to "the introduction of the Gospel into Britain from direct Apostolic sources and under the highest secular auspices in the kingdom," also on the character of Claudia tending "Paul the aged" at his martyrdom—"no instance more striking of the manner in which God, who bringeth good out of evil, overrules temporal calamities into agencies of eternal salvation."[3]

And more, for out of Saul's persecution came Joseph of Arimathea to Glastonbury; and his mission there to be blessed by Paul.

¶ 6. Senega, a Stoic philosopher, and in practice a grinding usurer, whose personal capital was fifteen millions, and Nero's adviser, had loaned two million pounds to the Iceni on the security of their buildings. Orders from Rome were sent to Suetonius to exterminate at any cost the chief seat of Druidism among the Cymry. Senega now demanded repayment at exorbitant interest.

[1] Cornelius à Lapide, quoted by Morgan, with comments, p. 183.
[2] Morgan, p. 184.
[3] R. W. Morgan, *History of Britain*, p. 117.

The King of the Iceni had died, and left his possessions to his two daughters and Nero. On the Icenic Senate demurring, Caius Decius, præfect at Caistor, claimed the whole as public property, stormed the palace with his legionaries; after outraging the household, carried off the treasures and confiscated the estates of the Icenic nobles, thus violating the Claudian treaty.

The Iceni send to Arviragus and with the Coritani abjure the Roman protectorate.

Meanwhile Suetonius by forced marches along the Wyddelian road reached the banks of the Menai, on either side of which extended the colleges and cemeteries of the Druids, the graves of 1500 years of chiefs and Archdruids with the hoary wisdom of the East. The massacre of the inmates and the burning of the colleges, which for several nights illuminated the waters of the Menai, are described by Tacitus.

Now all Britain joined in a religious war, and the Romans trembled. The statue of victory at Colchester fell and was shattered; a Pythoness with the spirit of divination ran through its streets crying: "Death is at hand"; the Straits of Dover at high tide were as red as blood, and, the waters receding, impressions on the sands were of dead men laid out for burial.

In the Senate at Rome the Britons' war-cry: "Trâ Môr, Trâ Brython!"[1] rang out, and cleared the chamber in panic, but the janitor who sought the cause found no one.

The Queen of the Iceni, Vuddig (Victoria) or Boeddig (Boadicea), harangued an army of 120,000 at Leicester. Colchester fell at the first assault, Verulam was stormed and burnt down, the Roman legions swept aside at Cocci Collis and Ambresbury. Her army, swollen to 230,000 men, advanced on London. Those who could escaped to Regnum and Ruputium, the rest took refuge in the Prætorium; its buildings were reduced to ashes, its walls levelled, 80,000 Roman citizens were slain. Then Boadicea swept west to intercept Suetonius, and in the fiercest battle ever fought in Britain, near Newmarket, where for the

[1] "As long as there is Sea, so long will there be Britons."

THE ROMAN WAR

whole day Dion Cassius describes the *mêlée* interlocked, the Britons were driven back and Boadicea either killed or died by poison.

This was A.D. 60, but the war was continued by Arviragus and in the north by Glaeneeg (or Gwallog). Suetonius, harassed like Ostorius, resigned his command to Petronius Turpilianus.

General followed general, Trebellius Maximus, Vectius Bolanus, Cerealis, Julius Frontinus, Julius Agricola, Sallustius Lucullus.

"*Magni duces, egregii exercitus,*"[1] writes Tacitus (lib. ii, c. 24).

"The series of invasions and sanguinary conflicts between the Romans and the Britons have no parallel in any age or country."[2] Sixty pitched battles were fought between A.D. 43–86.

In A.D. 83, Agricola had pushed the Roman arms into the north as far as the moors in Strathearn at the foot of the Grampian Hills, but in A.D. 85 he was recalled by the suspicious tyrant, Domitian. Arviragus reorganised the confederacy of the Cymry, Cymric Picts, the Caledonii of Western Albyn and the Brigantiaid, and renewed the attack until, in A.D. 86, the Plautian fortresses were stormed, London reoccupied, the Romans under Lucullus expelled from Kent, all their monuments destroyed, and a victorious peace concluded.

"Britain which was considered at last effectually conquered was lost in an instant," declares Tacitus.

¶ 7. In A.D. 114, Marius concluded a treaty with Trajan, wherein Britain agreed to become part of the Roman Empire and contribute 3000 lbs. weight of silver, but could choose her own kings, abide under her own laws and customs, no Briton to be disturbed in his estates, three legions only to be kept in Britain, at Caerleon, Chester and York, these to be recruited exclusively from Britons, and only for home service.

The Romans thus gained the assurance of being safe

---

[1] "Rome's greatest generals and picked armies."
[2] Smith, *Ancient Religions*, p. 457.

from British attack in the future, and called the island "Ferox Provincia" (the untameable province).

From the partial story furnished by the invaders themselves, conquest was never more dearly attempted than in the case of Britain by the Romans. By no people was every inch of country at any age contested with more bravery and surrendered more stubbornly. . . . They had settled laws and institutions, were distinguished for an ardent love of liberty, in defence of which the highest degree of valour and self-devotion were on all occasions manifested . . . a man must be a barbarian himself to suppose such a nation could be barbarous.[1]

In A.D. 125, Lucius, great-grandson of Caradoc, educated and baptized by a cousin, St Timotheus, succeeded Marius as king, and married his granddaughter, Gladys.

From the church at Avalon and the Cor-Eurgain had gone forth missionaries, among others:

(1) Beatus, a British noble, who converted Switzerland, died A.D. 96, his cell shown at Vuterseen, on Lake Thun.

(2) Mansuetus, who, with St Clement, preached in France, founded the Church in Lorraine, and was martyred in Illyria, east of the Adriatic (A.D. 89 or 110).

(3) Marcellus, who founded the Archbishopric of Treves, which for centuries was filled by Britons, and head of the Church in Gaul, was martyred A.D. 166.

(4) St Cadval, who founded the Church of Tarentum, in Italy, A.D. 170.

From the Triads:

There are three perpetual Choirs of the Isle of Britain, namely, Great Bangor in the forest of Maelor, Caer-Salog, and the Chrystal Isle in Avalon. In each of these are 2400 servants of Christ, singing night and day without intermission, a hundred every hour in rotation; so that the praises of God are sung without ceasing from year's end to year's end.

In A.D. 155, finding the British people prepared to support him, at a national council at Winchester, Lucius established Christianity as the national religion instead of Druidism.

The Gorseddan (high Druidic courts in each tribe) became episcopal sees; and those of the Archdruids at London, York

---

[1] Richardson, *Historian*, p. 10.

BRITISH CHRISTIANITY

and Caerleon became archbishops'. . . . In commemoration of this eventful change, Lucius endowed four churches from the royal estates, Winchester, Llandaff, St Peter's in Cornhill, St Martin's at Canterbury . . . and the Abbey of Bangor. . . . This British Church retained its national independence from A.D. 155 to 1203, when in defiance of the repeated protests of its clergy, it was incorporated with the Roman Catholic Church introduced into Saxondom by Augustine, A.D. 596. . . .

Lucius next enacted that everyone who made public confession of Christianity became entitled to all the rights of a native Briton. Multitudes of the persecuted faithful on the Continent found thus not merely a temporary refuge, but a free home in Britain.[1]

In A.D. 178, Lucius sent Elvan and Medwin, bishops of London and Llandaff, to Eleutherius, bishop of Rome, to obtain authentic copies of the Roman code of laws. Eleutherius with great wisdom urged him, as the sole vicar of God over his people, to make the New Testament the secular, as he had already made it the ecclesiastical basis of British legislation.[2]

They returned with the missionaries Dyfan and Fagan (both in name Britons).[3]

Cressy tells us . . . in his *Church History of Brittany* that in company with his sister, St Emerita, King Lucius finally went as a missionary through Bavaria, Rhœtia and Vindelicia, and was martyred near Curia in Germany.[4]

Perhaps his granddaughter, Frea, accompanied them on these journeys, through countries in the track of the Asar migration under Odin from Asgard. Evidently here, Frea and Odin met and married.

Their children, through Wecta, Skiold, Waegdaeg, Seaxnot, Baeldaeg and Niordr, originated most of the

---

[1] Morgan, *History of Britain*, pp. 132, 134.
[2] *Ibid.*, p. 135. *Cf.* the codification of English Law (first by Brute of Troy, *circa* 1100 B.C., then by Dunwal Moelmud, *circa* 500 B.C.) by King Alfred, called "Alfred's Dooms," which begins with the actual words of Exodus xx, 1, 2, and parallel throughout Exodus xx–xxiii. This record in the British Museum is summarised by Judge Thomas Hughes, *Alfred the Great* (Covenant Publishing Co.), also, W. Pascoe Goard, *The Law of the Lord, the Common Law*.
[3] L. S. Lewis, *St Joseph of Arimathea*, pp. 13, 14, quoting Bishop Urban, John of Teignmouth, Capgrave, Ussher, William of Malmesbury.
[4] *Ibid.*, p. 15.

Royal Houses of Europe (as per W. H. Milner's chart in *The Royal House of Britain*, see p. 194).

In A.D. 260, Constantius, aged twenty-five, was sent to arbitrate in certain differences between King Coel and the Duke of Cornwall, and four years later married Helen, Coel's daughter, heiress to the throne of Britain. In A.D. 265 their son was born, later to become Constantine the Great.

In A.D. 287, Diocletian, the reigning Emperor, nominated Constantius Cæsar.

> None of the first nine persecutions of the Christians extended to Britain. The tenth, under Diocletian, which raged for eighteen years over the rest of the Empire, was put an end to in Britain in A.D. 300, in less than a year, at the risk of civil war with his colleagues, by Constantius.[1]

Amongst the victims are enumerated archbishops and bishops of Britain, with ten to fifteen thousand communicants in different classes of society.

On his death, in A.D. 306, "the British legionaries elevated his son Constantine on their shields and proclaimed him Emperor . . . his pagan competitors, Maximian, Maxentius and Licinius, succumbed in succession to his victorious arms with the help of his legionaries chiefly selected in Britain." [2]

Till his death, in A.D. 337, Constantine extended Christianity throughout the whole Roman Empire, with only two short visits to his capital, Rome; and his mother, Helen, spent her last years seeking the sacred sites of Jerusalem, Bethlehem and Galilee and thereon erecting churches. He stated in a public edict the object of his life: "We call God to witness, the Saviour of all men, that in assuming the reins of Government, we have never been influenced by other than these two considerations—the uniting of all our dominions in one faith, and restoring peace to a world torn to pieces by the madness of religious persecution."

He presided over the Councils of Arles and Nice (A.D.

---

[1] Morgan, *History of Britain*, pp. 136, 138.   [2] *Ibid.*, p. 139.

BRITISH CHRISTIANITY

314 and 325). At the former, Eborius (Bishop of York), Restitutus (Bishop of London), Adelfius (Bishop of Caerleon), alone of all the bishops present, in proud independence refused the journey allowance and maintenance offered: at the latter, of 318 bishops present, there were only ten from the Latin-speaking Church.

The man must be mad who in the face of universal antiquity refuses to believe that Constantine and his mother were Britons, born in Britain (Baronius).

It is well known the great Constantine received his Christian education in Britain.[1]

Helen was undoubtedly a British princess.[2]

Constantine, born in Britain, of a British mother, proclaimed Emperor in Britain beyond doubt, made his native soil a participator in his glory.[3]

Until the reign of Constantine the Roman Christians had no other church than the Titulus to worship in.[4]

The Pope, it is well known, claims the sovereignty of the States of the Church by right of the decree of the British Emperor Constantine making them over in free gift to the Bishop of Rome.

That this decree was a forgery no one doubted; it was, however, confirmed by Pepin. By the Papal Church's own shewing, it is infinitely more indebted to the ancient British Church and sovereigns than they were to it. Without the benefactions of the Claudian family and Constantine, it would never have risen above the character given it by Pius the First, the brother of Hermas Pastor,—"Pauper Senatus Christi." For its earthly aggrandisement it is mainly indebted to ancient British liberality.[5]

Britain in the reign of Constantine had become the seat of a flourishing and extensive Church.[6]

---

[1] Sozomen, *Eccles. Hist.*, lib. i, c. v.
[2] Melancthon, *Epistola*, p. 189.
[3] Polydore Vergil, *Historia Brit.*, p. 381.
[4] Bale, *Scriptores Britan.*, p. 17.
[5] Morgan, *St Paul in Britain, or the Origin of British as opposed to Papal Christianity*, p. 164.
[6] Soames, *Anglo-Saxon Church* (Introd.), p. 29.

We can have no doubt that Christianity had taken root and flourished in Britain in the middle of the second century.[1]

Britain, partly through Joseph of Arimathea, partly through Fugatus and Damianus, was of all kingdoms the first that received the Gospel.[2]

The Christian religion began in Britain within fifty years of Christ's Ascension.[3]

Our forefathers were not generally converted, as many would fain represent, by Roman missionaries. The heralds of salvation who planted Christianity in most parts of England were trained in British Schools of theology, and were firmly attached to those national usages which had descended to them from the most venerable antiquity.[4]

From the *Triads of the Cymry*:

The three priorities of the Cymry:
  (1) Priority as the first colonisers of Britain.
  (2) Priority of government and civilisation.
  (3) Priority as the first Christians of Britain.

The glory of Britain consists not only in this, that she was the first country which in a national capacity professed herself Christian, but she made this confession when the Roman Empire itself was Pagan, and a cruel persecutor of Christianity" (Genebrard).

¶ 8. To this British Church then came Augustine in A.D. 596, and four years later in a report to the Pope he wrote:

In the western confines of Britain, there is a certain royal island of large extent, abounding in all the beauties of nature and necessaries of life. In it the first neophytes of the Catholic law, God beforehand acquainting them, found a church constructed by no human art, but by the hands of Christ Himself, for the salvation of His people. The Almighty has made it manifest by many miracles and mysterious visitations that He continues to watch over it as sacred to Himself, and to Mary, the mother of God.

---

[1] Cardwell, *Ancient History*, 1837, p. 18.
[2] Polydore Vergil, lib. ii.
[3] Robert Parsons, *Three Conversions of England*, vol. i, p. 26.
[4] Soames, *Bampton Lectures*, pp. 112-257.

Rome found here a Church older than herself, ramifications of which struck into the very heart of the Continent, the missionary triumphs of which in Italy itself in the life of Augustine were greater than his own among the British Saxons [1] —Cadval, at Tarentum; Columba, among the Lombards.

During the latter part of the sixth century there had been a continuous feud between the two churches, which represented the east and west divisions of the Roman Empire, the two feet of Nebuchadnezzar's image. John the Faster, Patriarch at Constantinople, had tried to assume the title of Universal Bishop of all Christendom; Gregory, Bishop of Rome, affirmed that whosoever took the title of Universal Bishop "doth forerun Antichrist"; to which John the Faster replied: "By this pride of his, what thing else is signified but that the time of Antichrist is now at hand?"

In A.D. 607, three years after Gregory's death, Emperor Phocas conferred the title of Universal Bishop or Pope on his successor, Boniface III.

This squabble as to "who should be first" in the Church penetrated as far as Britain, for Augustine (after introducing the Roman Church among the pagan Saxons, and being given St Martin's, Canterbury, by Ethelbert's Christian wife) was told to demand the submission of the British Church to the Pope at Rome. He was met with one voice:

> We have nothing to do with Rome, we know nothing of the Bishop of Rome in his new character of the Pope; we are the British Church, the Archbishop of which is accountable to God alone, having no superior on earth.[2]

Two conferences were held at Augustine's Oak in Herefordshire, A.D. 607, and ended by the following declaration of the Abbot of Bangor and seven bishops:—

> Be it known and declared to you, that we all, individually and collectively, are in all humility prepared to defer to the Church of God, and to the Pope of Rome, and to every sincere

---

[1] Morgan, *St Paul in Britain*, p. 155.
[2] *Ibid.*, p. 154; *vide* also Sir Henry Spellman.

and godly Christian, so as to love everyone according to his degree, in perfect charity, and to assist them all by word and deed in becoming the children of God. But as for further obedience, we know of none that he whom you term the Pope, or Bishop of Bishops, can claim or demand.

The deference which we have mentioned we are ever ready to pay to him as to every other Christian; but in all other respects our obedience is due to the jurisdiction of the Bishop of Caerleon, who is alone under God our ruler to keep us right in the way of salvation.[1]

"We cannot depart from our ancient customs without the consent and leave of our people," said the British bishops.[2]

Abbot Dionathus, of Bangor, treated Augustine with contempt.[3]

We have found the Scotch Bishops even worse than the British. Dagon, who lately came here, being a bishop of the Scots, refused so much as to eat at the same table, or sleep one night under the same roof with us,

writes Lamentius, Augustine's successor, to the Pope. Dagon's reason may be found in the sequela.

For on the break-up of the conference Augustine threatened the Cymry, as they would not accept peace from their brethren, they should have war from their enemies. At Augustine's persuasions Ethelbert instigated Edelfrid, the pagan king of Northumbria, to invade the territories of Brochwel, Prince of Powys, who had supported the bishops.

In the midst of the battle at Chester, Edelfrid observed on a hill 1200 priests of the University of Bangor, unarmed, praying for the success of the Christian arms. He led his force against them, killed as many as had not fled, and consigned the churches and colleges of Bangor to the flames. Thus was fulfilled, exclaims the pious Bede, the prediction of the blessed Augustine. Edelfrid was repulsed by Brochwel on the Dee, and a few days later was routed by Cadvan with the loss of 10,000 men, driven

---

[1] Hengurt MSS.; Selright MSS.; Humphrey Llwyd.
[2] Bede, *Eccl. Hist.*, lib. ii, c. 4 (writing A.D. 740).
[3] Nicholas Trivet.

BRITISH CHRISTIANITY

back to the Humber and besieged at York, where he surrendered.[1]

In A.D. 660, when the whole Heptarchy except Sussex had been converted, Wini, Bishop of Winchester (who had bought his first bishopric of London from Wulfhere, King of Mercia), was the only bishop of Roman origin, all the rest were British—such as Maelwyn or Patrick; Ninian, of the southern Picts; Finan, of the East Angles; Chad, of the Mercians; Aidan, of the Northumbrians; Columba, of the Scotch, who said:

> Except what has been declared by the Law, the prophets, the evangelists, and apostles, a profound silence ought to be observed by all others on the subject of the Trinity (Bede, iii, c. 4).

A.D. 666, Damianus was the last successor of Augustine in Canterbury and Rochester; afterwards Saxon Christianity was kept alive by the British Church.

The ancient British Church, by whomsoever planted, was a stranger to the Bishop of Rome and all his pretended authorities.[2]

The Britons told Augustine they would not be subject to him, nor let him pervert the ancient laws of their Church. This was their resolution and they were as good as their word, for they maintained the liberty of their Church 500 years after his time, and were the last of all the churches of Europe that gave up their power to the Roman Beast; and in the person of Henry VIII, that came of their blood by Owen Tudor, the first that took that power away again.[3]

It was the uniform practice of the Christians from the earliest times to read the Scriptures in the vulgar tongue, and it was not till the period of Charlemagne that Latin became the language of the Church services (*vide* Ussher, *Historia Dogmatica*). No two causes contributed so much to the declension of Christianity and the progress of Mohammedanism, as the suppression by the Church of Rome of the vernacular Scriptures, and her adoption of image-worship.[4]

---

[1] See Morgan, *History of Britain*, pp. 165–168.
[2] Blackstone, vol. iv, p. 105.
[3] Bacon, *Government of England*.
[4] Morgan, *St Paul in Britain*, p. 179.

Polydore Vergil, in Henry VII's time, affirmed, and Cardinal Pole (A.D. 1555) in his address to Philip and Mary (both rigid Roman Catholics): "Britain was the first of all countries to receive the Christian faith."

Only once was the priority questioned—and then on political grounds by the ambassadors of France, and Spain, at the Council of Pisa, A.D. 1417, when the Council confirmed it.

On appeal to the Council of Constance, 1419, it was again confirmed. A third time it was confirmed at the Council of Sena, when all acquiesced in the decision laid down, that the Churches of France and Spain were bound to give way in the points of antiquity and precedence to the Church of Britain, which was founded by Joseph of Arimathea "immediately after the Passion of Christ."

How truer to her God and more united would have been the Church of Britain to-day if she had traced her religious origins through the Stone Kingdom of the Bible and the Seth tradition of monotheism, Druidic direct to Christian, instead of being drunken with the Babylonian cup of Cainite mythology, and Græco-Roman arguments of the metaphysical that bred Agnosticism!

Thus to this blessed isle, washed by the Gulf Stream, *"a stone cut out of the mountains without hands . . . the place appointed . . . their fold"* prepared, came the House of Israel (the Angles, the Saxons, the Jutes, the Danes, the Normans)—*"scattered in the cloudy and dark day . . . wandering through all the mountains . . . sifted among all nations . . . gathered out of all countries . . . their captivity caused to return . . . brought again to their fold . . . upon the high mountain of Israel . . . to dwell safely . . . with My servant, the House of David, a prince among them . . . was lost and is found—to bring forth the fruits of the Kingdom of God."*

"Thus shall they know that I the Lord their God am with them, and that they, even the House of Israel, are My people, saith the Lord God; and ye, My flock, the flock of My pasture, are **men***, and I am your God, saith the Lord God."*

CHAPTER X

¶ 1. The Descending Passage of the Great Pyramid, at an incline when limestone rolls on limestone, under the view of the Dragon constellation, but watched overhead by the "sweet influences of the Pleiades"—the whole passage system being displaced 286·1 P" to the east, or negatively from the central vertical plane of the Pyramid—symbolises the spiritual history of the Adamic or white race from the deviation of Freewill of its ancestor, and under Cain-Bel influence from the first civilisation, the Sumerian, which brewed *"the golden cup of Babylon . . . of which all the nations have drunk and are mad,"* during the successsion of Nebuchadnezzar's kingdoms, the Times of the Gentiles, 603 (to 585) B.C.—A.D. 1918 (to 1936).

The Ascending Passage, at the same angle, 26° 18' 10", but upwards, with the commencement date 1486 B.C., symbolises the mode of spiritual ascent, directing to Bethlehem; in that year His chosen nation, Israel, was drawn out of Cain-Ham environment, to undergo spiritual training under Divinely given commandments, to become His servant nation to *"be a blessing to all the families of the earth,"* and the nucleus of His Kingdom on earth at His Second Coming.

The Ascending Passage leads into the Grand Gallery at the dating A.D. 30, His Crucifixion and Resurrection, when the height of 47·3", necessitating a stooping position *"under the Yoke of the Law, the Way of Truth in Darkness,"* is raised 286·1 P"[1]—*"the New and Living Way, the Way of Truth in Light, symbolising the Gospel Era."*

The preparation of *"the appointed place"* sees Joseph of Arimathea at Glastonbury with *"the light,"* and Britain

---

[1] *". . . will raise up by His own Power"* (1 Cor. vi, 14).

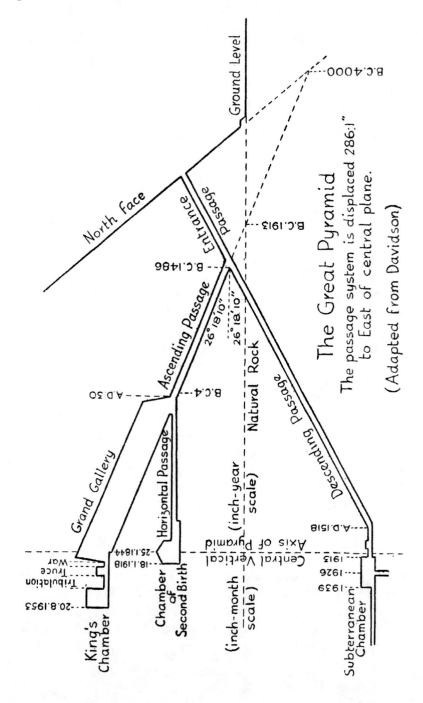

PYRAMID SYMBOLISM

proclaiming itself Christian (A.D. 155); then is seen the coming of the tribes, the purging of pre-Captivity stock and of the ten tribes from contamination of race stock and associations during their Continental migrations, in the fierce battles of Britons, Saxons, Angles, Danes, and Normans, and the Black Plague of A.D. 670-682, whereby within a few centuries after A.D. 1066 the whole should fuse into the United Kingdom.

The final moulding of His servant nation, Israel (*ruling with God*), is taking place in the first and second Low Passages (1914-1918; and 1928-1936)—the time of tribulation, for His Purpose which opens in the King's Chamber. Let Davidson explain:

¶ 2. Descent continued meant an increase in everything negative to spiritual progress. Thus, the chronological baseline of the Descending Passage symbolism is the roof-line. It is this line which defines the points 2500, 2520, and 3000 A.K. The roof-line is thus mathematically defined as the negative floor —thus symbolising the inversion of the conditions of spiritual man. . . . Its passing below the base-level (into natural rock) is negative, passing from the influence of the realm of Divine Love and its Law of mutual Love, into the complete influence of the realm of spiritual negation and its Law of mutual distrust, natural passion and internecine strife. Thus the passage continues to its abrupt end, where accelerated descent meets with the impact of the narrow and still lower horizontal passage leading to the Subterranean Chamber . . . symbolising a sudden arresting shock effecting rupture or schism . . . a change not necessarily in spiritual conditions, but in the material relations immediately affecting those spiritual conditions.

This Descent in natural rock symbolises the spiritual in man subordinate to the material in man, as the Pyramid— and its ascending passages and chambers constructed above —symbolise the material in man subordinate to the external in man (*vide* 1 Peter ii, 4-6).

In the Subterranean Chamber, the ceiling is smooth (= a negative floor), and the floor rough (= a negative ceiling), the opposite of a natural cavern. The Descending Passage is therefore symbolic of the "upside-downness" in relation to spiritual progress, the Subterranean Chamber symbolising the

302 THE PILLAR IN THE WILDERNESS

hall of Chaos, the only exit from which is the "Dead-End" passage ending in the natural rock—symbolic of annihilation.[1]

The datings of the junction of Descending Passage and the lower horizontal passage, for floor, axis and roof, are A.D. 1506–1518–1529 (the initial period of the Reformation in Europe).

An Apocalyptic pamphlet of 1508 shows on its cover the Church upside down. . . . It was into this seething mass of discontent that the spark of religious protest fell—the one thing needed to fire the train and kindle the social conflagration (*Ency. Brit.*, 13th ed., vol. xxiii, p. 10).

Anybody reading an unbiassed historical account of the conditions leading to and shaping the Reformation in Europe cannot but admit that the symbolism and the historical phase of the period thus defined by the Pyramid's chronology completely agree.[1]

At the end of 1517, Luther published his *Ninety-five Theses*. This point marks the low spiritual level to which had arrived every European country throughout Christendom, then under the undisputed sway of the Papacy, at that time as immoral as the Court of any pagan Emperor, and unfit as a spiritual guide.[2]

Thus it necessitated in the history of mankind the direct action of the Messiah, the risen Christ, to create a "field of influence"—the clear intention of this is indicated in the Grand Gallery's structural symbolism, with its Rhomboid of Displacement (1558–1844):

$$
\begin{array}{ll}
\text{A.D.} \ 27 & \text{—the first year of His ministry.} \\
1530 & \text{—} 153 \times 10.[3] \\
\hline
\text{A.D.} \ 1557 \ (\text{Dec.}) & \text{—the initial date.} \\
286 \cdot 1 & \text{—the Rhomboid of Displacement.[4]} \\
\hline
1844 \ (\text{Jan.}) & \text{—the Grand Step Epoch.}
\end{array}
$$

---

[1] Davidson, *The Great Pyramid*, pp. 374–376.
[2] Basil Stewart, *The Witness of the Great Pyramid*.
[3] 153 signifies "*the elect*," the number of fish in Peter's haul (John xxi, 11; also Matt. iv, 19; xiii, 47; Luke v, 10). 10 is symbolic of worldly completion without Christ.
[4] The Rhomboid—286·1″ marked off along roof of the Grand Gallery, by the height to which roof of Grand Gallery raised.

PYRAMID SYMBOLISM

The initial date of this "field of influence" is defined in relation to the Pyramid's displacement factor. This symbolises the "influence" of the Messiah, who *"purchased"* (Exod. xv, 16) the right to exercise this "influence" by this *"displacement"* or *"cutting off"* (Dan. ix, 26).[1]

The symbolic and astronomical indications depict the *"hastening"* (Dan. xii, 4) of man's material development to effect God's purpose, through Christ, in relation to spiritual development. In the scientific sense, the creation of a new field of influence brings into being a high potential relative to the former field, which has a disturbing effect on the media or entities in this field until they become attuned to the new field.

This will lead to the breaking up of old coalitions or groupings—temporary chaos—until readjustment is effected as the result of experimental groupings. When the regrouping process is complete the new field is a field of equilibrium, and its controlling units are attuned.

As the initial year of "the field of influence" in the Grand Gallery overhead is about to be passed under, in the Descending Passage the symbolism indicates arresting shock, rupture, or schism, and change of direction—chaos and blind following of the line of least resistance pouring towards the Subterranean Chamber.[2]

It is a remarkable fact that the initial date—ushering in the reign of Elizabeth in England—should be identified by Whewell, in his *History of the Inductive Sciences*, with what he defines as "the inductive epoch of Copernicus, beginning the history of formal Astronomy after the stationary period." It is further remarkable that prior to the Grand Step epoch (1844) gravitational astronomy was not sufficiently advanced in accuracy to check the translation of the Pyramid's co-ordinates with the degree of precision now possible . . . and since 1844 the attention of astronomers has been repeatedly directed to the solution of the Pyramid's astronomical purpose.[3]

---

[1] See Davidson, pp. 220, 221, 230–232.
[2] *Ibid.*, p. 375.
[3] *Ibid.*, p. 221. As the research of Professor Piazzi Smyth, Astronomer Royal for Scotland (*Our Inheritance in the Great Pyramid*), 1865–1880: and of John Taylor.

¶ 3. From 1557 (December), the initial date of this "field of influence" exerted by the risen Christ [1] to hasten man's knowledge to effect God's purpose at the times foreordained, there arises "a new Era of Inductive Thought . . . the creation by Higher Induction of a field of Intellectual Influence, analogous to the creation, by induction, of the field of Physical influence. Intellectual effort in the new field attunes the mind to synchronise with the laws of influence within the field. Synchronisation of one's will and effort with a stream of energy from the Godhead —whether realised or not—begets knowledge, and knowledge begets discovery of the law." [2]

All the outstanding scientific discoveries (as of Kepler, Newton, Einstein) came as instantaneous rays of light to their discoverers. Any research worker knows the process and its instantaneousness.[3]

Thus, Kepler (as he worked out the mathematical harmonies of his astronomical laws) says, "Almighty God, these are Thy thoughts, I am thinking after Thee," and goes on to say that our human minds think successively, advancing laboriously from conclusion to conclusion, God knows through simple intuition. It is to this he attributes the difficulty with which the human mind comes to understand the processes which are accomplished with so much ease in Nature, whether it be the ripening of a bunch of grapes in the sun, or the deeper and subtler processes, as the care of individual souls in an immense universe by a constant omnipresent spiritual Light—"*Pleni sunt cœli et terra gloria tua.*"[4]

> *Immediate are the acts of God, more swift*
> *Than time or motion ; but to human ears*
> *Cannot without process of speech be told,*
> *So told as earthly notion can receive* (Milton).

---

[1] "*In whom are hid all the treasures of wisdom and knowledge*" (Col. ii, 3), and by whom God "*created all things*" (Eph. iii, 9); "*the head of the corner*" (1 Peter ii, 7); the Apex Stone displaced, "*the Stone which the builders rejected*" (Ps. cxviii, 22; Matt. xxi, 42, etc.).

[2] Davidson, p. 232.

[3] *Ibid.*, p. 231.

[4] Alfred Noyes, *The Unknown God*, pp. 95, 258, 259, also following passages quoted, pp. 207, 126, 79, 68.

THE ERA OF INDUCTIVE SCIENCE

And Newton,[1] working on his laws, says: "The whole diversity of natural things can have arisen from nothing but the ideas and will of one necessarily existing Being, who is always and everywhere God supreme, infinite, omnipotent, omniscient, absolutely perfect."

Thus "intellectual effort with mind attuned to synchronise with a stream of energy from the Godhead, whether realised or not," produces:

> The whole of nature bespeaks an intelligent Author, and no rational enquirer can, after serious reflection, suspend his belief for a moment with regard to the primary principles of genuine Theism (David Hume, *Natural History of Religion*).

And Darwin, in *The Origin of Species*: "May we not believe that a living optical instrument might thus be formed, as superior to one of glass as the works of the Creator are to those of man?" And in *The Descent of Man*: "This grand sequence of events the mind refuses to accept as the result of blind chance. The understanding revolts from such a conclusion."

And Haeckel:

> Our cosmology knows only one sole God, and this Almighty God rules the whole of Nature without exception. We contemplate His operation in all phenomena of every description. . . . If each body *in vacuo* falls 15 feet in the first second, if 3 atoms of oxygen to 1 of sulphur always produce sulphuric acid . . . then these phenomena are the immediate operations of God, equally with the blossom of plants, the movements of animals, the thoughts of Mankind. We all exist by God's grace, the stone as well as the water, the radiolarian as well as the pine-tree, the gorilla as much as the Emperor of China. . . .
>
> This cosmology which contemplates God's spirit and power in all natural phenomena is alone worthy of His all-comprehensive greatness; only when we refer all forces and all phenomena of movement, all forms and properties of matter, to God, as the Author of all things, do we attain to that human

---

[1] When Newton saw the apple fall—and as many another had—in a flash of Intuition the conception of his theory came, to be later worked out and proved by Reason. So also Einstein, who got the idea of his theory when aged eighteen. And so is all real discovery, call it revelation or not (see p. 10).

intuition of God, and veneration of God, which really befits His immeasurable greatness. For "*in Him we live and move and have our being.*" . . . God is Almighty; He is the sole Author, the prime cause of all things, that is, in other words, God is the universal causal Law; God is absolutely perfect, He can never act otherwise than perfectly rightly. . . .

God is the sum of all forces, so also, therefore, of all matter. Every conception of God which separates Him from matter opposes to Him a sum of forces which are not of divine nature; every such conception leads to amphitheism, consequently to polytheism.

Since monism demonstrates the unity of the whole of Nature, it proves likewise that only one God exists and that this God manifests Himself in the collective phenomena of Nature. Since monism generates the collective phenomena of organic and inorganic Nature in the universal causal Law, and displays them as the effects of "active causes," it shows at the same time, that God is the necessary Cause of all things, and is the Law itself. Since monism acknowledges no other than the divine forces in Nature, since it recognises all laws of Nature as divine, it raises itself to the greatest and most elevated conception of which man, as the most perfect of all animals, is capable, to the conception of the unity of God and Nature.[1]

"*God is Light*" (1 John i, 5) in "the direct intuitions which lift untaught genius to the heights of philosophy and beyond them; the sudden illumination which reveals a universal law to the scientific discoverer, and shows him more in a moment than he had learnt in years of labour (as when Darwin read a paragraph of Malthus and saw his whole theory in a flash); the ecstatic vision of the contemplatives; the sudden splendour of inspiration in poet and prophet."[2]

"*All things were made by Him, and without Him was not anything made that was made. In Him was life, and the life was the light of men. And the light shineth in darkness; and the darkness comprehended it not*" (John i, 3–5).

---

[1] Alfred Noyes, *The Unknown God*, quoting—and commenting: "Eliminate from the passage the idea of a living, personal God . . . the whole passage collapses at once into an incoherent muddle. Thus, the philosophy of Nature becomes in fact theology."

[2] *Ibid.*, p. 263.

THE ERA OF INDUCTIVE SCIENCE

"Except a man be born again, he cannot see the Kingdom of God . . . for that which is born of the Spirit is Spirit" (John iii, 3, 6).

¶ 4. The increase of scientific knowledge to 1844 and its acceleration since, which should have been considered as the gift of Christ and administered by His teaching to be a blessing to mankind, has been turned by man, *"out of the imaginations of his heart"* and still under Cain-Bel influence, to the accumulation of power and wealth, individual and national,[1] in the present economic system.[2]

Mechanical achievements in an age of machinery had brought about the factory system, and a great monotony of existence for the multitude, and as a consequence, widespread discontent.[3]

This, again, was used by the Cainite idea to produce further discontent, and slaughter, by the deliberate creation of want and disorder, by propaganda of malice and envy, in a holocaust of revolutions.

The development of modern inductive science in current times has been the misdirection of scientific knowledge. Man has applied his discoveries in science to the creation of machinery of destruction. He has set in motion relentless energies that he cannot now control. He has made of his boasted world of civilisation a chaotic shambles. He is devising the means of protecting himself from the energies already released, and in thus devising, is releasing new and more terrible energies for his own destruction.

And quoting Professor F. Soddy, *Matter and Energy* (1911):

Civilisation . . . reaps what it has not sown, and exhausts, so far, without replenishing . . . a turning-point is being reached in the upward progress which has hitherto kept pace with the advancement of knowledge. . . . So far, Science has been a fair-weather friend. It has been generally misunderstood as creating the wealth that has followed the

---

[1] Tubalcain was the first organiser of industry, employing the pre-Adamite aborigines in the Sumerian civilisation to work the mines and for the smelting of brass and iron. [2] See pp. 337–341, 354–357, 363–365.
[3] Basil Stewart, *The Witness of the Great Pyramid*.

application of knowledge. Modern science, however, and its synonym, modern civilisation, create nothing except Knowledge. After a hand-to-mouth period of existence, it has come in for, and has learned how to spend, an inheritance it can never hope to restore. . . . The world is great enough and rich enough to supply human aspirations and ambitions beyond all present dreams. But the human intellect must keep pace in its development with the expanding vision of natural abundance.[1]

A few years later, the War burst on this civilisation and accelerated its spending, and human intellect lags behind in the conservation, development and distribution of natural abundance.

As Sir Oliver Lodge, October 1926, puts it:

The present epoch is . . . a danger. For it is an age of rapid material development, which is not at all the same thing as an age of true progress; and Science, not willingly, but incidentally . . . has provided men with new powers over the forces of nature, which may be used—and indeed have been used—purely for destruction. The industrial nations are completely equipped with all the requisite material for destruction. Are they adequately furnished with protection against the misuse of their powers?

If man has used for unhappiness and destruction the knowledge granted to him during this era of Inductive Thought, how much more necessary that He has reserved His greater gift of supreme Spiritual dominion over Matter —which was meant for man when the Adamic race was created—with the secrets of the utilisation of cosmic energy, of the origin of life in Nature, of the spiritual control after the death of the material body, to the gift of an *"incorruptible body"*; or, as Paul puts it: *"For I reckon that the sufferings of this present time are not worthy to be compared with the glory which shall be revealed in us. For the earnest expectation of the creature waiteth for the manifestation of the sons of God"* [2]—to be restored only to

---

[1] Davidson, pp. 234, 235.
[2] Romans viii, 18, 19; also the idea carried on in verses 20–23, *"waiting for the adoption, to wit, the redemption of our body."*

THE ERA OF INDUCTIVE SCIENCE

Freewill attuned to operate according to the One Divine Law of Love, through belief in Him.[1]

This all confirms that Chaos must precede Redemption. The joint symbolism of the Grand Gallery and the Descending Passage indicates that the arrested Descent in the latter is effected at the expense of an ultimate lowering (Rom. xi, 11–15) of the Spiritual "Potential" in the former . . . as the spreading or distributing of spiritual energy throughout the world (*i.e.* among the non-Adamic races also).[2]

"How long shall it be to the end of these wonders? . . . it shall be for a time, times and a half, and when he shall have accomplished to scatter the power of the holy people, all these things shall be finished . . . the words are closed up and sealed until the time of the end.[3]

"That time shall not come except there come a falling away first" (2 Thess. ii. 3).

"And the gospel must first be published among all nations. . . . For in those days shall be affliction" (Mark xiii, 10, 19).

"This Gospel of the Kingdom shall be preached in all the world for a witness unto all nations; then shall the end come" (Matt. xxiv, 14).

"And Jerusalem shall be trodden down of the Gentiles, until the times of the Gentiles be fulfilled . . ." (604 B.C. Jerusalem first taken by Nebuchadnezzar, 2520 years as the Times of the Gentiles bring to A.D. 1917, when Jerusalem delivered by Britain.)

"And there shall be signs . . . upon the earth distress of nations with perplexity . . . men's hearts failing them for fear, and for looking after those things which are coming on the earth" (Luke xxi, 25, 26).

¶ 5. Returning to the fulfilment of prophecy as to Israel-Britain's history to the present time (which was also "hastened"[4] during the Rhomboid of Displacement)

---

[1] John iii, 16.
[2] See Davidson, pp. 225, 226, "The Law of Contract."
[3] Dan. xii, 6–9: "*the time of the end*" = 360 years (A.D. 1558–1918), with Britain, as Israel, scattered over the earth.
[4] "*A little one shall become a thousand, and a small nation a strong nation. I, the Lord, will hasten it in His time*" (Isa. lx, 22).

as the Stone Kingdom, to be His servant nation, against Cain-Bel world influence:

"*And in the days of these kings shall the God of Heaven set up a kingdom, which shall never be destroyed; and the kingdom shall not be left to other people, but it shall break in pieces and consume all these kingdoms, and it shall stand for ever.*

"*Forasmuch as thou sawest that the stone was cut out of the mountain without hands, and that it break in pieces the iron, the brass, the clay, the silver and the gold; the great God hath made known to the king what shall come to pass hereafter; and the dream is certain, and the interpretation thereof sure*" (Dan. ii, 44, 45).

"*Therefore say I unto you, The Kingdom of God shall be taken from you* [the Jews], *and given to a nation bringing forth the fruits thereof.*[1] *And whosoever shall fall on this stone shall be broken; but on whomsoever it shall fall, it will grind him to powder*" (Matt. xxi, 43, 44).

"*In righteousness shalt thou be established, thou shalt be far from oppression, for thou shalt not fear . . . terror, it shall not come near thee. Behold, thine enemies shall surely gather together, but not by Me; whosoever shall gather together against thee shall fall for thy sake . . . no weapon that is formed against thee shall prosper*" (Isa. liv, 14–17).

"*He shall cause them that come of Jacob to take root: Israel shall blossom and bud, and fill the face of the earth with fruit*" (Isa. xxvii, 6).

The House of David "*shall have dominion from sea to sea, and from the river unto the ends of the earth*" (Ps. lxxii, 8). "*I will set his* [David's] *hand also in the sea, and his right hand in the rivers*" (Ps. lxxxix, 25).

"*He hath shewed his people the power of his works, that he may give them the heritage of the heathen*" [e.g. India] (Ps. cxi, 6).

"*In the last days it shall come to pass that the mountain of the House of the Lord* [Israel as the British Empire] *shall be established in the top of the mountains* [great nations];

[1] Britain, under Lucius, establishes Christianity in place of Druidism, as its national religion—the first nation to do so. And later, Constantine, a British king, enforces Christianity on the Roman Empire.

THE EMERGENCE OF ISRAEL-BRITAIN

and it shall be exalted above the hills [little nations]"[1] (Mic. iv, 1).

"*And the remnant of Jacob shall be in the midst of many people as a dew from the Lord, as the showers upon the grass, that tarrieth not for man, nor waiteth for the sons of men*" (Mic. v, 7).

Babylon, 539 B.C., destroyed by Cyrus, with Massagetæ in vanguard—the gates opened by captive Jews.

Medo-Persia, 528 B.C., Cyrus killed in battle with the Massagetæ.

Greece, 334 B.C., under Alexander, attacked the Getæ: fifty years later Lysimachus attacked them, and was taken prisoner. Greece worn out in fighting the Scythians, the Getæ.

Rome, 111 B.C., Marcius attacked the Getæ, Trajan's campaigns followed. The wars with Britain under Caswallon, Caradoc, Arviragus (55 B.C. to A.D. 86).

A.D. 451, the Ostrogoths with the Huns at battle of Chalons.

A.D. 493, Theodoric, of the Ostrogoths, rules Italy.

"*Thou art My battle-axe, and weapons of war, for with thee I will break in pieces the nations, and with thee will I destroy kingdoms*" (Jer. li, 20).

"*And the stone that was smiting the image* [the above four kingdoms] *became a great mountain, and filled the whole earth*" (Dan. ii, 35).

And followed "*smitings*" on the western and eastern divisions of the fourth kingdom, the former split into the "*toes of the feet*" (ii, 42) under the power of the Papacy as "*the little horn*," and the latter under the Greek Orthodox Church, "*the little horn*" of Dan. vii, 8, and viii, 9.

In the Middle Ages, Israel-Britain wars with France to regain ascendancy over the Continental Cymry and Celts, but by the Divine hand turned back from the Continent to the Island, at the death of Joan of Arc (1431), but still held its "*gate*," Calais, until 1558, the initial date of the time of the end.

Spain, 1588—"*the weapon*" of the Armada dispersed.

---

[1] In Biblical symbolism, mountain was a great nation and hill a small nation.

312 THE PILLAR IN THE WILDERNESS

A silver medal was struck by Elizabeth in commemoration. On the obverse, a Church founded on a rock, with "*I am assailed but not injured*," in Latin; on the reverse, Spanish fleet dispersed; above, sun-rays breaking through clouds; and at the top, the sacred name of Jehovah in Hebrew, with "*He blew with his winds, and they were scattered*," in Latin.

"*Thy seed shall take possession of the door of his enemies*" (Dillman's translation of "*gate*," Gen. xxii, 17)—

Quebec led to possession of the whole land, Canada.

Madras, Calcutta, Bombay, led to all India.

There followed "*gates*"—Shanghai, Hong Kong, Wei-hai-Wei, Singapore, Malacca, Penang, the Samoan Islands: Colombo, Aden, Perim, Suez Canal: Zanzibar, Mombasa, Durban, Cape Town, Sierra Leone: Port Darwin to Perth in Australia, Wellington to Dunedin in New Zealand: Tasmania: Falkland Islands, Jamaica, Bahamas, Bermuda, Newfoundland: Cyprus, Malta, Gibraltar, the Channel Islands.

When Britain first at Heaven's command
Arose from out the azure main,
This was the charter, the charter of the land,
And guardian angels sang this strain,
Rule, Britannia! Britannia, rule the waves.

¶ 6. And comparing the 2520 years' interval [1] of Divine time-rate:

B.C.		B.C.	
4000	Special creative selection of the Adamic race, endowed with Divine consciousness and Freewill, among the evolved races of men.	1480	Selection of Israel, endowed with Divine Commands, to be His *chosen*, servant Nation, among Cain-Bel civilisation.
2343	The family of Noah, *a just man*, saved from the Flood which destroyed Adamic civilisation in isolated, landlocked Turkestan, for His purpose: *with thee will I establish My Covenant* (Gen. vi, 18).	A.D. 178	Britain, having declared Christianity, sea-girt, saved from the Babylonian kingdoms, to become *a blessing to all the families of the earth* and the nucleus of His Kingdom (Matt. xxi, 43).

[1] In passing from B.C. to A.D. in calculation of years, add 1, as no year B.C. or A.D. 0.

THE EMERGENCE OF ISRAEL-BRITAIN

B.C.		A.D.	
1916 1913	Call of Abraham. The Covenant confirmed.	605 608	The British Church refuses any intermediary, as the Pope, between itself and the Voice of God.
1447	Balaam's prophecy [1] (Num. xxiii, xxiv) before Canaan entered.	1074	His prophecy being fulfilled after Britain entered, all the tribes of Israel have come to their *appointed place* (2 Sam. vii, 10): England settled under William I, who first introduced and settled the post-Captivity Jews in England (see Green's *English People*, pp. 86-87).
1007 (−1000)	The founding of Solomon's Temple. Preceded by the feud between Saul and David, and seven years of civil war until David crowned king over all Israel (1043 B.C.).	1514 (−1521)	The Reformation begins, which refounds the British Church to become the true "Temple of God." Preceded by the Wars of the Roses, and with Scotland.
975	The split in Solomon's kingdom. The divorce of Israel under Jeroboam, who put away the word of God and His prophets, and established idol worship under a new priestly order (*cf.* Hosea). For next 250 years the splitting of Israel further from their God in seeking idolatry, despite the warnings of the prophets. Israel wars with Judah to overthrow the House of David, and finally is punished for idolatry by the Assyrian Captivity (721 B.C.).	1546	Reversed by Henry VIII, who divorces the English Church from Rome, and separates Oxford and Cambridge from Continental scholarship, removal of foreign order of priests and Dissolution of the Monasteries, establishing the Bible as the final authority in British worship.[2] For 250 years Britain seeks to return towards a truer worship of God, with the preaching of Milton, Bunyan, Wesley—the Evangelical Movement, the founding of missionary societies, Sunday schools, crusade against the slave trade, prison reform.

[1] "*Lo, the people shall dwell alone, and shall not be reckoned among the nations. How goodly are thy tents, O Israel. Behold, the people shall rise up as a great lion; he hath as it were the strength of a unicorn: he shall eat up the nations his enemies: . . . in the latter days there shall come a Star out of Jacob, and a Sceptre shall arise out of Israel, and Israel shall do valiantly. Out of Jacob shall come he that shall have dominion, and shall destroy him that remaineth of the city.*"

[2] 1538. English Bible issued (1611, Authorised Version). 1548. Book of Common Prayer issued (1661, revised).

B.C.		A.D.	
721		1800	Israel free, and the Union of England, Wales, Scotland and Ireland completed under one king of the lineage of David.

And the *seven times* (2520 years) punishment on Israel.		The effect on Britain as the *seven times* ended.	
B.C.		A.D.	
771	The first captivity, by Pul (1 Chron. v, 26), of Assyria, half tribe of Manasseh separated.	1750	1739, War with Spain; 1745, the Pretender in Scotland. Clive in India (1757, Plassey); Wolfe at Quebec (1759). The separation of Manasseh as the American colonies (1776).
740	The second captivity, by Tiglath-Pileser, of all Galilee and Naphtali (2 Kings xv, 29).	1781	End of wars in America and India. The rise of manufactures (iron and coal) and of roads and canals. 1763, Potteries; 1764, Spinning Jenny; 1765, Steam engine; 1768, Spinning machine; 1771, English journals begun; 1773, War with American States; 1778, joined by France and Spain. 1793–1815, War with France, in life-and-death struggle with Napoleon, who aimed at World Dominion.
721 to 717	The third captivity, by Shalmanezer, against Samaria (2 Kings xvii, 6).	1800 to 1804	1800, United Kingdom formed; war against a world in arms. 1801, against Denmark; 1801, against the Dutch, Cape of Good Hope and Ceylon. 1805, Trafalgar; 1815, Waterloo. Followed by Crimean War (1854); Indian Mutiny (1857); War with China (1859); with Abyssinia (1867); Afghan (1879); Boer War (1881); Gordon in Egypt (1885); Kitchener in Sudan (1897); Boer War (1899), and the founding of the British Empire.[1]

[1] In war, Ephraim wears the red tabs, issuing reams of necessary and

THE EMERGENCE OF ISRAEL-BRITAIN

And the *seven times* (2520 years) punishment on Israel.		The effect on Britain as the *seven times* ended.	
B.C. 604	First captivity of Judah: Nebuchadnezzar takes Jerusalem.	A.D. 1917	Jerusalem freed by Britain; fall of Turkey as Edom (Luke xxi, 24; Ezek. xxv, 14; Obadiah, Dan. xii, 12; Rev. xvi, 12).
603	To Nebuchadnezzar, *Thou art this head of gold*.	1918	Fall of Babylonian succession of kingdoms—under Papal influence—Hohenzollerns, Hapsburgs; and Greek Orthodox influence—Czar of Russia.
595	Second captivity of Judah.	1926	The General Strike—subversive propaganda attack by Bolshevism against the British Church, State and King.
590	The Departure of Shekinah. The Image of Gold set up by Nebuchadnezzar (60 × 6 cubits).	1931	The return of the Jews into Israel territory (Jer. iii, 18). Questions in Press and broadcast, Whither Britain? The Way to God. Britain off the Gold Standard. In 1933, the attempt to bolster it up at World Economic Conference of 60 nations in London wrecked by Manasseh.

unnecessary red tape; while the Angle and the Saxon, irritated, await the great adventure.

In peace, Ephraim loves the procedure of Parliament, the issuing of law, the readings and the Committee stages; while the rest of the tribes still wonder at the need of so many, as D.O.R.A., and wish more of Ephraim exported elsewhere to the Empire, so there to exercise his birthright.

When the soap-box Socialist with blatant voice on his "toiling masses," as if their sole guardian, reaches Parliament, in contact with Zarah-Pharez and Ephraim, his irresponsibility tones down, and he may become a worthy member, which seems never to be understood by the "proletarian."

The Levites of the legal profession in carrying out the law delight in its ceremony and orderly routine with becoming slowness and levy of adequate tribute.

While the Levites of the Church love its creeds and ceremonies, and with diminishing tithe, which, as the relic of a legal levy, falls in these times on one part of the community and not the whole (see Malachi iii, 8–11).

And the *seven times* (2520 years) punishment on Israel.		The effect on Britain as the *seven times* ended.
B.C. 585	The third and last captivity.	A.D. 1936[1] — The entrance to the King's Chamber,— The times of the Gentiles ending, with the final war of Antigod element in the world, the Cain-Bel influence; *the restitution of all things : the consummation of the age*; so that nothing prevents His Second Coming, at His time.

¶ 7. The *"falling away"* or lowering of Spiritual Potential is symbolised by the overlapping descent, in seven stages, of the south end wall of the Grand Gallery, bringing the gallery roof down to the former 47·3″ level of the roof of the First Low Passage, which begins at the dating 5th August 1914.

The duty of the Church of Britain was clearly to speak with no uncertain voice to man as to how to direct the use of this scientific knowledge.[2]

Yet at the time of the Grand Step epoch (1844), the *"falling away"* or lowering of Spiritual Potential from even the standard of the A.D. 155–607 era, when the British Church and the government of the State were attuned, is seen by the questions then asked:

"What is the Church of England? Is it merely a human institution, or is it the Church of Christ, and what is its origin?"

Newman maintained that the English Church was a part

---

[1] The attempted subversion of England by Papal agents, 1558–1576, is paralleled, after the 360 years of *"the time of the end,"* by the subversive propaganda of Communism, 1918–1936.
And as the former reached its climax by the attempted invasion of the Spanish Armada, so in the near future will come the attempt of Soviet Russia on British possessions in the Near East to cut and finally disrupt the Empire—in each case frustrated and to be frustrated by Divine intervention.
"I will yet for this be enquired of, by the whole House of Israel, to do it for them" (Ezek. xxxvi, 37).

[2] "Glance with me for a few moments at English history, and you will see at once that the English Church and the English Nation are and always have been one" (Dean Farrar, 1881).

NINETEENTH-CENTURY RELIGION

of, and had an unbroken connection with, the Roman Church!

And even as late as 1935, G. K. Chesterton states (in a broadcast talk on "Freedom," 11th June 1935) that Catholicism is the foundation of British liberty and the origin of its Common Law.

Does he assert that Britain would have been freer if she had been conquered by the Armada and returned to the Roman fold?

Three hundred years before the city of Rome was founded, our laws were being formed, to be codified six hundred years before the Church of Rome was even established—to become the Common Law of the present time, on one of which even the safety of our roads to-day can repose (see pp. 80, 83, 291).

Centuries before monks elected their abbot, and bishops wore mitres, the Irish kings elected a Heremon, and the British a Pendragon (pp. 191, 280).

Before the Roman Church was founded—and endowed by a British royal family (pp. 284, 293)—there was a Glastonbury (pp. 277–279); and nobles of Rome and the Continent had sent their sons to a University in these isles at Tara; later, one of the largest, that of Bangor, was destroyed at Rome's instigation (p. 296).

The rule of Ephraim, even in Elizabethan days, was freer than the inquisition of the Papacy, when freedom was licence and spiritual descent met its arresting shock (p. 302); and but for His action during the Rhomboid of Displacement era (p. 303) the sun might still be revolving round the earth, and the earth round Rome.

What is liberty in material things worth—as the roof over your own house, the potato-patch in your own field your own shop to run, and beer to drink ("*for after all these things do the Gentiles seek*"), compared with freedom in spiritual expression (and this includes the scientific, as Galileo's)?

Britain has yet to learn that she is descended from Abraham, and founded at Sinai, and yet to stand in awe of His Purpose, that before even man on earth appeared she was a nation, the nucleus of His Kingdom, "*to bring forth the fruits thereof.*"

And not yet even has British scholarship dared to look beyond Julius Cæsar's own story.

In 1844 was born Julius Wellhausen, who first advanced Modernism in its present form with Higher Criticism.

In 1740–1836 there arose Rationalism, whose first great leader, J. S. Semler, another German,

> held that true religion springs from the individual soul and attacked the authority of the Bible in a comprehensive spirit of criticism. Rationalism in modern thought ... declines to accept the authority of the Bible as the infallible record of a divine revelation and is practically synonymous with free thinking. This type of rationalism is based largely upon the results of modern historical and archæological investigation (*Ency. Brit.*, 13th ed., p. 916).

Cain has been called the first free-thinker—he it was who instituted world history with his polytheism in the earliest Sumerian civilisation. And modern man, having drunk of "*the golden cup of Babylon,*" looks here for his religion!

> Rationalism may be defined as the mental attitude which unreservedly accepts the supremacy of reason, and aims at establishing a system of science, philosophy and ethics, consistent with intellectual honesty, verifiable by experience, and independent of all arbitrary assumptions or authority (Rationalist Press Association).

Cain would be well pleased with this from the Adamic race nearly 6000 years later—they have swallowed his dictum, "*Ye are as gods, knowing good and evil . . . your eyes are opened. . . . Hath God said* [anything at all]*?*" (Gen. iii, 1–5).

To each philosopher his own God, named according to his particular conception of one attribute of the Most High—in fact, in modern civilisation as it was in the Sumerian, when out of his own ancestry Cain created his polytheistic galaxy.

> Huxley's agnosticism was a natural consequence of the intellectual and philosophical conditions of the sixties, when

NINETEENTH-CENTURY RELIGION

clerical intolerance was trying to excommunicate scientific discovery because it appeared to clash with the Book of Genesis (*Ency. Brit.*, 13th ed., vol. i, p. 379).

The Church having clashed with Darwin's theory of Evolution, and Huxley's Agnosticism that followed, is so afraid to clash again with Science that it has even adopted the analytical or scientific method of Higher Criticism to its own Bible, hoping that by Reason and not by Spirit it will come unto the Truth.

Whereas it was by the Spirit (of Christ "*in whom are hid all the treasures of wisdom and knowledge*") that, during the Rhomboid of Displacement, Inductive Science originated in human spirit "attuned," seeking in the Ether the source of all energy.

"Almighty God, these are Thy thoughts, I am thinking after Thee" (Kepler). And in this spirit alone, recognising the Logos in Oneness in all Vital Evolution—call it by any name it likes—will Science make its discoveries; the more perfectly "attuned" the spirit, the clearer the Light that inspires the solution.

Science, by "accepting no creed that conflicts with the truth," was the first to discard the crude Græco-Roman mythology, the polytheistic source of Creation invented by Cain-Bel: thus Socrates, Plato and Aristotle searched towards the Unknown God of the Cosmos.

And Science, having savoured the aroma of "*the golden cup*," and spilt it out on the ground as noxious, is often nearer to God than the Church, which having drunk stands on uncertain ground and speaks with differing voice on the cause and the redemption of Man's displacement.[1]

And all nations, having drunk of "*the golden cup*" and become mad, too befuzzled to recognise Divine revelation, and to seek the One God, have applied scientific knowledge to their own destruction, to promote their own ambition for wealth and power, instead of creating harmony on earth, "*peace and goodwill among men*," according to the One Law of the universe—the Law of Divine Love.

[1] The Roman Church condemning Galileo, the English Church Darwin.

¶ 8. As the eleventh hour is about to strike, the Pyramid's ascending plane changes to the horizontal, and its time-scale of the inch-year to that of the inch-month.

The horizontal plane of the floor (in the Osirian texts "one day counts for a month") defines a chronological plane of the "*restitution of all things, the consummation of the age*," in Bible prophecy. If ever a Divine Message was prepared for the human race, it was prepared for the current time (see Davidson, pp. 384–407).

The graph is magnified, the picture enlarged, twelve times, so that in it the Anglo-Saxon race, holding the same inch as standard of measurement, may find the answer to its present perplexities—"The Way to God," "Whither Britain?" Nay more, it is the direct warning-message of the House of Seth, to the heirs of the Covenant at the times of "*the consummation*," the House of Britain, urgent with appeal to prepare them for some event—the redemption of Freewill of their fathers and of their race, deviated from Divine Will, and its restoration to Divine Harmony—

(1) His Incarnation, which Israel-Britain missed.

(2) His Second Coming.

And the message is embodied in universal scientific laws, and in stone, which has survived more than forty centuries, so that there shall be no quibble as to the words, so that it shall be as fresh now as when first imprinted.

The ascending floor of the Grand Gallery produced strikes the floor of the horizontal plane at A, the perpendicular from B to this line gives the dating of B as 3.54 A.M., 5th August 1914 (the dawn of the day of War); and by retrospective dating on the scale of the inch-month on the horizontal plane the point C is dated 2nd August 1909, and the dating forwards on same scale gives D as 8.28 P.M., 10th November 1918 (the flight of the Kaiser to Holland, and eve of the ending of War). The overhang of Grand Gallery before falling in seven overlaps dates 28th October 1912.[1]

In the rock below, the descending passage system enters

---

[1] Davidson, pp. 384, 386b, 387.

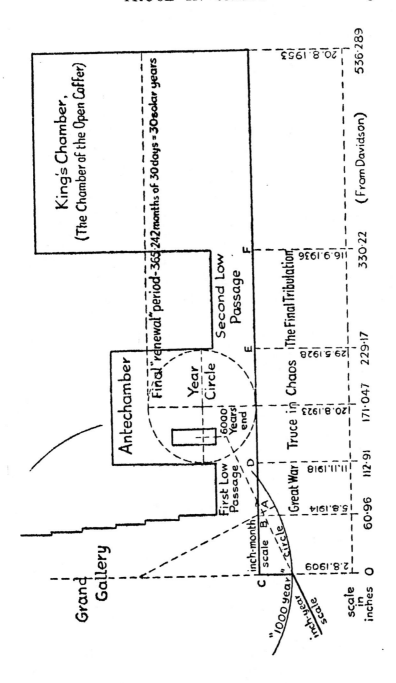

the Subterranean Chamber, as the hall of Chaos, 12th March 1913; and the drop to the floor of the Subterranean Chamber, 11th August 1913; and these dates:

2nd August 1909. King Edward at Cowes meets the Czar of Russia, who had just visited the French President.
28th October 1912. The defeat of the Turks at Lule Burgas.
12th March 1913. Austro-Russian crisis on the Balkan Wars.
11th August 1913. Treaty of Bucharest ending second Balkan War.

Of the last, J. L. Garvin wrote: "They had made Armageddon inevitable at a second and not distant remove . . . within a year the world was in flames."

The dates of the first Low Passage are those of the Great War, the entrance at the date when Britain involved.

Man's artificial fabric of civilisation collapsed, and until the entrance to the King's Chamber man seeks to restore it.

The first Low Passage ends at the Armistice—Lord Fisher said that the sudden cessation of the Great European War was the result of Divine intervention. Certainly the men in France saw no chance of it ending until within three months of the close, and then dared not hope till the last days.

To accomplish this at "*the appointed time,*" the United States, the west half of the tribe of Manasseh, as the daughter nation, springing from Britain as Israel but independent of her (Gen. xlviii, 19; Isa. xlix, 20), was by Divine Plan reserved, to throw in her weight at this moment and for this purpose—not so much by material arms as by the power of her monetary might, of her natural resources, of her fresh enthusiasm on the side of war-wearied Britain. For still against Britain were the strongest defences and the main concentrations of the enemy.

In Palestine, little progress had been made until "*the appointed time,*" when Ephraim-Britain alone, with no Allied help but that of the descendants of Ishmael, "the Arabs," of whom was promised "*he shall dwell in the presence of all his brethren*" (Gen. xvi, 12), then under

TRUCE IN CHAOS

Allenby marched through the Turks as "*fire through stubble*" (Obadiah 2, 11, 18, 20, 21), and took Jerusalem without firing a shot into her, in December 1917.

"*And I will lay My vengeance upon Edom by the hand of My people Israel*" (Ezekiel xxv, 14).

"*And Jerusalem shall be trodden down of the Gentiles, until the times of the Gentiles be fulfilled*" (Luke xxi, 24).

604 B.C.—2520 years, as the times of the Gentiles—A.D. 1917.

A.D. 27 (His Ministry begins)—1530 (= 10 times the number 153 signifying the elect) = 1557 A.D.: 360 years as "*time of the end*" = A.D. 1917.

¶ 9. The combined length of the first and second Low Passages (51·95″ and 101·05″ respectively) is 153″, and interposed between them is the Antechamber, the length of which is 116·26″, as a period of "truce in chaos," the vertical height being temporarily raised; the whole system indicating the time of *tribulation*, before the entrance to the King's Chamber.

The number 153, when indicated with Messianic symbolism, relates to Christ's elect, their trials, and their mission in being prepared for, and in preparing for, the *restitution*. A period of chaos had to precede the *restitution*. Such a condition of chaos is symbolised by the stooping process necessary in the two Low Passages. These are symbolised as relating to a single historical phase of chaos by the associated measurement of 153 P″, representing 153 calendar months of 30 days; the Antechamber being defined as inserted to symbolise a truce, or temporary cessation of chaos, until some purpose, indicated by its symbolism, is served.[1]

"*And then shall be great tribulation, such as was not since the beginning of the world to this time, no, nor ever shall be. And except those days be shortened, there should no flesh be saved; but for the elect's sake those days shall be shortened*" (Matt. xxiv, 21, 22).

Here is emphasised (1) *flesh*—*i.e.* man as body, not soul;

---

[1] Davidson, p. 386, with pp. 354–358, 378–380.

(2) *elect*—*i.e.* Israel, His servant nation, chosen for His Purpose.

The passage means that the Great War would have dragged on, with economic stress and financial crises superimposed, from 4th August 1914 [1] till September 1936 (the end date of second Low Passage), and who could tell how many of friend and foe who fought, and the civilians from resulting intenser food-rationing, starvation diseases—as influenza, vitamin deficiency—would have survived this period? But for Britain's sake Christ Himself "*shortened the days*" by interposing a period of 116·26 months, before "*the final tribulation*" (the second Low Passage, 29th May 1928 to 16th September 1936).

Now, *re* the significance of 116·26″ length of Antechamber as symbolic of Christ's action:

> The period of Intervention is represented by the diameter, 116·26″, of the year-circle, 365·24″—this circle being the symbol of the "Lord of the Pyramid" and the "Lord of the Year," the promised Messiah of ancient prophecy.[2]
>
> The year-circle of the Antechamber symbolises the Intervention of the Sun of Righteousness to end the Great War, and to provide a period of Truce in Chaos, during which should be revealed, both from Biblical and Great Pyramid prophecy, the Plan of God for the redemption of humanity, and the purpose of His plan in relation to current history.
>
> In ancient chronology, the Great Year of 365·242 months of 30 days was known as the period of renewal or regeneration (=30 solar years, "a day stands for a month"). So the Antechamber year-circle of 365·242″ circumference unrolled along the floor symbolises the Renewal period, 20/21.8.1923 to 19/20.8.1953, during which the building race (the Stone Kingdom), under the spiritual propulsion or tuitional guidance of the Master-Builder, learns to discard its materialistic ideals of building, to substitute therefor the true spiritual principles of building (Davidson).
>
> At this date, 4th–5th August 1914, the great tribulation of Chaos began. Man's artificial fabric of civilisation collapsed,

---

[1] Davidson, p. 402. On the exact anniversary of the destruction of the Temple by the Romans, 4th August A.D. 70 (1844 years).
[2] *Ibid.*, p. 401, with pp. 370, 371, 381–385.

TRUCE IN CHAOS

and mankind was too stunned by the shock to realise the true significance of the catastrophe.

Here, the insertion of the Antechamber, as "truce in chaos," is therefore symbolised as a period of realisation, the final opportunity for man to learn of his error and the futility of his patchwork "reconstructions," prior to the final phase of "compulsion" being resumed.

The compulsion here is not symbolised as an "act of God," but as the consequence of God permitting man's artificial "law"—substituted for His Divine Law—to run its complete and natural course in effecting the collapse of civilisation; Divine Reconstruction being symbolised as forthcoming when the better part of mankind has learned the lesson intended and has asked, at a time appointed in God's foreknowledge, for His intervention.[1]

And in the times appointed in 1918 there fell the last kings of the Babylonian succession of Empires: of the western *foot*, the Hohenzollerns of Prussia, the Hapsburgs of Austria; of the eastern *foot*, the Czars of Russia.

"The ex-Kaiser recognised and professed his belief that Germany was Assyria, and Great Britain, Israel,"[2] and he declared in November 1914:

"If we gain, as we must, a new Empire will arise, more splendid than the world ever saw, a new Romano-German Empire which shall rule the world."

"*Thou sawest a stone cut out without hands smote the image upon his feet and brake them to pieces.*

"*Then was the iron, the clay, the brass, the silver and the gold, broken to pieces together, and became like the chaff of the summer threshing-floors; and the wind carried them away, that no place was found for them; and the stone that smote the image became a great mountain, and filled the whole earth.*"[3]

[1] *Ibid.*, p. 394.
[2] M. H. Gayer, *The Heritage of the Anglo-Saxon Race*, p. 47.
[3] Dan. ii, 34, 35.

CHAPTER XI

¶ 1. And at the eleventh hour, of the eleventh day, of the eleventh month, and the eleventh month after the delivery of Jerusalem and eleven days after signing the treaty with Edom (Turkey)—(The Armistice), the eleventh hour struck.

Britain, the stone kingdom, ended the War, man-wearied, but with prestige never higher, extending over the whole earth, and with the ideal of a new earth. Then Chaos. Why?

His-Story has shown that man's history in world affairs is but the expression of the spiritual problem—the Evil inherent in Creation, and diverged from Divine Harmony, influencing the Freewill of man: on the other, the Divine Purpose for the restoration of Freewill to operate to His One Law of Love. Reality is in Spirit, not in Matter, and in terms of Spirit only can the present chaos be understood.

Cain, expelled from a landlocked Adamic civilisation for Freewill succumbing to the natural law of force, arrived in the land of Sumer, and there, with the special knowledge and power endowed in the Adamic race, founded the world's first civilisation.

This land, later as Babylon [1] and the heritage of Adam's eldest son, with its succeeding kingdoms, was allowed by the Most High to be the ruling power in the world (1) until it proved itself inefficient, productive of world chaos as shown in the descent to the Subterranean Chamber, the end (or south) wall of which is dated 27th November

[1] 603 B.C., Nebuchadnezzar: as the first king of the Babylonian kingdoms.

3123 B.C., 2520 years previously, is the approximate date of the founding of the Ist Dynasty of Ur, of Chaldea, and the Ist Pharaoh Dynasty at Thinis. Thus as the civilisation of Cain divided into two "feet," so did that of Rome, the last Babylonian kingdom.

THE ELEVENTH HOUR

1939, when it is driven underground altogether; (2) until the seven times punishment meted out to Israel and Judah under the terms of the national Covenant (Lev. xxvi; Deut. v–vii, xxviii; 1 Chron. xxviii) should pass.

The institution of the Babylonian Empire was founded on the Laws of Hammurabi,[1] derived from Cain's Sumerian civilisation, and on his polytheistic galaxy of 6500 gods and goddesses, of whom Merodach (himself deified) was now chief, having put into second place those representative of his ancestry, Anu, Ea and Bel.

It was followed by the setting up of the Image of Gold (590 B.C.), 60 cubits × 6 cubits—"*To you it is commanded, O people, all nations and languages, that ye worship the golden image*"[2]—and the institution of the gold standard system, with its bankings and insurance, interest and bookkeeping systems, mortgage and joint-stock companies, public and private loans, as the recovered inscriptions of the period show.[3]

For an idea of the full measure of a man, unattuned to the One God, the complete carbon string of evolved protoplasm, 666 as his mark, with power over life and death of his fellow-men, his accumulated wealth and lust rampant, see the vivid pen-picture of the time in Madame Tabouis' *Nebuchadnezzar*.

"*Babylon hath been a golden cup in the Lord's hand, that made all the earth drunken, the nations have drunken of her wine, therefore the nations are mad*" (Jer. li, 7, 9).

This economic system based on gold was well developed in the times of Greece and Rome, whose empires succeeded, and with the growth of wealth in the last two centuries was intensified and worshipped as "the rock on which all stable government and civilisation is founded" (*The Times*, 1924).

And with this idea of wealth has come in the "golden

---

[1] Which have omitted the two greatest commandments, against idolatry and murder, and begin with a dedication to the quartet—Anu, Ea, Bel and Merodach (see p. 112).

[2] Dan. iii, 4, 5 (see p. 137).

[3] For their source, see C. L. Woolley, *The Sumerians*, pp. 91–129; and Delaporte, *Mesopotamia*, pp. 90–134.

cup," the idea of the man-governed State, the ambition for power and material success, Atheism and Agnosticism as the reaction of Reason to its former crude polytheism. Such then are the tares sown in the first place by Cain, and in every age by Freewill deviating from Divine Law of Love.

With the fall of the last Babylonian succession of kings, in 1918, followed the collapse of the Babylonian economic system, and amid the confusion temporary governments under dictators arose—"*their time is prolonged for a season.*"

¶ 2. Now this sequence is given in Revelation (which requires careful interpretation in view of xxii, 18, 19):

"*There was war in heaven: Michael and his angels fought against the dragon.*

"*And the great dragon was cast out, that old serpent, called the Devil, and Satan, which deceiveth the whole world, he was cast out into the earth, and his angels were cast out with him*" (Rev. xii, 7–9).

Thus the origin of evil was before and during the creation of the Universe, as "deviation of Freewill of Spiritual Being in operation from the Divine Law of Harmony."

Described by philosophers as—Good being positive, evil is a negative phase.

For perfect understanding, the question can be left until with immortal body, free of Matter, we come to knowledge.[1] Enough for mortal body and mind, endowed with Freewill, to know that evil was inherent as a spiritual force in Creation.

"*The secret things belong unto the Lord our God: but those things which are revealed belong unto us and to our children for ever, that we may do all the words of this law*" (Deut. xxix, 29).

"*Now is come salvation, and strength, and the kingdom of our God, and the power of His Christ*" (Rev. xii, 10)—

[1] "The ultimate realities of the universe are at present quite beyond the reach of science, and may be—and probably are—for ever beyond the comprehension of the human mind" (Jeans, *The Universe Around Us*, p. 356).

indicating God's foreknowledge and His Purpose in Christ as given in Chapter I.

"Woe to the inhabitants of the earth! for the devil has come down to you, having great wrath, because he knoweth that he hath but a short time . . . and he persecuted the woman which brought forth the man-child" (12, 13).

Showing the action of Evil on the Freewill of the Adamic race, as given in Gen. iii, the woman being the selected seed via Seth, Noah, Shem, Abraham, Isaac, Jacob, David, whose Freewill is attunable.

The *"woman clothed with the Sun"* (Rev. xii, 1, 2) is the mystical Eve, the mother of all living—*i.e.* of all that have alive within them the Spirit of God (Gen. ii, 7) . . . the maternal spirit of the latent *"living waters"* pent up in spiritually progressive humanity . . . from her is spiritually reborn the nation of the Stone or Pyramid Kingdom, destined to bring forth the Spiritual fruits (Matt. xxi, 42, 44) . . . the mystical mother of spiritual Freedom . . . the true spiritual Britannia (literally the Covenant One). The woman in travail signifies the expectant spirit of humanity . . . gives birth to the *"man-child,"* the Early Church of the Spirit, the infant *"body of Christ"* which, when it grows *"unto a perfect man, unto the measure of the stature of the fulness of Christ,"* shall *"rule all nations with a rod of iron"* (Davidson).

"And to the woman were given two wings of a great eagle, that she might fly into the wilderness, into her place, where she is nourished for a time, times and a half,[1] *from the face of the serpent"* (Rev. xii, 14).

This seed in every age has always had two wings—Israel and Judah; Getæ and Massagetæ; Britain and America. And for 1260 years the identity of Israel was hid from the Babylonian Empires, and their existence was carried on until the ten tribes arrived in Britain on the outskirts of that Græco-Roman civilisation to which our scholarship bows.

Circa 608 B.C.—escape from Assyrian Captivity—1260 years—A.D. 653—when came the final disintegration of the Roman Empire and the *"woman"* coming *"into her*

[1] A time, times and a half = $1 + 2 + \frac{1}{2} = 3\frac{1}{2} \times 360$ years—*i.e.* half of the seven times.

place," Britain,[1] with the arrival of the Angles and Saxons.

"*And the dragon was wroth with the woman, and went to make war with the remnant of her seed, which keep the commandments of God, and have the testimony of Jesus Christ*" (17).

Before A.D. 653, Britain had received "*the testimony of Jesus Christ*" first from Joseph of Arimathea, and had declared Christianity as its national religion.

In A.D. 655 died Penda, King of Mercia, the last heathen king of England.

And ever since that time the opposing Babylonian religious *cum* economic system has been handed to her, sometimes by force of arms, sometimes by subversive propaganda.

"*A beast rose up like unto a leopard and his feet were as the feet of a bear, and his mouth as the mouth of a lion, and the dragon gave him his power and his seat and great authority. And all the world wondered after the beast. And they worshipped the dragon . . . and the beast, saying, Who is able to make war with him?*" (Rev. xiii, 1–4).

This identifies the beast as that of Daniel's, the kingdoms of Babylon, Persia, Greece and Rome, whose civilisations adopted the Cain-Bel standards of religion, power and wealth, and whose constant wars continually upset the world.

"*And there was given unto him a mouth speaking great things and blasphemies; and power was given unto him to continue forty and two months.*[2] *And it was given unto him to make war with the saints and to overcome them. And all that dwell upon the earth shall worship him, whose names are not written in the book of life of the Lamb slain from the foundation of the world*" (5–8).

The first free-thinking was done by Cain in his perverted story of the Creation; on it was based Græco-Roman mythology and their mystery religions; there followed the man-made philosophies based on Reason as

---

[1] Brith, in Hebrew, means Covenant; -ish = man; -ain = land.
[2] $3\frac{1}{2} \times 360$ years, verifying the 1260 years above.

APOCALYPTIC SIGNIFICANCE

a reaction to such crude conceptions; and farther east, the cult of the children of the Sun and the heathen debasements of Buddhism and Zoroastrianism; mingled with them were the false accretions on Christ's teaching.

There came the persecutions of the early Christians as "saints"; but the meaning, saints, as His elected nation, Israel, is evidenced in the actual wars on the Getæ and Massagetæ, Ostrogoths, Angles, Saxons, from which they did not escape unscathed, for in Deut. xxviii list of punishments:

"The Lord shall cause thee to be smitten before thine enemies; thou shalt be oppressed and crushed; ye shall be left few in number whereas ye were as the stars of heaven for multitude; the Lord shall scatter thee among all people and there shalt thou serve other gods which neither thou nor thy fathers have known, even wood and stone."

And so from Israel, worship of Baal and Ashtoreth, the wood and stone worship of the Messagetæ, to the Odin religion of the Asar men and of the Angles, Saxons and Danes—the Cain-Bel system spiritually, assailed them, as it did all the earth, to cause "*all that dwell upon earth to worship him; . . . whose names are not written in the book of life*"=indicating continued deviation of Freewill from Harmony is death, annihilation of such spirit before the "*second world*" when Divine Law of Harmony is restored and God is All in All.

"*If any man have an ear, let him hear*" (9). Placed in this context is deliberately emphatic to each Freewill.

"*Here is the patience and the faith of the saints*" (10). Yes, those who believe Him. But also a message of hope to Britain and associated nations as His Israel in the present Chaos.

"*Another beast coming out of the earth, and he had two horns like a lamb, and he spake as a dragon; and he exerciseth all the power of the first beast before him . . . and deceiveth them that dwell on the earth, that they should make an image to the beast*" (11–15)—indicating the influence of the Babylonian succession of religious, economic, political ideas, from A.D. 653+1260 years, to A.D. 1913, the entrance date to the Subterranean Chamber, the hall of Chaos; then follows the result.

"*And he causeth all, both small and great, rich and poor, free and bond, to receive a mark in their right hand or in their foreheads; that no man might buy or sell save he that had the mark, or the name of the beast, or the number of his name. Here is wisdom, Let him that hath understanding count the number of the beast, for it is the number of a man, and his number is 666*" (the completeness of the carbon string in the acme of protoplasm—Man) (Rev. xiii, 16–18).

"*For before these days* [that is, the days before the Temple was rebuilt[1]] *there was no hire for man, nor any hire for beast, neither was there any peace to him that went out or came in because of the tribulation: for I set all men every one against his neighbour*" (Zech. viii, 10).

"*Behold, victuals shall be so good, cheap upon earth, that they shall think themselves in good case* [so it was in 1913] *and even then shall evils grow upon the earth, sword, famine and great confusion. For many of them that dwell upon earth shall perish of famine, and the other that escapes the famine, shall the sword destroy*" (2 Esdras xvi, 21, 22).[2]

And in Russia, where the Babylon idea has been carried to its logical conclusion, the State has taken over the production and rationed the consumption, with the resultant famine and sword, and yet further in the future, for the effect is not yet ended.

"*The great whore, with whom the kings and the inhabitants of the earth have been made drunk with her wine . . . a woman sitting upon a scarlet-coloured beast, full of names of blasphemy, decked with gold and having a golden cup in her hand, full of abominations*" (Rev. xvii, 1–4)—the same "*golden cup*" of Jer. li, 7, brewed by Bel-Cain.

"*And upon her forehead was a name written, Mystery, Babylon the Great, the mother of harlots, and abominations of the earth*" (Rev. xvii, 5).

---

[1] And so in these days before the Sanctuary in Palestine is built.

[2] V. 74 follows, to Israel (and her associated nations): "*Behold the days of tribulation are at hand* [1914–1936], *and I will deliver you from them* [after 1936]. *Therefore be ye not afraid, neither doubt, for God is your guide.*"

Mystery, because the connection has never been realised —its origin in Cain-Bel, its influence throughout the Babylonian kingdoms, and the resulting religious *cum* economic *cum* political system in these present times.

She, as *"Babylon the Great, the mother of harlots, who saith in her heart, I sit a queen, and am no widow . . . and lived deliciously"* (Rev. xviii, 7)—

on the material fruits of the earth, with its materialistic kingdoms as lovers . . . is the seductive spirit of materialistic building, the spirit of greed, luxury and gainful intercourse, that stimulates to false and unhealthy activity the natural system of commerce, and prostitutes the gifts of God's providence in usurious barter and exchange. She is the Spirit of the World's sociological system that dominates all kingdoms and races on the earth (Davidson).

"Drunken with the blood of the saints and with the blood of the martyrs of Jesus" (Rev. xvii, 6).

Throughout the ages, it has warred with Israel, and persecuted Christianity. Rome and that part of her Church, which elected political power on the decree of the Emperor Phocas, have been always in antagonism with early British Christianity, Judaism and Protestantism.

"The ten horns are ten kings, which have received no kingdom as yet but receive power as kings one hour with the beast; these shall hate the whore, and give their kingdom unto the beast, until the words of God be fulfilled. And the woman which thou sawest is that great city which reigneth over the kings of the earth" (12, 18).

Ten in symbolism denotes worldly completion without Christ: one hour,[1] as fifteen years, the horns thus indicating "dictators" of states. And they have hated the previous religious systems,[2] and subjected their conception of State to the Cain-Bel idea (as Babylon), which as the woman governed the rule of the previous kings.[3]

---

[1] Twenty-fourth part of a time $=\dfrac{360}{24}$ years.

[2] Roman Catholic and Greek Orthodox Churches.

[3] Thus the autocracy of the Czars is succeeded by the autocracy of the Kremlin; the national pride in might of the Kaiser and the Prussian, by the same pride in the race to dominate, of the Nazis.

"*These shall make war with the Lamb, and the Lamb shall overcome them*" (14).

These have rejected Christ's teachings—but further, in the next paragraph:

"*They shall have one mind; they shall give their power and strength to the beast*" (17).

They shall agree as to what constitutes the State.

¶ 3. The Divine teaching given to the Adamic race was: the attuning of man's individual Freewill to the Universal Will was to be the free act of each man as a Spiritual personality; so that in the community each person acts (or gives) in mutual service, by the Divine Law of Harmony (Love God and thy neighbour): that finally in the next world God is All in All.

Cain's perversion of this in the Sumerian civilisation (and so in those of the succeeding Assyrian and Babylonian), when his higher spiritual and intellectual powers mastered the inferior-evolved mind of the pre-Adamite aborigines, was that in the nation the king took the place of God. Man had no life as an individual, no value except as the property of the king, who had power of life and death over him—a family was punished for the act of the father, a whole tribe massacred for the deed of one individual.

The king alone was a Personality; his will was the government of the State, to decide its wars, build its temples, dictate its "culture," and on his death his aborigine subjects were also entombed to give him further service.

In modern times [1] nations which have lost their king, having drunk of the golden cup of Babylon, translate it thus:

The State or nation alone matters, the individual matters not—he belongs to it, not to himself. So the State may make demand on the person, his religion, his working capacity, his life; to advance its own power and wealth (the 666 standard), and with increase demanding expansion, then aggression on the next state, in war if necessary.

---

[1] In the Zarah-Pharez House of Britain every Prince of Wales is brought up on his motto "*Ich dien*," and, more than any man, the King and his son in service each pulls his weight.

THE TOTALITARIAN STATE

Of what is the State, its particular culture or merit, is decided upon some human conception, some human standard.

(The Divine standard, explained by Christ in person, of mutual service in Love, is not considered, "for the nations have drunken of the golden cup of Babylon, therefore the nations are mad.")

Liberty, Equality, Fraternity—are perverted by the Cain idea from his earliest associations.

Liberty to all in the Adamic world—Cain used it by killing his brother.

Equality, also—but Cain resented being a tiller of the ground, while his brother, Abel, was a shepherd.

Fraternity, certainly—as long as the other fellow, Abel, has not something more acceptable than yourself.

On such great catchwords revolutions [1] started, to build some standard of State.

King, Church and Government have destroyed Liberty, Property has destroyed Equality, therefore down with King, Church, Government and Property.

"Cause want, govern opinions, and you will overthrow all existing systems," is the basis of all revolutionary propaganda.

Where in all this the happiness [2] of the individual comes in, perhaps Cain-Bel only knows, certainly not his dupes, the "proletariat" and "their leaders."

Though some high-sounding ethics is usually attached, Freewill attuned to Harmony can nowhere be found in it.

The State model of Socialism varies according to the conception of the particular "dictator," influenced by some previous race ancestry—whether Communism, Fascism, the National Socialism of the Nazis, or the Socialism of Cripps.

---

[1] As Esdras prophesied (2 Esdras xv): "*There shall be sedition among men; and waxing strong one against another, they shall not regard their king, nor the chief of their great ones, in their might. For a man shall desire to go into a city, and shall not be able; for because of their pride the cities shall be troubled, the houses shall be destroyed, and man shall be afraid.*

"*A man shall have no pity upon his neighbour, but shall make an assault on their houses with the sword, and spoil their goods, because of the lack of bread, and for great tribulation.*"

[2] Happiness—defined, p. 269.

They all agree, having drunk out of the same "cup," that the individual Freewill has to be subordinated to the State.

In Russia, recognising no God but the State, Antigod and Atheism have displaced the Greek Orthodox Church and all religion.

In Nazi Germany has arisen a neo-paganism [1] based on the polytheistic galazy of its Hittite origin and assumed migrations (having the same emblem of the Vulture on their standards).

And the intentions of Mussolini are hourly becoming clearer.

They all agree—to have "arms in keeping with their right as a great nation"—anything from half to one million or more to stand in uniform.

---

[1] "Our highest ideal is not Christ the King, but the German people," amid tumultuous applause (at Munich, 29th March 1935, from a daily paper).

"The A B C of the German heathen"—numbering one million (from daily paper, 4th April 1935): "The word heathen is for us no insult but a title of honour. We are proud of our German faith, of our Nordic paganism.

"We believe no more in the Holy Spirit, we believe in the Holy Blood. The essence of Jewish Christianity is Sin and Absolution; the essence of Nordic paganism is, Blood and Honour. Everlasting consciousness of sin is a symptom of racial cross-breeding and consequent degeneration. The German people needs no Bible. The Edda and the Sagas, Master Eckehart and Frederick the Great, Goethe and Schiller, Hoelderlin and Nietzsche, and many other great Germans were no Christians; they believed in the life in Nature, and in the power of the German soul."

Again: "Our Bible is the German soul and its merit! Our symbol is the ancient heathen Swastika! Our future is Germany!" (*Blitz*, quoted from daily paper, 15th April 1935).

And of what is the swastika a symbol?

The swastika in earliest times consisted of a Greek cross, either enclosed in a circle, as ⊗, or with its arms bent back 卍, believed to represent the sun, in the nature religions from Scandinavia to India, and in the later worship of the sun-gods (Apollo, Odin). Similar devices are found in monumental remains of the ancient Mexicans and in burial-mounds in the United States.

It was the symbol of the "Children of the Sun" (or Bel)—see pp. 63, 66—which spread the influence of Cain-Bel civilisation from Sumer to the furthest east, and via Ham-Naphtuhim to Mexico (see p. 109).

Thus it is the sign of Cain-Bel; nay more, it may even have been "the mark"—"*And the Lord set a mark upon Cain, lest any finding him should kill him*" (Gen. iv, 15).

Reading the above from *Blitz*, can one doubt the influence that has frothed to the top as Nazi Germany—their origin from an Assyro-Hittite race? Now under a thin veneer of "Kultur" they are back to their "Jotun" origin (see p. 172), a civilisation discarded by Israel 1500 years ago.

THE BABYLONIAN ECONOMIC SYSTEM

"*These shall make war with the Lamb*" (Rev. xvii, 14).

On whose side stands Britain?—or the implication does she understand?

¶ 4. Rev. xviii indicates the fall of the Babylonian economic system.

"*Babylon the great has fallen* [2]. *For all nations have drunken of her wine, and the merchants of the earth are waxed rich through the abundance of her delicacies.*

"*Alas, alas, that great city Babylon, that mighty city! for in one hour is thy judgment come* [10]. *And the merchants of the earth shall weep and mourn for her, for no man buyeth their merchandise any more* [11]. *And the fruits that thy soul lusted after are departed from thee, and all things which are dainty and goodly are departed from thee*" (14).

"*The merchants of these things, which were made rich by her, shall stand afar off weeping and wailing*" (15).

"*For in one hour so great riches is come to nought. And every shipmaster, and all the company in ships, and as many as trade by sea, stood afar off*[1] [17]. *Alas, that great city! wherein were made rich all that had ships in the sea by reason of her costliness! for in one hour is she made desolate* [19]. *For thy merchants were the great men of the earth; for by thy sorceries were all nations deceived*" (23).

Thus the suicides of the finance magnates [2]—Kreuger (Sweden), with liability of 25 million pounds and more; G. Eastman (U.S.A.), Dr Jucht (Munich), Stavisky (France): and in England the Hatry mess; in America the unprecedented orgy of speculation on Wall Street (6th November 1929, stocks of 1000 million pounds in paper value lost in a few hours), and, in the years 1930–1932, 4665 bank failures, with estimated loss of 666 million pounds.

---

[1] "The trade of the world is practically stopped, and only foodstuffs, mainly grain, sufficient to feed human beings and livestock are being carried overseas, and then only to those countries that are not self-supporting" (Philip Haldin in *The Times*, 16th April 1932).

[2] "*The love of money is the root of all evil: which while some coveted after, they have erred from faith, and pierced themselves through with many sorrows*" (1 Tim. vi, 10).

"*... Silver and gold, the peculiar treasure of kings and princes, is vanity and vexation of spirit, and there is no profit under the sun*" (Eccles. ii, 8, 11).

"*Go to now, ye rich men, weep and howl for your miseries that shall come upon you. Your riches are corrupted, your gold and silver is cankered, and the rust of them shall be a witness against you, and shall eat your flesh as it were fire. Ye have heaped up treasure together for the last days. Behold, the hire of the labourers who have reaped down your field, which is of you kept back by fraud, crieth: and the cries of them which have reaped are entered into the ears of the Lord of sabaoth. Ye have lived in pleasure on the earth and been wanton*" (James v, 1–6).[1]

In 1931 (2520 years after the Image of Gold 60 × 6 cubits set up, 590 B.C.) the Babylonian economic system was crashing and Britain came off the Gold Standard. In 1933 (in "one hour"—fifteen years after the fall of Babylonian succession kings, 1918) there sat in London the World Economic Conference, at which some sixty nations were represented, to bolster up the system, and failed, mainly due to Manasseh (U.S.A.).

In Pyramid symbology, "The ceiling of the Antechamber (datings of which are 11.11.1918 to 29.5.1928) is circumscribed horizontally by a circle of 66·6" ... denoting the circumscribing influence (during this period) of 666, the fullness of human labour and achievement ... circumscribed, limited or bound by the value of gold" (*cf*. Rev. xiii, 18).

Through the gate of Tyre there came to Solomon "*666 talents of gold*" annually (1 Kings x, 14); also she provided the gold of Nebuchadnezzar, "*she is a mart of nations ... whose traffickers are the honourable of the earth*" (Isa. xxiii, 3, 8).

But for 2520 years this qualifying condition is placed as a curb upon human cupidity—"*her merchandise and her hire shall be holiness to the Lord: it shall not be treasured nor laid*

---

[1] There follows, v. 8, "*stablish your hearts, for the coming of the Lord draweth nigh.*"

THE BABYLONIAN ECONOMIC SYSTEM

up ; for her merchandise shall be for them that dwell before the Lord, to eat sufficiently and for durable clothing" (Isa. xxiii, 18).

Corners in food and clothing . . . purity and durability, are Divinely controlled within limits; and the mass-restriction or control of necessary world commodities is Divinely prevented until the appointed time for the withdrawal of Divine Restraint.

The Restraining Power (*cf.* 2 Thess. ii) hitherto operating to prevent the assumption of world power by ultra-international financial organisations (*cf.* Rev. xiii) was withdrawn from 29.5.1928 [1]; and as a consequence the current period of economic depression is the period during which is being fulfilled the condition that "*no man might buy or sell, save he that had the mark or the name of the beast, or the number of his name—666*" (Rev. xiii, 17, 18).

The initial period of the inception of disorganisation that follows from 29.5.1928 and reaching its crisis at 23.8.1931, when the financial crisis forced the resignation of the British Labour Cabinet, clearly indicates the gradual withdrawal of Restraint [2] . . . leading to cessation of world exchange and shipping (*cf.* Rev. xviii; Isa. xxiv).[3]

The present monetary system had evolved, when forced by the expansion of Industry (from about 1776,[4] with scientific discoveries), to meet the increased wealth of the community, but restricted by the quantity of gold cover in the world. What was the banker to do? He adopted the idea of the early Babylonian goldsmiths, who found it "safe" for them to issue banknotes or promises-to-pay gold on demand many times in excess of the gold they possessed —until he deemed a 10–15 per cent. gold cover sufficient for modern financial safety, and governments permitted it.

In prosperity, with wealth [5] produced and used, with confidence in honesty, when the demand in trading was

[1] "A decline in commodity prices is actually an increase in the price of gold" (Walter Runciman); such began on the London Market from the end of May 1928.

[2] This is symbolically shown as a gradual process by the four vertical channels on the south or end wall of the Antechamber running from the ceiling to over the entrance doorway of the second Low Passage.

[3] Davidson, *The Great Pyramid's Prophecy concerning the British Empire and America*, pp. 17, 19.

[4] For date significance, see pp. 309, 314, 353, 354.

[5] "*But thou shalt remember the Lord thy God, for it is He that giveth thee power to get wealth, that He may establish His Covenant which He sware unto thy fathers*" (Deut. viii, 18).

not for gold but for credit, which was easily repaid by goods and service, all went well—until the War broke. Then people in panic for the safety of the real wealth accumulated demanded gold, and governments came to rescue the banks from bankruptcy by declaring moratoriums.

Now the monetary system still attempts to bind increasing production to a gold cover (varying in each country, to the confusion of trade), even when the Real Wealth of the community many times exceeds the gold available; and the folly of present-day finance is seen in digging gold out of the earth of South Africa, worshipping it as wealth, and burying it again in the Kentucky mountains of the U.S.A.—on a par with a child raising a castle out of sand, calling it "my castle," and knocking it down to sand again.

An echo over the ages is heard—"*a golden cup, the nations have drunken of her wine, therefore the nations are mad.*"

Why should a higher standard of living, the increasing wealth (due to Science) and the happiness of man be bound to a yellow metal dug out of the ground, the quantity of which is limited, and hoarded from circulation?[1]

The creation of new money, freed from greed, Socialist and Capitalist, to meet increasing production and consumption, and regulated in amount to keep price-levels constant, is to be based on (1) scientific laws according to (*a*) the capacity of the community to produce goods and give service, and (*b*) the need of the community for its consumption of food and clothing; and on (2) natural law that each unit gives or puts in more than he takes out.[2] The balance covers the needs of the young, the sick and the aged.

---

[1] Total bank deposits (in 1935) at Bank of England, £2023 millions: covered by £50 millions in silver and bronze, £200 millions in gold, +the 1928 fiduciary issue of £250 millions Bank of England "promise to pay" notes. Instead of one-fourth, why not wholly cover the deposits (money surely, and Britons', as good as American gold but for international wangling) and banks to hold pound for pound (printed to mean "Britain owes you £1 of goods or services")? And the Bank doles out a paltry £20 million of new note issue at Christmas because of a few extra toys in the shop window which it expects consumers might buy (meanwhile, palpitating with fear lest such should be inflation), until after Christmas it breathes again with the notes safely returned by the shops to the Bank.

[2] Refer pp. 355, 365, 369, the Mosaic Covenant, and Luke vi, 38.

THE BABYLONIAN ECONOMIC SYSTEM

The creation of new money is not to be given into the hands of international financial and banking groups, "circumscribed" by a gold standard to restrict output and raise or lower price-levels, whereby to amass limited wealth and to create want and discord.

Professor Sir F. Soddy [1] sums up the present monetary system as

> defeating technological progress by turning it into the channels of destruction, and challenging the autonomy not of one nation but of all alike, so that now the original authorities constituted for the preservation of that autonomy needs must fawn upon it to rule at all. Hampered by national frontiers, nothing can satisfy it till the whole world is made safe for banking, that its fundamental insolvency may defy exposure. Under the specious guise of a unification of humanity, it aims at absolute dictatorship under which none shall be allowed to live save by its favour and for the advancement of its transcendent whims. ... It prefers the dark in times when all men seek the light; it is sowing the seeds of hatred and war in a world weary to death of strife; it is poisoning the wells of Western civilisation.

Compare this quotation with Revelation xiii, 16–18 (given p. 332). The words scientific and apocalyptic may differ, but the meaning identical and attuned.

Let us "*come out of her*"—as put in Revelation xviii, 4 (given p. 342)—or again by Professor Sir F. Soddy:

> The British Way.[2] Let us not, as other countries have done in the grip of these anti-social innovations, discard a peculiarly native growth, the freedom of the individual and personal life, or be goaded into paroxysms of futile despair under this new absolutism. Let us see it for what it is, deriving its power from the loan of licences to live ... its loans being fictitious, its pawn-tickets can never afterwards be redeemed. Let us not enslave men that pretenders may rule, but take back our sovereign powers over money in order that men may be free. It is a road Britons have trod before.

---

[1] Professor Sir F. Soddy, *The Rôle of Money*, concluding pages, 218–220 (George Routledge & Sons).

[2] Read the wording of a Petition to His Majesty to set up a select committee to review the monetary system, now being circulated by the League to Abolish Poverty.

The command to Britain as His Israel is to come out of Babylon. It has been repeated three times:

(1) To our ancestor, Abraham: *"Get thee out of the country"* (Gen. xii, 1, Ur of the Chaldees); and he did.

(2) To Israel at the time of Assyrian Captivity: *"Israel hath not been forsaken. . . . Flee out of the midst of Babylon, and deliver every man his soul, be not cut off in her iniquity"* (Jer. li, 5, 6, 45).

"Go ye forth of Babylon, flee ye from the Chaldeans, with a voice of singing declare ye, tell this, utter it even to the end of the earth; say ye, the Lord hath redeemed His servant Jacob" (Isa. xlviii, 20); and they did.

(3) And here in Revelation xviii, 4, 20, the risen Christ to-day to Britain:

"Come out of her, My people, that ye be not partakers of her sins, and that ye receive not of her plagues."

Will Britain?

"And what agreement hath the temple of God with idols? . . . Wherefore come out from among them and be ye separate" (2 Cor. vi, 15–18, quoting Isa. lii, 11).

¶ 5. Thus the problem—on one side, Evil as a spiritual force inherent in Creation and manifest in the material development of the civilisations, from the Sumerian, through the Babylonian and its succeeding kingdoms to the present times, as manifest on the Continent and infecting Britain and America.

On the other side, God had, with foreknowledge of the descent to Chaos caused by deviation of Freewill, reserved from that seed of Adam, whose Freewill was attunable, through Seth, Noah, *"a just man and perfect in his generations,"* Abraham, *"because thou hast obeyed my voice,"* Isaac and Jacob, a Nation, chosen, "elected" to be the nucleus of His Kingdom on earth, *"to be a blessing to all the families of the earth."*

The difficulty in spiritually training this people and moulding them as "*a potter*" for His Purpose, within and yet to be apart from the above civilisations, is the theme

THE GENESIS OF BRYTHON

of the Old Testament,[1]—difficult, but not an impossible task for the Most High to be given up at the Assyrian Captivity.

The Divine Covenant, mentioned to Noah, emphasised to Abraham, Isaac and Jacob, was everlasting, the promise to the seed of Abraham, of nationhood, kingship, territory, multitude of seed, and the great mission of carrying the blessing of God to all the families of the earth. The Mosaic Covenant [2] to the twelve sons of Jacob was conditional; and the two Covenants are not incompatible.

If Israel would obey the Voice of the Lord her God, and keep His Covenant, then she should enjoy the given schedule of blessings, which would assure to her and her associated nations, "*the strangers within the gate*," health, peace, abundance and prosperity of every kind.

And as a token of such she should keep the Sabbath holy.

"*That ye may know that I am the Lord that doth sanctify you . . . the children of Israel shall keep the Sabbath, for it is holy unto you . . . to observe the Sabbath throughout their generations for a perpetual Covenant. It is a sign between me and the children of Israel for ever*" (Exod. xxxi, 12-17).

But if as a nation she refused to obey the Voice of the Lord her God, and turned away from His Covenant, then she should be punished with a given schedule of evils, including four sets of seven times punishment which may be found in our past and present catalogue of evils.[3]

"*See I have set before thee this day life and good, or death and evil; blessing and cursing; therefore choose*" (Deut. xxx, 15, 19, 20).

And it was to be told and taught to her children, and her children's children, to each successive generation, even to this day.

"*And yet for all that, when they be in the land of their enemies, I will not cast them away; neither will I abhor*

---

[1] "*O Israel, this day thou art become the people of the Lord thy God*" (Deut. xxvii, 9).
[2] See pp. 118-123.
[3] See Pascoe Goard, *The Empire in Solution*.

them to destroy them utterly, and to break My Covenant with them ; for I am the Lord their God. But I will for their sakes remember the Covenant of their ancestors, that I might be their God ; I am the Lord" (Deut. vi, 4–9).

This means, if it means anything at all, that even in the Assyrian Captivity of Israel her seed was not destroyed. If to-day a part of the tribe of Judah as Jews persists,[1] in spite of all their vicissitudes, then Israel. And Christ Himself was definite on this:

"Therefore I say unto you [the Jews], *the Kingdom of God shall be taken from you, and given unto a nation bringing forth the fruits thereof"* (Matt. xxi, 43), and v. 44 following connects the nation with the Stone Kingdom.

And with the passing of the "seven times" punishment (1804), Israel-Britain with a back-to-the-wall fight with a world in arms emerged and miraculously expanded; followed in 1805 Trafalgar, in 1815 Waterloo (Chapter X).

The population of the Celto-Saxon race was

in 1558—6 millions,

in 1800—16 millions,

in 1933—160 millions,

—increased by ten times in a century and a third, which averages four children per family.

And Britain has forgotten her origin.

"And it shall come to pass, when ye be multiplied and increased in the land, in those days, saith the Lord, they shall say no more, The ark of the Covenant of the Lord, neither shall it come to mind, neither shall they remember it, neither shall they visit it" (Jer. iii, 16), and yet, unconsciously and gropingly uncertain, Britain begins to fulfil His Mission.

Possessions came to her, often unwillingly, but for a Purpose. What would Germany do with such possessions?

---

[1] When Herr J. Streicher (as reported 30th July 1935) introduced the new-style academic oratory before 600 intellectual *élite* at Munich University, which included, in a subject-matter of four and a half hours— "You old men with beards and gold-rimmed glasses, with your scientific faces, are really worth next to nothing" (imagination faints if such were said by a British Cabinet Minister before a meeting of the Royal Society!) —he is typically Hun-unconscious of Truth in his aim to plaster the wall of every German school with "The Jewish Question explains all World History."

She would, as on the 1913 road to commercial supremacy, organise their resources with ruthless efficiency, and create a world of Krupps and chemicals, orderliness and etiquette. Would the free spirit of man be happier?

¶ 6. Over that sea, in calm majesty, lies the proud island whose existence consoles me for a thousand Continental crimes, and vindicates for me the goodness of Providence. Yes, yes, proud Britain, thou art justly proud of thy colossal strength —more justly of thy God-like repose. . . . He awaits his hour but counts not the hours between. He knows that it is rolling up through the mystic hand of destiny. Dare I murmur the mists will clear for me? That I shall hear the rumbling wheels of the chariot of the hour of Britain? It will come, it is coming, it has come. The whole world, aroused as if by some mighty galvanism, suddenly raises a wild cry of love and admiration, and throws itself into the bounteous bosom of Britain. Henceforth there are no nations, no people—but one and indivisible will be the world, and the world will be one Britain. Her virtues and her patience have triumphed. The lamp of her faith, kindled at the apostolic altar, burns as a beacon to mankind. Her example has regenerated the erring. Her mildness has rebuked the rebellious, her gentleness has enchanted the good. Her type and her Temple shall be the Mecca and Jerusalem of a renewed Universe (Victor Hugo).

And Professor Wilhelm Dibelius (Berlin)[1]:

England has always possessed a skill superior to that of any other country to direct economic and intellectual forces. Practically every war between 1700 and 1918 ended with a victory for Britain. The fact remains that Britain is the solitary Great Power which has never injured the vital interest of another European Power by annexation: the single country where patriotism does not represent a threat or challenge to

---

[1] Also another German, Professor Muth, of Munich, says: "A World Empire, such as the British, is the work of God, it was not created by human plans. But no other nation on earth can destroy it, for it is necessary for the defence of the liberty of the world."

And Bismarck (unconsciously confirming later British action in 1914), from his newspaper: "England is the Deus ex Machina, the arbiter of the situation from every point of view, and that the only remedy for the present ills would be for England to assume the position which she took up in 1815, and declare that the nation which first took up arms would be her mortal enemy, and would be obliged to withdraw, or fight against two."

the rest of the world, the single country that invariably summons the most progressive, idealistic and efficient forces in other nations to co-operate with it.

But even more important than all economic influence has been England's rôle as champion of spiritual movements in the world. Under the purely humane inspiration of Wilberforce and Clarkson it eradicated slavery everywhere. Throughout the world, England is the political ally of every great religious force. The world's holy places are almost all under British protection. Wherever any International question arises, whether inside the Protestant Churches, the Labour Movement, the women's question, the war on alcohol, the Youth Movement or Mission Field, England will be behind it, sympathetic, disinterested.

Britain is the solitary Great Power with a National programme, which while egotistic through and through, at the same time promises to the world something the world passionately desires—order, progress and eternal peace. England alone knew how to arouse something like an ethical feeling all over the world. And for the people of Europe who saw after 1800 [1] that English freedom, unlike the freedom of the French Revolution, kept order at home and allowed other nations to live their own lives in all essentials, for them, the slogan seemed a veritable gospel.

The Imperial Conference in London, December 1926, gave to the Empire a kind of new constitution . . . into a Union of British States or British Commonwealth of nations under the nominal headship of a common Monarch. . . . Through the very freedom and looseness of the new form of British Commonwealth, is it not possible to hope that, on the basis of complete independence of detail but the pursuance of common aims in most large questions, America may too be brought into a kind of all-British Confederation? Is it not possible that an Empire of this character, which while compelling none, influences all and gives to those within it the guarantee of complete peace, may exert a certain attractive force over other nations? Is the hope quite excluded that one day a British [2] League of Nations may stand over against the League of Nations at Geneva?

---

[1] See the date significance, p. 314.
[2] For British here read Brith-ish (men of the Covenant), to include with our Empire associated nations—America; the "remnant" of Israel

THE GENESIS OF BRYTHON

The Anglo-Saxons [say Israel [1]] have built a dominion that more nearly than any other approaches the ideal of a self-sustaining, self-sufficient State; whose centre has surrounded itself with a series of Daughter States essentially like it; an empire only force could have founded, but force cannot maintain, held together by the free voice of all its members, partly from political and economic reasons, partly from the force of the Anglo-Saxon idea; which rests upon the premise that every citizen who recognises its power is a free man—can express any opinion he likes, in the Press or on the platform; can profess any religion he chooses; free from conscription; can move from place to place without any police regulation, no official interferes with his private life; where the State makes no demand on him, merely invites him to co-operate with it.

Despite all this freedom, the Empire holds together, a fact which proves that this State, unlike any other in the world, has a moral basis.[2] It is an advantage for any nation to belong to it. There is but one condition attached—the new entrant to the community must recognise the rule of the Imperium [3]; apart from that he may keep his peculiarities, his language, and his religion.

It is the strenuous but glorious task of Anglo-Saxondom to stand for freedom all over the world, and to draw the sword in the cause of small and oppressed nations; the development of the world will one day bring it about that the entire world will be filled with the Anglo-Saxon [say Israelite] idea, when Anglo-Saxon (with the U.S.A.) dominion will transform itself into a League of Free Nations [say His Kingdom on earth].

This dominates the feeling of every Briton with the force of a Gospel.

The Anglo-Saxon idea can only be understood as the confession of faith of a Community, organised like a Church. Nothing in England is so hard for the Continental observer to

---

stock on the Continent (the Scandinavian countries; the Protestant, peace-loving element of Germany—now, if any; the Belgæ, the Celts of old Gaul), to be the commencing nucleus of His Kingdom on earth at His Second Coming, to take over the Throne of David (as promised, Luke i, 32, 33).

[1] Words in brackets inserted.
[2] Compare the idea of the Totalitarian State, previously in chapter.
[3] This applies to the Sein Feiners of Ireland, and the discontent Congress Party in India.

understand as the Anglo-Saxon attitude to religion and the Established Church. The German Protestant goes to church because he seeks to derive some spiritual benefit from the service of God: the Englishmen, in order that God may receive from him His due service, honour and gratitude.[1]

In August 1920 the Lambeth Conference addressed a great manifesto to the world . . . for the erection of a World Church. The issue of such an invitation by the head of the English Church at a time when the shouts of victory in war were resounding in the air constitutes a landmark in the history of civilisation.

Christianity, for a people with such a passion for deeds and want of comprehension for dogma, is an action and an organisation dedicated to the glory of God.

Is not the present epoch, when naked brutality of economic interests seems to rule the world, expressly designed to call for the resistance of spirit against matter, of the eternal against the temporal, of Christendom against the international Cæsaro-papism of capital [say the "golden cup of Babylon"]? Such a line of thought is widespread among the intellectual leaders of English Christians everywhere [then it wants to be more outspoken].

From the Old Testament comes the doctrine peculiar to all forms of English piety and tinging them all with what the Continental observer feels as a repulsive hypocrisy [not when he realises its Israel source]—the idea of the English as the Chosen People.[2]

[Of Post-War England]—The moral looseness which swept through the world like a destructive plague on the wings of war was felt in its full force in England. The discipline of the Church disappeared, the observance of Sunday died out, Puritan rigour in sex relations was replaced by a licence bordering on shamelessness. Now self-expression is the cry, the old moral code is scorned as prudery. There is a deterioration of manners everywhere, the Church has nothing for the rising generation, everything Puritanic is dismissed as ridiculous hypocrisy. For the first time in history, it would seem as though England has broken with Christianity. England itself is the scene of a struggle for mastery between profiteering

---

[1] Compare Exodus xxxi, 12–17—His Sabbath to be holy, a sign of a perpetual Covenant.

[2] Here Dibelius goes on to speak of the British-Israel World Federation and Puritanism.

capitalism and the crude egotism of the working class—with the spectre of Moscow brooding in the background. The hand of Moscow is discerned in the Nationalist agitation in India, and even at home the English soul is infected with Bolshevism.

Has England become powerless? Is it moving towards dissolution? The answer to this can only be a decisive, No! England's greatness does not rest with its thinking minority, but with the great mass of men of instinctive action and powerful will.

What England is now experiencing is a grave crisis in which a new state and a new type of society are coming to birth. The old has to a large extent lost its justification for existence, and must now yield place to the new. To a nation with its ultimate powers unimpaired, as with England, any new development is possible out of the present political and economic chaos of Europe.

A country which is to fight Bolshevism with spiritual weapons must have first cast off from its own economic system the fetters of a predatory civilisation—a policy sufficiently long-sighted and strong to have substituted a genuine system of European peace for the regimen of force.[1]

¶ 7. Thus is seen—Evil as a spiritual force inherent in Creation and manifest in the material world civilisations; on the other side, the Divine Purpose for "the restitution of all things," to be accomplished finally by Christ.

The clash between the two forces in man will come to a climax at some time known as "the last days"[2]—the time by Divine foreknowledge embodied in the Pyramid's internal structure.

"Woe unto them that shall be left in those days! and much more woe unto them that are not left! for they that are not left shall be in heaviness, understanding the things that are laid up in the latter days, but not attaining unto them.

"But woe unto them that are left, for they shall see great perils and many necessities. Yet is it better for one to be in peril and come unto these things, than to pass away as a

[1] Extracts from *England*, by Professor Wilhelm Dibelius, of Berlin, critically reviewed by Nelson Heaver in *The English People* (The Covenant Publishing Co.).

[2] See pp. 394–398.

cloud out of the world, and not see the things that shall happen in the last days" (2 Esdras xiii).

The present is a better age in which to have passed through an earthly life than any other, if mind is spiritually directed and Freewill attuned; but if mind looking on earthly things is loaded with the cares of this world, it is . . . Chaos.

And Esdras asks: *"What shall be the parting asunder of the times? or when shall be the end of the first and the beginning of it that followeth? From Abraham unto Abraham, inasmuch as Jacob and Esau were born of him, for Jacob's hand held the heel of Esau from the beginning. For Esau is the end of this world and Jacob is the beginning of it that followeth. The beginning of a man is his hand, and the end of a man is his heel; between the heel and the hand seek thou nought else, Esdras"* (2 Esdras vi).

Esau in the Bible is revealed as Edom, Dumah, the river Euphrates, and historically as Turkey.

> The destruction of the power and independence of the Ottoman Empire should be as a trumpet blast to Christendom, proclaiming that the day of Christ is at hand.[1]

From 1917, by the delivery of Jerusalem, the Treaty of Sèvres, and the fall of the Caliphate,[2] the Ottoman Empire slowly dried up like a lake under a hot sun and shrivelled from 613,000 to 175,000 square miles.

From the drying up of the River Euphrates, Revelation xvi, 12–21, gives the sequence of future events.

"And the sixth angel poured out his vial upon the great river Euphrates; and the water thereof was dried up, that the way of the kings of the east might be prepared. And I saw three unclean spirits like frogs come out of the mouth of the dragon, and out of the mouth of the beast, and out of the mouth of the false prophet. For they are the spirits of devils, working miracles, which go forth unto the kings of the earth

---

[1] Dr Grattan Guinness, *The Approaching End of the Age*, p. 474, 1880.
[2] A.D. 634—Oman set up his Mosque on the site of Solomon's Temple, *"the abomination that maketh desolate,"* and succeeded to the Caliphate.
1290 solar years of Dan. xii, 11 = 1924, the fall of the Caliphate.

and of the whole world, to gather them to the battle of that great day of God Almighty" (Rev. xvi, 12–14).[1]

The number three is symbolic of individual completion; and in this context applied to the Babylon system of Cain-Bel teaching defines the resultant forces within man based solely on Mind and Reason, and without spiritual intuition of the Divine—*"ye are as gods."*

Out of the mouth of the dragon (as original evil, Bel) emerges the Antigod idea, Atheism, Bolshevism.

Out of the mouth of the beast (as the religious system and pride of power in the Babylon kingdoms) emerges the neopaganism and race pride of the Nazis; the desire again for world dominion by material aims on the part of the Babylon kingdoms; and includes any political interest of the Papacy behind the Italian resuscitation of arms to restore its Temporal Power.

Out of the mouth of the false prophet (as Mind and Reason without spiritual intuition) emerges philosophic thought which denies Divine revelation, denies that the Creator can speak directly to man, and asserts that man alone works out his evolutionary progress on this earth, as in Rationalism [2] and Modernism.

"For they are the spirits of devils, working miracles, which go forth unto the whole world to gather them to battle."

It is peculiarly suggestive that post-War events seem to have followed the exact lines of the *Protocols of the*

[1] Follows: *"Behold I come as a thief. Blessed is he that watcheth"* (xvi, 15).

[2] In a recent broadcast it was stated that the world to-day was witnessing a struggle between God's religion and the social religion of Rationalism, described as a "humanist religion"; that Reason was the instrument for producing progress, and unless the religious impulse (and "speaking with tongues" now considered an emotional manifestation of low religious value) was prepared to subordinate itself to Reason, then it was harmful; that God as an hypothesis had been useful in the past, but now such idea must give way to Science, which was controlling disease and by psychology was explaining man's impulses.

Science has postulated in the evolution of man "two opposite and complementary kinds of knowledge, expressive of two diverse faculties, Instinct-Intuition and Intelligence-Reason." Has man then for 6000 years been developing only one half of his cortical area? Or is the other half the seat of hyperplasia, tending with age to malignancy and thus requiring immediate radical operation?

Cain-Bel intended man to think so, and smiles cynically, well pleased at the "Utopia" to which it leads.

Learned Elders of Zion (copy dated 1905 in the British Museum).

After boasting of instilling class hatred into the people, the protocol continues:

> This hatred will be magnified by the effect of an economic crisis, which will stop dealings on the exchange and bring industry to a standstill.
>
> We shall create by all the secret subterranean methods open to us, and with the aid of gold, which is all in our hands, a universal economic crisis, whereby we shall throw upon the streets whole mobs of workers simultaneously in all the countries of Europe. . . .
>
> It is indispensable for us to undermine all faith, tear out the very principle of Godhead and the Spirit, and to put in its place arithmetical calculations and material needs.
>
> In place of the rulers of to-day we shall set up a bogey which will be called the Super-Government Administration. Its tentacles will reach out in all directions, and its organisation will be so colossal as to subdue all the nations of the world [suggesting the League of Nations].

Out of some secret society of Continental Freemasonry, which inspired the protocols, Germany produced Lenin to conspire the fall of Russia, but it came to overpower Germany also, and still directs the policies of both countries.

The conspiracy seems directed mainly to undermine the British Empire and her industrial supremacy. There, the Cain-Bel influence, which unites material force, subversive and anti-Biblical propaganda, and international finance, opposes Brython; and there the danger which awaits after the Italo-Abyssinian dispute (see p. 401).

"*Come out of her, My people, that ye be not partakers of her sins*" (Rev. xviii, 4).

¶ 8. In these times, we have come to the change-over, under compulsion of Divine force exerted by the Logos —on to the plane of "the consummation" of the age, when all these things are to be "restituted." And the crucial date when these things are accelerated is given by the point where the ascending plane of the Grand Gallery produced meets the horizontal plane, 2nd December 1924.

Davidson has shown the importance of this date in Pyramid symbology.

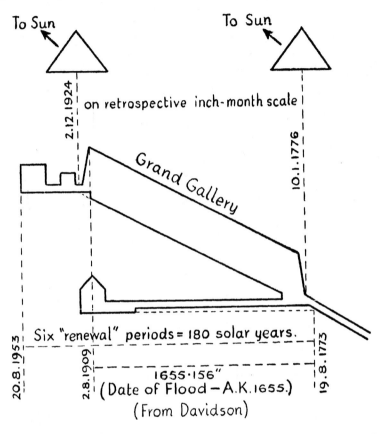

(From Davidson)

On the following three dates the noon reflexion of the Sun on the South face of the Pyramid pointed direct to the Sun:—

        (1) 10th–11th January 2625 B.C.
        (2) 10th January A.D. 1776.
        (3) 2nd December A.D. 1924.

(1) This date is defined by the floor-point of the entrance doorway on the *built* North face; whereas as defined by the *designed* North face the date is 2644 B.C.

The symbolism of "Light returned to its Source" defines the date, 10th–11th January 2625 B.C., at which the builders of the Great Pyramid consummated the error that led to their rejection of the Apex Stone ("the headstone of the corner").[1]

(2) On 7th April (Julian) A.D. 30, the date of His Crucifixion, the passage height raised 286·1" symbolises that "the spiritual barriers to the rectification of Displacement in the Spiritual environment of our visible Universe were removed by Our Lord's atonement for all the errors of the mystical builders." [1]

This point taken on the retrospective inch-month scale is dated 10th January 1776.

On this date was published Thomas Paine's pamphlet, *Common Sense*, that gave momentum to the American Revolution (*i.e.* "displacement" from the potential Commonwealth of the British Empire), and which actually contained the essential elements of the Declaration of Independence (signed 4th July 1776).

In 1776 was also published Adam Smith's *Wealth of Nations*, which gave direction and momentum to the Industrial Revolution.

Thus in both countries was happening a spiritual displacement—the harnessing of the Puritan spirit to the yoke of material achievement.

It was at this stage that the Anglo-Saxon race built its own casing of Industrial Bondage, in the construction of which they rejected the Guidance of the Master. . . . The error in building the false casing of Christian civilisation has been discovered as the false casing approaches apparent completion (4th June 1932 to 5th December 1935). So now it is in process of demolition, to prepare the pyramid of Christian civilisation for the construction of its perfect casing under His guidance [1] (16th September 1936 to 19th August 1953).

(3) Again, on 2nd December 1924, the symbolism of "Light returned to its Source" signifies that "the reason for the Unfinished Pyramid of British and American

---

[1] Davidson, *The Great Pyramid's Prophecy concerning the British Empire and America*, pp. 21–24.
For the significance of these dates, see pp. 86, 89, 97.

civilisation is the rejection of Divine guidance in the work of planning and founding their sociological structure."

Just as at the Great Step dating (25th January 1844) so now, it was the duty of the Church of Britain to direct the Spiritual way to the State: "See, I have set before you, this day, Christ or Chaos, therefore choose ye."

A new Parliament had just assembled (having been displaced for 286 days by a Socialist Ministry) after fighting a General Election on the Communist danger, and it returned immediately to the gold standard as "the rock on which all stable government and civilisation is founded" (*The Times*).

And on (1) Communism and (2) Gold, Britain nearly foundered.

The Divine system of economics commanded to Israel is—Produce goods, give service, gold and silver to be merely a medium of exchange.[1] For the accumulation of these, in spite of all his wisdom, Solomon fell. So the British sterling is to be based on the exchange of goods and services,[2] and not on the value of gold.

The troubles of Manasseh (the U.S.A.) in accumulating gold instead of circulating it as monetary exchange are due to the disregard of this command. Now she has begun the same on silver she is laying up more trouble for herself and the rest of the world.

(1) The infection of Communism and irreligion, insuf-

---

[1] In *Farming and Money* (J. Jenks and J. T. Peddie, 1935, Williams & Norgate) the troubles of agriculture traced to our present monetary system—as:

(1) Agricultural producers throughout the world form the bulk of the world's markets for manufactured goods.

(2) Eighty per cent. of the world's population is engaged in the production of primary products.

(3) Most of the real difficulties to trade due to currencies being linked to gold at varying stages of valuation.

Then follows, industrial depression due to gold, or the lack of it (gold having become a commodity and not mere exchange is taken out of circulation and hoarded as wealth).

[2] Emphasised much more by Christ as:

"*Give, and it shall be given unto you, good measure, pressed down and shaken together, and running over*" (Luke vi, 38).

"*Lay not up for yourselves treasures upon earth, where moth and rust doth corrupt, and where thieves break through and steal. For where your treasure is, there will your heart be also*" (Matt. vi, 19–21).

ficiently regarded, led to the General Strike of 1926. At this time Britain was passing over the downward shaft in the floor of the Subterranean Chamber.[1] If she had succumbed to Bel influence, in a few years she would have been eliminated from the World Powers, absorbed and disintegrated in the coming and final destruction of Babylon. She skated over the crisis guarded by the "sweet influences" above, and by weight of her transgressions from Divine Purpose lessened by the "services given" over and above their pay (often not without grousing when its value compared with the easy wealth of business) by her men of good will in mission field, churches, hospitals, philanthropy, Services (Civil, Colonial and other) and the Professions (Scientific, Medical—and here must we include the Legal?).

(2) Deviation from Divine Purpose with the burden of the gold standard and the crashing of the Babylonian economic system forced her into "the final tribulation" on the entrance of the second Low Passage, dating 29th May 1928.

The bankers now state that the fall in prices started then. Since the summer of 1928 signs of credit limitation were

---

[1] The central date of the Subterranean Chamber is 20th July 1926, when occurred—

    (i) the reorientation of French financial policy;
    (ii) the reorientation of Soviet Russia's economic policy.

(i) The Herriot Ministry fell, with the franc at its lowest value.
A new financial policy was inaugurated which, in consequence of the flight of French funds abroad, and their subsequent repatriation, jointly with the movements of War Debts' and Reparations' payments, has been mainly responsible for the sterilisation of the world's gold stocks, the consequent increase in the value of gold, and the resulting relative decline in commodity prices.

(ii) At the plenary conference of the Central Committee of the Communist Party, which was held to settle vital questions of policy, "fraught with the most serious historical consequences" . . . violent disagreement developed amongst the leaders . . . Dzerzhinsky, the arch-terrorist, suddenly died . . . Zinovieff and other prominent leaders expelled. . . .

The crisis was attributed to the failure of the General Strike in England and to the related failure of Communistic propaganda. . . . The policy developed from this conference led to the Soviet Five-Year Plan, put into execution from October 1928, and to increased Communistic activities in India and China (Davidson, *The Great Pyramid's Prophecy*, p. 39).

evident. In the process of stabilising the French franc at the end of May, 1928, it was obvious that drastic measures must be taken.

Sir Arthur Balfour (11th March 1929) stated: "I am convinced that before we get through we shall go through a harder period than we have yet seen and one that will affect everyone. In the end you cannot take more out of production than you put into it, and we have been trying to do this for some time. The facts have been hidden by our large accumulation of capital at home and abroad."

Robbery with and without violence and motor banditry are becoming daily events.

Unsettled movements are seen in many phases of life to-day—Financial, as in the Gold Standard; Industrial, as in unemployment; Political, as in the relations of France, Italy, Germany, Russia.

All these effects are caused by the spiritual powers of Evil in a last gigantic attempt to overthrow Christian civilisation. This is being allowed by God to operate to such extent only as is necessary to purge Christian civilisation of its evils. He is gathering and will gather out of His Kingdom all things that offend and them that do iniquity.[1]

"*This know also, that in the last days perilous times shall come. For men shall be lovers of their own selves, covetous, boasters, proud, blasphemous, disobedient to parents, unthankful, unholy; without natural affection; truce-breakers, false accusers, incontinent, fierce, despisers of those that are good; traitors, heady, hard-minded, lovers of pleasures more than lovers of God; having a form of godliness, but denying the power thereof. For of this sort are they which creep into houses, and lead captive silly women laden with sins, led away with divers lusts; ever learning and never able to come to the knowledge of the truth*" (2 Tim. iii, 1–7).

"*In that day, as for my people, children are their oppressors, and women rule over them. O my people, they which lead thee cause thee to err, and destroy the way of thy paths. Because the daughters of Zion are haughty, and walk with stretched forth necks and deceiving with their eyes; walking and mincing as they walk, and making a tinkling with their feet.*

[1] Clive Kendrick, *The True Economic System*, pp. 72, 73.

"Thy men shall fall by the sword, and thy mighty in the war" (Isa. iii, 12, 16, 25).

"And in that day, seven women shall take hold of one man, saying, We will eat our bread, and wear our own apparel; only let us be called by thy name, to take away our reproach. Woe unto them that rise up early in the morning, that they may follow strong drink; that continue until night, till wine inflame them! And the harp, and the viol, the tabret and pipe, and wine are in their feasts; but they regard not the work of the Lord, neither consider the operation of his hands.

"Woe unto them that are mighty to drink wine, and men of strength to mingle strong drink!

"Woe unto them that say, Let him make speed and hasten his work, that we may see it, and let the counsel of the Holy One of Israel draw nigh and come, that we may know it!

"He will hiss unto the nations from the ends of the earth, and behold, they shall come with speed swiftly, their horses' hoofs shall be counted like flint, and their wheels like a whirlwind. And in that day they shall roar against them like the roaring of the sea, and behold, darkness and distress, and the light is darkened in the heavens thereof" (Isa. iv, 1; v, 11, 12, 22, 19, 26, 28, 30).

"For as in the days before the flood, they were eating and drinking, marrying and giving in marriage, until the day that Noah entered into the Ark, and knew not until the flood came and took them all away." [1]

In the modern world it is usually said that there is no such thing as sin.[2] Re-define it as deviation of freewill from harmony, both Divine and in natural law and in human affairs, and it is seen rampant—in the petty squabbles in the family circle, the constant nagging that

---

[1] Then follows, *"so shall also the coming of the Son of Man be"* (Matt. xxiv, 38, 39).

[2] The modern definition of sin as the living of a selfish life is too limited; also it is unconvincing, being founded on the moral code of society, and not on a deeper Reality; from the view of society, a hermit's life is thus sinful, but in terms of Reality it is not necessarily so. Such a view forces men to say of it: "As a matter of fact the higher man of to-day is not worrying about his sins at all, still less about their punishment; his mission, if he is good for anything, is to be up and doing" (Oliver Lodge, *Man and the Universe*, p. 220).

finishes love,[1] the half-true gossip at a tea-fight, the worthless wrangle over some absurd garment at a bargain sale, the mess that comes up before the Law Courts and reported in the Sunday papers, the ill-health due to deviation from natural law that fills a doctor's consulting-room—"nerves,"[2] demanding constant sympathy and produced by discord in living with their fellow-beings; the waste of time in Parliament when one party opposes another because it has to, in everything whether good or bad; the discordant note of Hyde Park oratory, urging class war, that pernicious invention of the Babylon system, and never meant for Britons, born not of Spirit but of Discord; the deviation from righteousness in the dealings of one nation with another that produces war.

Even in war, within the destruction of the trenches, when petty discords are eliminated in a common purpose, there emerges comradeship of man with man.

¶ 9. Let Davidson sum up the meaning of the present chaos:

> What is happening now is that Britain is bearing the initial shock of the passing from the Old Order of things to the New. In passing through it we are purged and purified, the first amongst the races of mankind, to emerge spiritually regenerated into the new economic order of things that constitute the living

---

[1] "*Whom God hath joined, let not man put asunder.*"
The moral code of society requires a State licence and registration for marriage, and if certain things fulfilled the State dissolves it. Let the State make and break, as it wishes.
But does the Church dare presume to say "*Whom God hath joined*" where it makes no preliminary investigation but the babbling of three Sundays' banns—in those cases which "*join*" by natural passion or as a business proposition—and in many wholesale cases of Bank Holiday weddings? Does it expect respect from such for obliging them? By the progeny of such Christ's task of salvation becomes harder.
By sane advice and sympathetic sifting (before, and not after) much of later discord may be saved. By making marriage and divorce easier in the one and harder in the other, by stressing the difference between the two requirements, man's thoughts are led higher—having passed the intermediate, he likes to take the degree!
"*Render unto Cæsar the things that are Cæsar's.*"

[2] *The sound of a shaken leaf shall chase them; they shall flee, when none pursueth*" (Lev. xxvi, 36).
"*In the morning thou shalt say, Would God it were even! And at even thou shalt say, Would God it were morning! for the fear of thine heart*" (Deut. xxviii, 67).

conditions of the Kingdom of Heaven on earth. In spite of all our material efforts and strivings we shall ultimately be compelled by circumstances to ask God "to do it for us"— not as isolated individuals nor from the lips in formal prayer, but racially in spontaneous prayer from the stricken spirit of every individual of the race. We have yet to learn how this spontaneous prayer of conviction can come from the mass of the race, and to learn this we have yet to experience many failures in material action. We have His promise that, when the spiritual consciousness of the nation is thoroughly awakened, our prayers shall be answered by our being guided safely through our difficulties.[1]

"Thus saith the Lord God, I do not this for your sakes, O House of Israel, but for Mine holy name's sake, which ye have profaned among the heathen whither ye went. For I will take you from among the heathen, and gather ye out of all countries and will bring you into your own land. Then will I sprinkle clean water upon you, and ye shall be clean, from all your filthiness, and from all your idols will I cleanse you. A new heart also will I give you and a new spirit will I put within you; and I will take away the stony heart out of your flesh, and I will give you a heart of flesh. And I will put My Spirit within you, and cause you to walk in My Statutes, and ye shall keep My judgments and do them.

"Thus said the Lord God, I will yet for this be enquired of by the House of Israel, to do it for them" (Ezek. xxxvi, 22–37).

How can the Anglo-Saxon race attune each individual Freewill in common prayer to meet the Spiritual compulsion now exerted?

For amid the wilderness of opinions expressed in books and speeches, wherein is Righteousness?

The present practice in public speeches and of broadcast prophets to the prolongation of phrases and the protraction of pauses wherewith to give painful birth to a pettifogging platitude or a "polysyllabic petrifaction," to impress the big British public, makes plain men protest against this pomposity of verbosity.

[1] In *The National Message*, p. 24.

THE PATH TO BRITAIN

So did the Pharisees delight to be seen in long prayers.

It needed Christ to use the fewest words to express a symbol, "*God is Love,*" for Man's Reason, "*born again*" to spiritual intuition, to interpret the concept.

"*Seek ye first the Kingdom of God and His Righteousness.*"

"*Six days shall thou labour and do all thy work. But ye shall keep My Sabbath for it is holy unto you, a perpetual Covenant, a sign between Me and the children of Israel for ever, that ye may know that I am the Lord that doth sanctify you.*"

It is for Britain as Israel to maintain the Sabbath,[1] whereon to think on these things—nay, in these days when work and thought are inextricably mixed, to pause at times and think on Righteousness: to go back to the Church, there to find again the Early Church of the Spirit displaced by formal Christianity, even if the sermon is not attuned; with mind on these things to reinterpret the readings of the Scriptures nationally.

And why was Joshua i, 7–9, read instinctively at the Jubilee Service?

For to Israel-Britain to-day,—

"*Only be thou strong and very courageous, that thou mayest observe to do according to all the law, which Moses, My servant, commanded thee; turn not from it to the right hand or to the left that thou mayest prosper whithersoever thou goest.*

"*This book of the law shall not depart out of thy mouth; but thou shalt meditate therein day and night, that thou mayest observe to do according to all that is written therein; for then thou shalt make thy way prosperous, and then thou shalt have good success.*

"*Have I not commanded thee? Be strong and of a good courage; be not afraid, neither be thou dismayed: for the Lord thy God is with thee whithersoever thou goest.*"

---

[1] On such lines are the broadcast programmes of Britain on Sundays, in contrast with the Continental; their daily morning and mid-week services. And, on New Year's Eve, 1935 was ushered in over the ether by the service from Winchester Cathedral, and with the hymn, *Praise, my Soul, the King of Heaven.*

So also to "see" the hymns such as—

> *O God, our help in ages past,*
> *Our hope for the year to come.*

And

> *O God of Bethel, by Whose Hand,*
> *Thy people still are fed—*

with vision of Jacob's Stone, raised at Bethel, on its journeyings, following the line of David, to its resting-place in Westminster Abbey.

> *God of our fathers, be the God*
> *Of their succeeding race.*

And another, instinct with our origin, which will yet be our cry, *Great King of Nations, hear our Prayer*:

> *Our fathers' sins were manifold,*
> *And ours no less we own,*
> *Yet wondrously from age to age*
> *Thy goodness hath been shown;*
> *When dangers like a stormy sea*
> *Beset our country round,*
> *To Thee we looked, to Thee we cried,*
> *And help in Thee was found.*

And more, it is for us to absorb the English Prayer Book, unique in all Christendom, as ours (and expressed in perfect language), wherein our origin rises up to strike us.[1]

"*We have erred and strayed from Thy ways like lost sheep*" (Liturgy).

"*For He is the Lord our God, and we are the people of His pasture, and the sheep of His hand.*"

"*Forty years long was I grieved with this generation, and said, It is a people that do err in their hearts, for they have not known My ways, . . . and our fathers grieved Him in the wilderness*" (Venite).

"*Wherefore let us beseech Him to grant us true repentance*" (Absolution).

"*It is He that has made us, and not we ourselves; we are His people, the sheep of His pasture*" (Jubilate).

---

[1] "We make assertions and advance claims in our Prayer Book that have nothing whatever to do with our race unless those people are correct who assert that the English nation is descended from the lost Ten Tribes of Israel" (Mrs Annie Besant).

"We magnify the Lord . . . He remembering His mercy hath holpen His servant Israel, as He promised to our forefathers, Abraham and his seed, for ever" (Magnificat).

"Blessed be the Lord God of Israel, for He hath visited and redeemed His people, that we should be saved from our enemies . . . to perform the mercies promised to our forefathers, and to remember His holy Covenant: to perform the oath which He sware to our forefather, Abraham, . . . that we being delivered out of the hands of our enemies might serve Him all the days of our life" (Benedictus).

"For mine eyes have seen Thy salvation, which Thou hast prepared before the face of all people, to be a light to lighten the Gentiles, and to be the glory of Thy people, Israel" (Nunc Dimittis).

"We believe that Thou shalt come to be our Judge. . . . O Lord, save Thy people, and bless Thine heritage: govern them, and lift them up for ever" (Te Deum).

Nay more, it is for us to be "*born again*" to see Jesus as the Christ, who is at hand to "restitute" out of Chaos —the longed-for hope of the world.

Even the cinema comes to the idea in the film *The World Moves On*, for after showing the destruction of man by man in war, and the sordid scramble of man on the Stock Exchange in pursuit of wealth, it ends with the figure of Christ on the Cross, fading into the sky to the strain of the Resurrection hymn.

¶ 10. Are we, His people, the sheep of His pasture, His servant nation amidst a world of nations, prepared to become the nucleus of His Kingdom on earth, which is now at hand, "*to bring forth the fruits thereof*"? What then is the future state of Britain to be?

(1) Britain is to adopt the Divine system of economics. Whereas the nineteenth century deviated in regarding (*a*) profits, not consumption, as the aim of production, and (*b*) wealth as consisting not in goods and services but in money as an interest-bearing commodity, wherewith to levy tribute on production and the national income,[1]—

[1] See Jeffrey Mark, *Analysis of Usury* (J. M. Dent & Sons).

such developed since 1776 by Britain to its logical conclusion of monetary chaos: the Divine command to the twentieth century is (a) to break such vicious circle, proclaimed by the Babylonian Nebuchadnezzar, and (b) to regard the production of goods as for the consumption of all, and money (gold and notes) as mere medium of exchange, not to be accumulated by nations nor by individuals, but circulated. And money, if interest-bearing, is to be limited to seven years, or arranged to cancel out with all debts in fifty years, so not to burden future generations.

For this, the first essential is to place the issue of money under the nation's statistical control, to maintain a constant price-level, such issue being freed from speculative gambling on international exchanges.[1]

"The earth is the Lord's, and the fulness thereof"—not man's, to levy tribute on his fellow-man.

The accumulation of wealth from an office chair without the equivalent of service given or goods produced [2] is denounced, as also the boosting-up of share values on a problematic dividend.

This is not an argument for Socialism, which unlike

[1] Sir F. Soddy, bringing a scientific mind to monetary muddle, writes, in *The Rôle of Money*, pp. 204–206, on "the signs of a new truth," as these signs "at first slight but cumulative and interwoven, by which a scientific investigator or pioneer into novel regions of thought knows when he is on sure ground, even when everyone else may think him mad. . . . One, certainly, of these signs is how what appears nothing so much as a jig-saw puzzle of disconnected events and conundrums suddenly seems to fit together into a picture, to be lost again in a haze of uncertainty, but always returning, each time a little more orderly and definite. . . .

"Another sign is the projection of the new view back into the past, and how, there also, it throws light upon what before was mysterious and inexplicable. . . .

"Another sign of a new and true idea is its extension from its immediate application to throw new light upon cognate problems."

Personal comment on this—words cannot better express the making straight of the path of "His-Story" these many months.

[2] *Vide* the Pepper Pool—report in daily paper (5th April 1935): "The purchase of 3000 tons of pepper would secure complete control of the market and allow the purchaser a handsome profit before the next season's crop was due." (Such to-day may be legitimate business, but is opposed to Divine economics and so productive of financial chaos.) Report continued: "At a subsequent meeting of the shareholders there was a deficiency . . . practically a million pounds 'gone west'" (because it represented no "real" values).

THE THEOCENTRIC STATE

Nature produces with niggardly hands and overpays for service given, while arousing no urge to give—it takes more out of production than it puts in. Capitalism is condemned when it limits production and destroys the fruits of Nature to keep up price-level.

The emphasis is on giving—good measure, pressed down, shaken together and running over:

Give all thou canst,
High Heaven rejects the lore
Of nicely calculated less or more.

It means that man is to give of his best without stint to the utmost of his talent, regardless whether the other fellow seems to acquire more than he has earned—the reversal of the present universal mentality of (getting) 9d. for (giving) 4d.

"*For what shall it profit a man if he gain the whole world, and lose his own soul.*"

Britain has again balanced its Budget, and feels secure "Babylonly." The shopkeeper rubs his hands in satisfaction, with money in the till at the year end, and looks with pride over his material possessions. Yet another bolt here, another bar there is needed, the risk taken last year must be remedied, for the fellow opposite does not look too good.

Has Britain balanced its Budget on the Divine Scales? Against service given and goods produced, is the fulfilment of His promise to Israel as this glorious country of Britain. The collection of the direct demand note to each British-born to attune his freewill to Divine Harmony is due, plus a spot of thanks.

(2) Britain is to found a League of Free Nations, on His economic system, and in which Christ is at the head of its constitution; to be based on peace and goodwill, between each no dread of war; in which the British Commonwealth of Nations is the model (see Dibelius, already quoted).

Therefore India and South Ireland [1] are to be retained

---

[1] The Sinn Feiner of Hamite-Phœnician origin (see p. 186) seeks to drag South Ireland out of the Empire, and involve it in Babylon's crash.

The Phœnicians' merciful escape from death after the Easter Rebellion

—for these are to be within His Kingdom on earth and not lost.

Ask any Briton in the pay of an Egyptian or Indian firm (or government) whether he has seen righteousness in rule in the land.

"*One land shall ask another, Is righteousness, is a man that doeth righteousness, gone through thee? And it shall say, No* (2 Esdras v).

And he will instance the custom of pettifogging difficulties and intrigues deliberately made to be smoothed away by backsheesh. Give service? No, buy it, and oil its palm with gold and silver.

Is Ephraim to hand over six million Indian Christians to the rule of unrighteousness? There is One, whose time is near, Who is preventing him.

Thus we see the Indian princes have accepted as king the House of Zarah-Pharez, and the birthright of Ephraim in India,[1] and rejected the spurious claim to such birthright by the sons of Abraham by Keturah (the Brahmins, see p. 110).

---

allows them to continue as "*the thorn in the side*" (Num. xxxiii, 55) of Israel. The solution suggests itself of the transfer and exchange of the Sein Feiners to Germany, union with whom they desired in the War; for "*the remnant*" our ancestors left behind on the Continent, the Lutheran element. With South Ireland repopulated with the spirit which induced the Christmas truce of 1914 between the trenches, and with the Protestant in the south, some working arrangement could be made with Ulster to the unity of Ireland. While with the rest of the Semites expelled from Germany, for the "glory" and unity of their "Aryan" (*sic*) race, the threatened split in Germany would be avoided, and the Prussian Nazi and the Sinn Feiner could lie down comfortably together, and within a century evolve again that galaxy of Assyrian gods in the new Nazi mythology.

The I.R.A. would shake down with the S.S.A. to form a common "playboy" army, the brown shirt and the green would tone easily to a jersey of some earthy hue, the ASSIRA.

In the "last days"—"*Thus said the Lord, When I shall have gathered the house of Israel from the people among whom they are scattered, and shall be sanctified in them in the sight of the heathen*" (Ezek. xxviii, 25); then said of Tyre: "*And there shall be no more a pricking brier unto the House of Israel, nor any grieving thorn of all that are round about them*" (24).

[1] Ranjitsinhji, Jam Sahib of Nawanagar, at Chamber of Princes, 25th March 1933 (his last words before rebuked by the Viceroy): "I confess that these words" (referring to the safeguards proposed, as valueless) "cause me profound disquiet, for they coincide with certain apprehensions arising from our studies as to the difficulty of the Crown retaining in future any effective sovereignty in India."

And Ephraim will yet be compelled (*vide* the difficulty to the passage of the India Bill) to continue to rule India with righteousness and not delegate the birthright to "heathen" politicians, who though Westernised by education know not yet what is "righteousness" in ruling.

The Continent may have looked on the pre-War Briton as, with cloak of righteousness trimmed with altruism, the shopkeeper who seeks to gather his shekels.

But the generation of the War, who entered it unprepared and for a Cause, and the younger one, are down to bedrock, ready to shake the hand of any truly righteous.

Yet when he sees a nation which

(i) teaches its children hymns of war [1];

(ii) beheads women on a block (in February 1935);

(iii) garlands in the ring the national winner of a boxing bout, amid 25,000 hands uplifted in salute and as many voices singing their national "*über alles*," while the vanquished, who fought with insane courage through two rounds that would never have been allowed in Britain, and where he would have received an ovation greater than the victor for such, received only the encouragement of his own manager (March 1935)—

Is it "superior" to feel superior to that? But, come, allow a trodden nation some expression of joy at the recovery of national pride.

Will a snake striking to kill understand why it is bruised and not crushed? And if it inflates its hood again, will you be eager to nurse it to your bosom?—or, defenceless, will you this time seek a gun?

---

[1] *Yet one day the world will tremble,*
Mankind will shiver with fear,
When we Germans rise up again,
For the fight for freedom and justice.
Now the hour is come,
When this rumbling is perceived.
The whole world already understands
That Hitler is mighty and leads.

We trot far afield, the banner waving in the wind,
Many thousands are by our side who have gone out
To ride into the enemy's land.
I die on foreign soil, well, that was to be.
Press on your horses, trot over my grave.
Hurrah, Victory.

(3) Canon Streeter [1] on co-operation as the basis of future world civilisation. And re-define co-operation (a word which Socialism has encased with its own meaning) as mutual service (the idea given to the word service, in deeds rather than words, by the Prince of Wales).

Civilisation is the creation of three activities—intelligence, energy and goodwill. This active Will to Good is in the last resort the regulative and directive principle behind all achievement which is really creative.

Science is the by-product of the disinterested love of truth, ... is the clearest proof of all that the spirit of disinterested constructiveness (active Will to Good) is the mainspring of progress.

Discovery, Invention and Organisation are possible only because of man's unique gift of Co-operation. Invention and Organisation have been too often used by men to compass the detriment and destruction of fellow-men. Advance (in such) has been due to the desire for personal profit or advantage, in legitimate but not ideal ways, such as belong to common human nature.

Co-operation on the grand scale is a new thing—the slow result of centuries of effort and of organisation continually improving. Before man could hold his own with Nature, the great State, the great Industry, World Trade, International Science had to rise.

The creation of great states and industries has been effected under the leadership of men whose outstanding characteristics have been intellect, energy and courage. These qualities lead to progress only in a society dominated by the ideals of Honesty and Justice. And they can be directed to purely selfish or to purely ideal ends ... creative forces when directed towards unselfish ends; destructive, when directed towards ends purely self-regarding.

Production and distribution are a necessary public service.

In trade, honesty is creative, dishonesty destructive; commercial prosperity depends on confidence. Commerce and industry thrive only where there exist security of person and property. But these depend entirely on the intelligent and impartial administration of Justice by the State. Their poor success in solving the problem of civil Justice has been the

---

[1] Canon B. H. Streeter, *Reality*, pp. 159–171.

main cause of the relative stagnation of civilisation of India and China (one might add, of Russia and Egypt also).

Laws, institutions and customs are intimately related to national character. And to what is reform of law and custom due? At once we come upon the creative function of the prophet, the martyr, the reformer.

Progress depends partly upon a growing keenness of perception for ethical ends, partly upon an advance in the art of reducing them to practice.

That growing capacity of Co-operation, which alone has made possible the advance of man, has in the last resort been due to generosity, the readiness to do more than one need, to give more and exact less than is strictly in the bond, to sympathise and to forgive.[1]

An act of temper which is to be spiritually creative must have an aspect of abandon (*debonnaire*); a flame from which men seek to light their torches must be a flare.

The prodigality of Nature is a true reflection of a necessary element in the highest spiritual life "*good measure, pressed down, and running over.*" Progress has depended on the direction of energy and intelligence by the still small voice which bids man stake all on his intuition of the highest.

Let British Socialism ponder on this. It is dead against the teachings of Karl Marx, the Communism of Soviet Russia, and all Continental Socialism; dead against the class war, which destroys all co-operation and buries in its ruins every creative element.

¶ 11. Britain is wearied of fighting elections on vote-catching schemes which produce nothing but more legislation, wearied of contending parties mouthing clap-trap and class war, "toiling masses" and gold-thirsty capitalists, wearied of a top-heavy superstructure of laws in language that no one understands. God has legislated, it is for man to obey.

And so Britain could, if united in a common purpose, with "*a new heart*" in her, of flesh and not of stone—"*a new spirit,*" to recognise herself as His servant nation,[2] elected to be "*a blessing to all the families of the earth.*"

---

[1] *Vide* Matt. v. 39-42.
[2] "*If thou shalt keep the commandments of the Lord Thy God and walk in His ways, all people of the earth shall see that thou art called by the name of the Lord, and they shall be afraid of thee*" (Deut. xxviii, 9-10).

Is it beyond Britain's intellect, with the resources of her God-given Empire, to fulfil His economics, and put every man back into work?

Britain is tired of the reformer who speaks of his convictions, elaborating proof, of hearing the bigoted experiences of him that has saved his own petty soul, tired of politicians who speak of enthusiasm for the cause and their duty to advocate it, tired of hearing arguments for and against—argument which originated in the analytic rationalistic categories of Greek philosophy and developed by modern education.

Britain wants to hear again the direct "*Thus saith the Lord.*"

The kingdom of heaven is like unto a Rugby field with straight white touch-lines. Freewill knowing the laws of the game can cross them, but as long as it stays beyond the lines it is not in the game but a mere spectator. The game is played with every bit of body, mind and spirit, in co-operation, to each his position, with talents trained and untrained, knowing the rules of fairness and harmony, in which you do to the other man only that which you would that he should do to you.

The rewards of the game are not in the plaudits of the crowd, but in a suffused feeling of well-being, of treading on air, when Freewill is in harmony with everything on earth.[1]

Socialism, of whichever Continental brand, has the touch-lines marked and bounded by the brick wall of the State, so that Freewill tends to play for safety towards the centre. As often will happen, hurled into touch by tackle, the body emerges bruised and mind bewildered, Freewill still has to remain in the game, unable to choose a better field until the referee blows his last whistle.

Britain is being asked to accept the brickfield, when it has been offered a better game.

For the future of Britain, the Finger points, not back-

---

[1] "*The Kingdom of God is not eating and drinking [i.e.* entailing strict training to attune Freewill and when having pulled one's weight, as] *but righteousness, and peace and joy in the Holy Spirit*" (Rom. xiv, 17).

wards to the dilapidations of the past, as Democracy or Dictatorship, Socialism or Capitalism, but forward to a Theocentric State.

"Arise, shine, for thy light is come, and the glory of the Lord is risen upon thee."

"Howbeit when the Son of Man cometh, shall He find faith on the earth?" (Luke xviii, 8).

CHAPTER XII

¶ 1. THE Kingdom of God is at hand!

"*At midnight, a cry! Behold, the bridegroom cometh.*

"*Then all those virgins arose, and trimmed their lamps.*

"*And the foolish said unto the wise, Give us of your oil, for our lamps are going out.*

"*But the wise answered, Go ye and buy.*

"*And while they went to buy, the bridegroom came; and they that were ready went in with him to the marriage, and the doors were shut*" (Matt. xxv, 1–13).[1]

From His many quotations from the Old Testament, confirming and explaining the meaning therein—by parables of the bride, of the marriage of the King's son, the ten virgins, etc.—Christian Israel is to be the bride.

From the seed of the woman (Gen. iii, 15), Israel was chosen at Sinai according to His Covenant with her forefathers, divorced for following after other gods (the parable of Hosea i and ii), and redeemed (by the effect of His Death and Resurrection on Matter, whereby a new force is created to divert the natural and inherent destruction to which Freewill, deviated from Divine Oneness, tends): Israel—scattered and regathered in "*the appointed place*," cleansed, coming to learn the commandments of His Kingdom, and "*bringing forth the fruits thereof*"—is to be remarried under a New Covenant (Jer. xxxi, 31–37), to found the Kingdom of God on earth.

This is "*the restitution of all things*," "*the consummation of the age.*"

Christ is the head; His true Church, of thousands of saints whose Freewill is attuned, is the body. At His Second Coming, to these are restored "*the incorruptible body*," with Spirit Dominion over matter.

---
[1] See Basil Stewart, *At Midnight, a Cry!*—and for interpretations of the parables, J. J. Morey, *The Parables of the Kingdom*.

THE MIDNIGHT CRY

Thus is symbolised the bridegroom.

The parable of the ten virgins follows His summary of A.D. history (Matt. xxiv, 1–35), the warning of His Second Coming to each Freewill (xxiv, 36–51), symbolised by the "faithful and unfaithful servant."

The ten virgins are, then, the ten-tribed Israel-Britain (ten signifying worldly completion, without their heavenly king), "*the lost sheep of the House of Israel.*" They possess the lamp of knowledge, Christian teaching from their childhood onwards, Christian atmosphere in Church and State organisations.

"*Five were wise, and five foolish. . . . While the bridegroom tarried, they all slumbered and slept.*"

In Pyramid symbology it has been seen that the overlap descent of the Grand Gallery ending signifies a "falling away"; the chaos of the two low passages as the result of Cain-Bel tares sown in the world affecting Israel-Britain, drunk with "*the golden cup of Babylon,*" befuzzled as to spiritual origins, and still imbued with Babylon ideas of State government and economic wealth.

As the eleventh hour (struck at Armistice, 1918) closes in 1933–1934, preachers, publicists and Press have gone forth to cry "Give us oil" — *The Way to God, Whither Britain?* — in discussion of religious, philosophic and scientific problems, the survival after death, propaganda for world peace, national and economic conditions being adjusted to some new and uncertain situation.

Many trust in the lamps—social standing, Church membership, Biblical knowledge, philanthropic and public-spirited service.

Christ emphasised the intention behind all such as "*the oil*"—an immanent "animus," "born again" (*i.e.* freed from mere morality as a code of social rules), which originates and brings forth an instinctive capacity for simple, direct thought and right action, which flares out as "light."

Dibelius writes of England (see pp. 348):

> The Church has nothing for the rising generation, everything Puritanic is dismissed as ridiculous hypocrisy. For the first

time in history, it would seem as though England has broken with Christianity.

England itself is the scene of a struggle for mastery between profiteering Capitalism and the crude egotism of the working classes. . . . What England is now experiencing is a grave crisis in which a new state and a new type of society are coming to birth.

The Puritan spirit of Britain harnessed to material achievement had erected a casing stone of Christian civilisation which refracted (or diverted) the Light of the Sun of Righteousness.

Thus has followed the intuitive rejection by "the rising generation" of formalism and dogma in Church service, the irritation at a ponderous mass of legislation, unintelligible and lengthy in procedure, and the dissatisfaction in political catchwords.

At Midnight, a cry has been heard—the Bridegroom comes to His Theocentric State,[1] and this generation of Britain searches for oil wherewith to light it.

"*And while they went to buy, the bridegroom came . . . and the door was shut.*"

"*Watch, therefore, for ye know neither the day nor the hour wherein the Son of Man cometh.*"

The parable of the marriage supper [2] is given in Matt. xxii, 1–14.

At the wedding feast, which was postponed, there was one who had not a wedding garment and was cast out—stressing the garment of His righteousness, "born again," and attuned to Harmony, as essential to each individual Freewill.

"*For many are called, but few are chosen*"—to many the Kingdom offered, but few choose to attune Freewill.

Vide 2 Esdras viii and ix:

"*The Most High hath made this world for many, but the world to come for few.*

"*There be many created, but few shall be saved.*

"*For as the husbandman soweth much seed upon the*

[1] For which three premises are given, pp. 363–369.
[2] For interpretation see J. J. Morey, *The Parables of the Kingdom*, pp. 207–211.

ground, yet not all that is sown shall come up in due season, even so they that are sown in the world shall not all be saved. . . . Thou comest far short that shouldest be able to love my creature more than I. . . . Therefore ask thou no more questions concerning the multitude of them that perish.

"*For when they had received liberty, they despised the Most High, forsook His ways, and said in their hearts, There is no God. For the Most High willed not that men should come to nought; but they which be created have themselves defiled the name of Him that made them, and were unthankful unto Him which prepared life for them.*

"*For there was a time in the world, even then when I was preparing for them that now live, before the world was made for them to dwell in. . . . So I considered my world, and lo, it was destroyed,*[1] *and my earth, and lo, it was in peril, because of the devices that were come into it. And I saw and spared them and saved me a grape out of a cluster, and a plant out of a great forest. . . . Let my grape be saved, and my plant: for with great labour have I made them perfect.*"

¶ 2. Matt. xxv, 14–30, gives the parable of the talents:

"*The Kingdom of Heaven is as a man travelling into a far country . . . after a long time the lord of those servants cometh and reckoneth with them.*"

Indicative of His passing from the earth, and of His return, not immediately as His disciples expected, but at a distant future, given in xxv, 31:

"*When the Son of man shall come in His glory, and all the holy angels with Him, then shall He sit upon the throne of His glory: and before Him shall be gathered all nations.*"

Thus fulfilling Luke i, 32, 33:

"*And the Lord God shall give unto Him the throne of His father David, and He shall reign over the House of Jacob for ever, and of His Kingdom there shall be no end.*"

Thus at His Second Coming, having dealt with Israel-Britain and associated nations under Christian rule, in the parable of the sheep and the goats (Matt. xxv, 31–46),

---

[1] Freewill in deviation bringing its own destruction.

He turns to the judgment of the nations of the world according to their action in history under Cain-Bel influence in the Babylonian succession of kingdoms, their resultant ideals and purposes—the Totalitarian State that now prepares war.

To the sheep—the name applied to Israel through all time, now used to these other nations who accept His ideal:

"*Come ye, inherit the kingdom prepared for you from the foundation of the world: for I was an hungred, and ye gave me meat; I was thirsty, and ye gave me drink; naked, and ye clothed me.*

"*Verily I say unto you, Inasmuch as ye have done it unto one of the least of these my brethren, ye have done it unto me.*"

Compare Deut. xxviii, 48—to Israel; in exile as the punishment of "seven times"—

"*Therefore shalt thou serve thine enemies which the Lord shall send against thee, in hunger, and in thirst, and in nakedness, and in want of all things, and he shall put a yoke of iron* [the Babylon economic and religious system] *upon thy neck.*"

Thus is envisaged the reaction of the Babylon kingdoms to the migrations of Israel through the midst of the races of Europe—which of the Divine commandments to Israel that these have accepted, which attacked; their reaction to His life and teaching on earth.

In the course of His-Story, civilisations rise and fall.

So it is said that the rise of the Anglo-Saxon will be followed by its fall.

In that do the Totalitarian States imagine a fallacy.[1]

They will be included in His Kingdom as "the sheep"—or will as "the goats" be destroyed.

There is yet time for the Totalitarian State to avoid "*the battle of that great day of God Almighty,*" by renouncing Cain-Bel, and to receive this His decision. Individuals may, but a race under Cain-Bel influence will not. Even

---

[1] "*The stone that smote the image became a great mountain and filled the whole earth . . . a kingdom . . . and it shall stand for ever*" (Dan. ii, 35, 44).

THE MEANING OF THE KING'S CHAMBER 377

if Stalin accepted Christ's teaching and pursued it in government, or if Hitler listened to the voice of His Church in Germany, either he would be "crucified" by his own people in Moscow or in Munich—a glorious Fate wherein to tread if but the freewill to choose it—or he would flee the country. With his disappearance, another deviating freewill would arise, to be given power by Cain-Bel, and elected to carry on what is foreshadowed.

¶ 3. But in these modern times, it will be said, surely His Second Coming is unbelievable.

From the early days of the Church, men expected it: the expectations have not been realised, so in these times there is a general disbelief in any such event—purely visionary; and if ever fulfilled then it was at Pentecost.

The Second Coming is "a delusive hope founded on a primitive fancy that has never so far been realised,"[1] and therefore never will be.

"*There shall come in the last days scoffers . . . saying, Where is the promise of His Coming? for since the fathers fell asleep, all things continue as they were from the beginning of the Creation*" (2 Peter iii, 3, 4).

"*For as a snare shall it come on all them that dwell on the face of the whole earth*" (Luke xxi, 35).

So, His Second Coming scientifically impossible and politically unthinkable?

Then so was His Incarnation in 4 B.C.

If you believe His Resurrection (and such physical explanation man's intelligence has not reached, maybe beyond its reach, so, stressing Faith, p. 247), then His Second Coming is no more difficult of belief—putting aside its political import.

Future and past, plus and minus, make and break, exit and entrance, are distinguished merely by a difference of sign; scientifically there is no essential difference between them.

And as to its political import—man's proud civilisation is in Chaos, with another war impending.

[1] The words of a British bishop.

> The universal expectation . . . exists of the coming of one who should put an end to the present state of things and establish a reign of universal peace and righteousness, even among Hindus, Buddhists and Moslems (Canon Horsefield, *Return of the King*).

Science and religion have not yet even begun to consider the effect of the action of Him "*by Whom all things were made, and all things consist,*" in the heart of the earth (words given as symbols).

At His Second Coming, the new force then created becomes manifest: with "*the restitution of all things*" and as "*the consummation of the age,*" a new and visible spirit dominion over matter, granted to the first Adam and lost, is restored by the Second Adam.

This has been put down in words, and talked out by argument bred of Cain-Bel.

But, such result being foreknown, this has also been put down in stone, in mathematical language in the Pyramid symbology of the King's Chamber, wherein it is more difficult to quibble.

Having come on to the horizontal plane of the restitution from August 1909, with the scale increased to the inch-month for more detailed interpretation, we are now stooping through the second low passage which ends in September 1936, and opens into the main chamber, the intentional destination of the internal passage symbology.

The entrance to the King's Chamber is dated 16th September 1936, and its end (at the south wall), 20th August 1953.

The importance of the King's Chamber is given by the architecture of the Chambers of Construction which form its protective roof. These consist of five rows of granite beams placed right across the Chamber, the largest being 27 × 7 ft. high × 5 ft. wide, weighing seventy tons. The topmost beams are sloped at an angle, like a house roof; the whole thus arranged to take the weight of the stone superstructure above, and protect the King's Chamber from the effects of subsidence and earthquake.

The measurements of the King's Chamber are: breadth

THE MEANING OF THE KING'S CHAMBER

(N.S.), 206·066″, length (E.W.), 412·132″, height (floor-ceiling), 230·38″.

And within it, placed towards the western end, is a stone coffer—hollowed out of a single block of granite by drilling and chiselling[1]—its cubic capacity being exactly four British quarters of wheat, and its measurements—height, 41·18″, length, 89·79″, breadth, 38·65″ (Piazzi Smyth).

(i) $\dfrac{\text{Length + height of King's Chamber}}{\text{its breadth}} = \pi = \dfrac{\text{length + breadth of Coffer}}{\text{its height}}$.

(ii) Area of Circle with diameter 412·132″ (length of K.C.) = area of a square with side of 365·242″.

(iii) Breadth of K.C. (206·066″) × square root of π (1·7742) = 365·242″.

(iv) Twice the length of the King's Chamber (412·132″) set off along the floor of the Grand Gallery (inclined at an angle of 26° 18′ 10″) will give in that distance a vertical rise of 365·242″.

So far the historical plane of the internal passages has been displaced 286·1 P″ negatively (*i.e.* to the east or left-hand) from the central vertical plane of the Pyramid. Not until he has entered the King's Chamber (dated 16th September 1936) can man turn positively (*i.e.* to the right-hand).

Now instead of stooping, as in the low passages, he can walk upright down its length to the west towards the central plane of the Pyramid.

And when his steps have led him "positively" exactly 286·1 P″ (the displacement factor), so that he stands in the central vertical plane, underneath the missing Apex Stone, his feet touch an open, lidless and empty coffer, down into which he looks.

Above is "*the Stone which the builders rejected, the chief cornerstone*"; in front and looking into, an empty Tomb —symbol of "*The Resurrection.*"

Thus, "*in these last days,*" Man's displacement in Creation is to be restituted.

---

[1] How was such done in 2600 B.C.; and how seventy-ton stone placed in position? Certainly by means known to a then civilisation and lost for centuries; also the mathematics and astronomy (built into stone and symbolising an intent) unknown to Pythagoras and Aristotle, the Greeks and Romans. Even after the Great War, one says, there has been no "fall of man"!

"And He shall bring forth the headstone thereof with shoutings, crying, Grace, grace unto it" (Zech. iv, 7).

The symbolism of the King's Chamber shows that between its datings is the process of restitution (Acts iii, 21): by Divine hand, it is to be a period of reconstruction for Britain, who will recognise her origin as Israel, when all Israel and Judah are reunited under her King of the line of David, and the whole world also will realise it. The Anglo-Saxon race is to be under Divine rule, so that after 1953 it will be reprepared to carry this government to the rest of the world in preparation for the Millennial era.

The symbolism of the Chambers of Construction shows that Britain-Israel is under Divine protection during drastic world changes.

"Thus saith the Lord God, I will yet for this be enquired of by the House of Israel to do it for them" (Ezek. xxxvi, 37).

"Watch ye therefore that ye may be accounted worthy to escape all these things that shall come to pass, and to stand before the Son of Man" (Luke xxi, 36).

The time of the Second Coming is hidden,[1] but the era has been defined—A.D. 1936–2001.

"But ye, brethren, are not in darkness, that that day should overtake you as a thief" (1 Thess. v, 4).

Nor can it be known how His actual presence is to be made manifest; but maybe premonitory to man before actual seeing (Rev. i, 7) will be an event (in the 1936–1953 period), during the coming crisis, which will prove to all beyond any manner of doubt this personal action in intervention, so definite that—

"Wherefore, if they shall say unto you, Behold, he is in the desert, go not forth; behold he is in the secret chambers, believe it not.

"For as the lightning cometh out of the east, and shineth even unto the west, so shall also the coming of the Son of Man be" (Matt. xxiv, 26, 27).

---

[1] *"But of that day and that hour knoweth no man, no, not the angels which are in heaven, neither the Son, but the Father"* (Mark xiii, 32).

The warning of the era is given by the House of Seth in a message embodied in Stone and in physical laws to the Anglo-Saxon race (pp. 100, 320)—to watch the King's Chamber period.

THE THEOCENTRIC STATE

The symbology of the Bible and of the Pyramid coincides. The "wedding" in the parable of the ten virgins is paralleled by the "mitring" of the rejected Apex Stone to the Perfect Pyramid as designed. Each interpretation leads to the Divine Purpose for our race to become the Theocentric State, the Stone Kingdom, the nucleus of His Kingdom on earth.

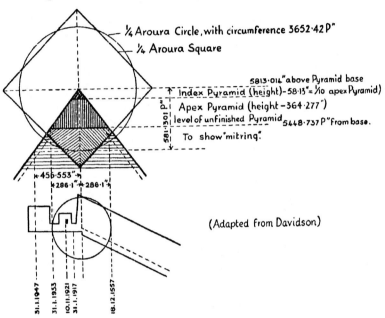

(Adapted from Davidson)

In the diagram,[1] the date 18th December 1557 signifies the rise of Israel in British sea-power, and during the next 286·1 years, when the 2520 years of punishment elapsed, the Stone Kingdom emerged as "a great mountain and filled the whole earth" to break in pieces and consume all these kingdoms.

So the date 31st January 1933 signifies "the displacement of material power, the end of the need for the punitive exercise of naval power," and the germination of a new

[1] In the Great Seal of the U.S.A. is represented a pyramid unfinished, awaiting its apex stone to be mitred thereon. To the U.S.A., Davidson shows, the dates 31st January 1917, 10th November 1921, 31st January 1947 are especially applicable (*The Great Pyramid's Prophecy concerning the British Empire and America*, pp. 28-31).

382 THE PILLAR IN THE WILDERNESS

idea of spiritual power, the founding of the new "body politic" (the State and Church of Britain and the U.S.A.) as the Theocentric [1] State.

And 286 days after (13th November 1933), when midnight had struck, such conception is seen intensified in the turmoil awakened to find "the Way to God" and to ask "Whither Britain?" And in her groping towards Peace and Goodwill among men—in her example of disarmament to the world "to the edge of risk," Britain again erects further false casing stones, trusting not in Divine guidance, but in a man-reasoned League of Nations, who "have not called upon the name of God" (Ps. liii, 4, with context of whole Psalm), but would welcome to its fold any wolf in sheep's clothing, an Antigod Soviet, a neo-pagan Germany, or a revived Rome.

Now Davidson [2] from Pyramid chronology actually dates the founding of the Theocentric State.

The horizontal line at the ceiling of the Chamber of Second Birth [3] (the Queen's Chamber)—with the symbolism of the year-circle and Second Birth influence—continued to the floor of the Grand Gallery defines the date, A.D. 385.

With such as central epoch, the displacement factor 286·1 P" as circle defines A.D. 98 "as the end of the Apostolic Age, the gradual passing of the Early Church of the Spirit, and its displacement by formal Christianity"[2]; and A.D. 671 "as the date at which Church and State jointly began to function in England as the Body Politic of the building race, or as 'the Two Witnesses' of Revelation xi."[2]

"Their period of 'witnessing' thus mystically clothed in the sackcloth of law and ritual (Rev. xi, 3) is defined as a period of 1260 solar years, beginning at 23rd August A.D. 671 and ending at 23rd August A.D. 1931."[2]

---

[1] The word "theocentric" is used where Davidson uses "theocratic." Theocratic meaning literally the rule of God, and so the Kingdom of Heaven on earth, yet is associated in these times with the idea of Church rule and the Papacy; theocentric conforms with the concentric way of thought (see Index), and with the idea of the year-circle, the Sun of Righteousness.

[2] *The Great Pyramid's Prophecy concerning the British Empire and America*, pp. 37, 38, 41-43.

[3] Its floor-level defines the birth of Christ at 4th October (Greg.) 4 B.C.

THE THEOCENTRIC STATE

The ensuing period may be more clearly shown as a table:

23.8.1931	286·1 days as displacement period.	The passing of State and Church from reliance upon human law and civil procedure, from formalism and dogma.
4.6.1932	*"three days and a half"* (Rev. xi, 9, 11).	The "mystical" death of the Body Politic: in the economic bondage of "mystical Egypt" (the Babylon religious *cum* economic world system): on 31.1.1933, the germination of the new idea of spiritual power in place of material power.
5.12.1935[1]	286·1 days as displacement period.	"Rebirth" of government in State and Church on the spiritual basis of the Theocentric State.
16.9.1936		

The significance of this last period lies in this—for the dates mark the effect of the General Election, and the new Parliament by 16th September 1936 will meet a Crisis, arising out of the Italo-Abyssinian War, in which some question will force the issue of the Theocentric State into the Lobby—"Ayes" or "Noes." On the M.P. of your election choice will depend the decision, emphasising personality (the freewill to choose) and cutting across the party Whip . . . (*"five were wise and five foolish"*).

By 16th September 1936 (the entrance date of the King's Chamber[2]), the regenerated State and Church of Britain have passed through their own tuitional Exodus, and have emerged therefrom, prepared as Moses and Aaron, to lead the building race, under Divine protection, guidance and tuition, into the unknown economic wilderness, that lies between the bondage

---

[1] Within a week of this date: the new Parliament sits (3rd December); the Naval Conference of Israel ("between each no threat of war"), France and Italy (9th December)—Manasseh, united and allied, will be required on the eastern flank; the League halts at the peace plan and oil sanction (12th December).

[2] *"And when these things begin to come to pass, then look up and lift up your heads, for your redemption draweth nigh"* (Luke xxi, 28).

of the world system [1] (mystical Egypt) and the promised land of the Kingdom of God (mystical Canaan).

Joined by the British Empire and the U.S.A. by 31st January 1947, the mass of the race is prepared and instructed to come under the Headship of the Body of Christ, and to await His Visible Presence.

On 20th August 1953, the Covenant race crosses the mystical Jordan—offering itself as individual "stones" rather than as individual "builders" to enable the "Master Builder" to complete the finished structure to the fulness of the Design for the Perfect Structure [2]—into the promised land of the Kingdom of God on earth, its true inheritance, the spiritual dominion of the earth.

The governmental centre of this inheritance is an actual New Jerusalem in the actual Palestine, at which centre after the site has been cleansed from the pollution of former occupations (Dan. viii), between A.D. 1953 and 2001, the true Sanctuary of God on earth is to be built.[3]

¶ 4. As the "seven times" punishment (2520 years) of the Mosaic Covenant finally closes in 1936, the Covenant race emerges into the "wilderness" again under Divine guidance to be prepared for a new Covenant (Jer. xxxi, 31–34; Hebrews, viii).

The King's Chamber (dated 16th September 1936 to 20th August 1953), the destination of the Pyramid internal system, symbolises the Chamber of Divine Assessment and Judgment of the Adamic or White Race. Such parallels the "consummation of the age" in Divine Purpose as to the Freewill of man, whether still (1) in Harmony, or (2) in Deviation.

Thus the references for this period run on parallel lines —(1) as being under Divine protection (symbolised by the Chambers of Construction roofing the King's Chamber), and (2) the round-up of the powers of Evil (symbolised by the Subterranean Chamber and dead-end passage).

(1) The symbolism of protection shows the "halt" of

---

[1] "*Come out of her, My people, that ye be not partakers of her sins, and that ye receive not of her plagues*" (Rev. xviii, 4). See pp. 342, 352.
[2] The mitring of the casing stones into the Apex Stone completed: "*the Bride hath made herself ready*" (Rev. xix, 7).
[3] *The Great Pyramid's Prophecy*, pp. 6, 38, abbreviated.

THE THEOCENTRIC STATE

Revelation vii, 1–7 (between the sixth and seventh seals), as after 16th September 1936 and before the seventh vial is poured out (Rev. xvi, 17).

"Hurt not the earth till we have sealed the servants of our God . . . and I heard the number which were sealed, an hundred and forty and four thousand of all the tribes of the children of Israel."

The number is symbolic (as all Biblical numbers), not actual, of

- 12 the twelve tribes, as the *"dry bones"* of Ezekiel that lived again as the Celts-Anglo-Saxons.
- ×12—as completion, 4×3 meaning the earth in union with Trinity—Christ's Millennial Kingdom.
- ×10—as the Virgins, His earthly Church-State without
- ×10   the Bridegroom-King.
- ×10—being repeated 3 times, as restored to Harmony with the Divine Will.

—thus meaning the individuals whose Freewill is restored by Faith to harmony with Divine Will, in the united Nation, State and Church of all Israel (Britain, the Empire and U.S.A.), prepared to become the nucleus of His Kingdom on Earth, sealed as *"the servants of our God,"* thus to bring about in the world,—

"Lo, a great multitude which no man could number, of all nations, and kindreds, and people, and tongues stood before the throne, and before the Lamb" (Rev. vii, 9).

The emphasis is on the responsibility of the individual components; in particular addressed in Christ's warnings of Matt. xxiv, 27, 36–44; Mark xiii, 32–37; Luke xxi, 34–36.

"Watch, lest coming suddenly He find you sleeping."

"For they are not all Israel, which are of Israel" (Rom. ix, 6).

"Lord, Lord, open unto us. . . . We have eaten and drunken in thy presence, and thou hast taught in our streets. . . . Verily, I say unto you, I know you not . . . depart from me all ye workers of iniquity." [1]

From the context, during the "halt" and extending to

---
[1] Matt. xxv, 11, 12; with Luke xiii, 25–28.

the time of His Second Coming, there is placed as the "escape from all these things that shall come to pass," [1]—

"*Then shall two be in a field; the one shall be taken and the other left*" (Matt. xxiv, 40).

The signs of the imminence of His Second Coming are to appear immediately after the Tribulation (Matt. xxiv, 29). Thus for these signs the injunction is given:

"*Watch, therefore . . . for in such an hour as ye think not the Son of Man cometh*" (Matt. xxiv, 42-44).

(2) The Tribulation, due to the present conditions on the earth of the "second woe" (Rev. xi, 14), extends into the King's Chamber period as the dawn of the "great day of God Almighty," and its termination seems to be indicated in the Subterranean Chamber as 29th November 1939.

The work of "rounding up" the powers of Evil is symbolised as a process of Divine Diversion in continuous operation from 12th March 1913 to 29th November 1939—the Subterranean Chamber period. . . . The symbolism of Divine Diversion indicates that by 20th August 1953 (King's Chamber ending) the powers of Evil in the world will have been diverted into the channels of God's purpose, harnessed, by the "shock-absorption" of the impacts of Evil, by exhausting thereby the energy of Evil, and by finally eliminating the Will to Evil. The direction of Diversion is towards the West . . . for the purpose of bringing humanity into the path of the going down of "the Sun of Righteousness."

Such elements of evil as escape the "round-up" towards the mystical West are symbolised as proceeding Southwards along the Dead-End Passage . . . therein to be overtaken and eliminated by Jesus Christ as the Index Pyramid." [2]

Thus, the seventh angel, of Revelation viii, 6; x, 1–7, who utters the mystical words, "*that there should be time*

---

[1] Such "escape" may have been seen as happening to our glorious dead in the War. Maybe it is operating now—as recently in a Dorset lane between two fields, when two were taken to hospital, and one was taken and the other left; one, freed from the Babylon concept of honours and wealth, whose quickness of eye, that had saved him in tighter corners, deliberately failed for a fraction of a second, and whose last conscious thought as the crash came must have been to endanger his own life rather than the other innocent party on a cycle. (And I begged T. E. Shaw to get himself safe to the Transjordan in 1936.)

[2] Davidson, *Ibid.*, p. 38.

no longer," heralds the closing of the Subterranean Chamber and "the dawn of the age in which time should cease to witness to the ruthless exploitation of the commodities of air, land and sea by man" (*cf.* Rev. x, 6)."[1]

"*And the seventh angel poured out his vial . . . and there came a great voice out of the temple of heaven, from the throne, saying, It is done.*

"*And there were voices, and thunders, and lightnings; and there was a great earthquake, such as was not since men were upon the earth, so mighty an earthquake, and so great.*

"*And the cities of the nations fell, and great Babylon came in remembrance before God, to give unto her the cup of the wine of the fierceness of His wrath.*

"*And there fell upon man a great hail out of heaven, every stone about the weight of a talent*" (Rev. xvi, 17–21).

Such ushers in the "third woe" (Rev. xi, 14), to begin the final drive of the great "round-up" of the powers of Evil—to the mystical West, or southwards into the dead-end passage to annihilation.

Before 20th August 1953, all Israel (the British Commonwealth of Nations with the U.S.A., the Jews and the Scandinavian countries) having been united into the Theocentric State, which under a new Covenant founds the new economic world system and establishes the law of righteousness, the division of the Gentile branches of the Adamic or White Race falls (in the parable of the sheep and the goats) as each by its choice elects to join this Kingdom, or still chooses deviation, thus to partake in the resulting "wipe-out" of the old 666 or Babylon system.

And when the seventh angel had sounded, "*there were great voices in heaven, saying, The Kingdoms of this world are become the Kingdoms of Our Lord, and of His Christ; and He shall reign for ever and ever*" (Rev. xi, 15).

¶ 5. Before considering the seventh seal, let us return to trace events, as "*the things that are still to come to pass*" in the "*final tribulation*" of the second low passage, denoting these present times.

[1] *Ibid.*, p. 19.

Chapter XI has given Israel-Britain's part, and the modern tendency of the remnant of Babylon nations as the Totalitarian States—

"*Three unclean spirits . . . which go forth unto the kings of the earth and of the whole world, to gather them to the battle of that great day of God Almighty*" (Rev. xvi, 13, 14).

In these days are seen these three Continental nations, with 2,000,000–3,000,000 men standing armed, with munitions piled up, ready to war at their opportune moment (*i.e.* presumably when their national pride, called "*heroic destiny*," gets upset).

Also there seems a mask over the "*Abmachungen*" (the military co-operation of the Soviet and Germany) since Hitler's régime—perhaps Goering knows what became of it.

Can Britain, seeking righteousness in international dealings and herself disarmed to ensue Peace, force the League of Nations to take concerted action on any such aggression? Diplomacy covers up the act of aggression until months later after several committees have sat; and in a dissonant League some will diplomatically withdraw from action in fear that they themselves will be "aggressed."

"Nothing will stop War save a Second Advent of Christ." [1]

As the old spirit [2] of Caradoc faced the whole might of the Roman Empire, so now Britain—hating war and refusing to build up excess of armament and gas-mask and waste of uniform—with a new Spirit and a new Faith in her "*five loaves and two fishes*" as sufficient for Spirit to act through matter to accomplish His purpose—will oppose the "*heroic destiny*" of aggression which seeks to bind universal Freewill to a false kingdom, the Totalitarian State.

To quell her spirit, they tried then the smell of elephants, yesterday poison gas, and to-morrow it will be some other frightfulness.

As then the fear of the Coritani viewing the material

---

[1] Sir Ian Hamilton on "War," in *Ency. Brit.*, 13th ed., vol. xxxi.
[2] "Trâ Môr, Trâ Brython" (As long as there are seas, so long will there be Britons), their battle-cry; see p. 288.

might of Rome gave birth to "*Safety First*," and hampered in council and action both Caswallon's and Caradoc's resistance; so now, their direct descendants, the "Conchie" and the maudlin Pacifist,[1] would sacrifice an unprepared and amateur army of Britain's best.

Thus they seek to save their own life; and thus, having survived but never having known action nor suffered sacrifice, they seek by talk to degenerate the spirit of the Israel race in Britain's youth.

(Mr Rudyard Kipling put it extremely well in his Jubilee Day speech to the Royal Society of St George.)

Not to these is this inheritance of Britain.

"*Blessed are the debonnaire, for they shall inherit the earth.*"

So it is, and so the history of Britain proves by the men who made her Empire.

And the prestige of Britain was never higher than when she risked war for "*a scrap of paper*," because she believed such righteous, and chose not the path of political expediency in the accumulation of wealth out of a world in misery (for which in the scale of real values the U.S.A. now suffers).

And Britain's prestige declines when she explores avenues

---

[1] From them, unattuned to Harmony, preaching class war, and disarmament for ten years with war in the eleventh, we shall soon be hearing the voice of Moscow to hamper our councils—

"Now is the time to defend Soviet Russia by defending Abyssinia" (Citrine, at the Trade Union Congress, 6th September 1935, supporting the force of sanctions, man-made, against Italy).

"We are willing to share the territories under our flag, and the markets we control, with the rest of the world" (Lansbury).

Then fear of material might handed Thanet over to Cæsar, now lack of trust in spiritual power would surrender some part of His Theocentric Kingdom to the rule of unrighteousness.

The raw materials and natural resources of our Empire, Divinely given to us in trust, belong to His Kingdom, and will be willingly shared with any nation which is attracted to, joins, and abides in the free, religious, economic and monetary union of His Theocentric State—with mutual service between each, without thought of war or of rebellion to His Rule of Righteousness.

These will not be shared with (but under certain conditions may be distributed to) any nation whose Freewill is bound to the State requirements of a dictator—not even at the dictate of a League of (666-ruled) Nations. And in the defence of our kingdom we are prepared to die (armed preferably than disarmed, please, pacifists), as always, until His Coming, for which all Brython seeks to prepare, and humbly awaits.

to accommodate the unrighteous, wherewith to seek to solve present trouble and thus to create future discord.

If a foreign fleet engaged in a war which France has not sought, and in which she had not been the aggressor, came down the English Channel and bombarded and battered the undefended coasts of France (the French fleet being at the time in the Mediterranean), we could not stand aside. If in a crisis time like this, we run away from these obligations of honour and interest as regards the Belgian Treaty, I doubt whether whatever material force we might have at the end, it would be of very much value in face of the respect we should have lost. Let every man look into his own heart and his own feelings (Sir Edward Grey, at 3 P.M. on 3rd August 1914).

Britain is still stooping through the low passage, and will have to meet a Great Crisis in which each man will be compelled to "look into his own heart," and "racially in spontaneous prayer from the stricken spirit of every individual" inquire of God "to do it for them," before she will be able to stand upright in the King's Chamber.

¶ 6. In Chapter VII, quoting Esdras, the sequence of events are clearly prophesied—

"When thou seest a certain part of the signs are past . . . then . . . the Most High will visit the world which was made by him. And when there shall be seen in the world earthquakes, disquietude of peoples, devices of nations, wavering of leaders, the beginnings are manifest in wonders and mighty works, and the end in effects and signs" (2 Esdras ix).

2 Esdras xi and xii give the vision of the eagle and lion, and its interpretation, confirming the visions of Daniel.

2 Esdras xiii continues in a further vision:

"Whereas thou sawest a man coming up from the midst of the sea, the same is he whom the Most High hath kept a great season, which by his own self shall deliver his creature, and he shall order them that are left behind.

"Behold, the days come, when the Most High will begin to deliver them that are upon the earth. And there shall be astonishment of mind upon them . . . and one shall think to make war against another, place against place, people against people, kingdom against kingdom.

"And it shall be when these things shall come to pass, and the signs shall happen which I shewed thee before, then shall My Son be revealed, whom thou sawest as a man ascending.

"And it shall be when all the nations hear his voice, every man shall leave his own land and the battle they have one against another. And an innumerable multitude shall be gathered together, desiring to come and fight against him.

"But he shall stand upon the top of Mount Sion. And Sion shall come and be shewed to all men, being prepared and builded, like as thou sawest the mountain graven without hands. And this My Son shall rebuke the nations. . . .

"And whereas thou sawest that he gathered unto him another multitude that was peaceable; these are the ten tribes, which were led away out of their own land in the time of Osea the King, whom Shalmanasar the King of the Assyrians led away captive."

(Going on to say that these tribes escaped from over the Euphrates and came to Arsareth—the region of the Danube—and identifies them again as)

". . . the multitude gathered together with peace. Those that be left behind of thy people are they that are found within my holy border.[1] *It shall be therefore when he shall destroy the multitude of the nations that are gathered together, he shall defend the people that remain; and then shall he shew them very many wonders."*

In these days the difference between the two "*multitudes*" is seen in the clash of two distinct and separate views of this world.

On the one side:

The general economic state of the world foreshadows a long period of heavy economic convulsions and acute crises, tending towards further proletarianism, pauperisation, and consequent development of class war.

The proletariat will not consent to any reduction in its standard of living, and this will lead to strikes and disorders, which in their turn will render economic reconstruction impossible.

---

[1] Thus to-day, as Israel-Britain, working for peace.

These problems are further complicated by the conflict for world domination now ripening. The late war was therefore merely a prelude to what will indeed be a World War, the war for single-power domination (*Bolshevist Politgramota—Catechism*—1921).

On the other side:

Do we realise that we alone to-day are those whom God hath chosen in the world's history as a people to whom is given incomparably the greatest trust and decision for the world's safety, that has been laid upon a great company of people? This is a trust upon the English-speaking peoples, but primarily upon Britain and America (Archbishop of Canterbury, 1927).[1]

The clash between the two outlooks, the Continental and Anglo-Saxon, is well brought out in a speech of Mr Baldwin.[2] (He also finds the dilapidations of the past—p. 371—are not the keys which fit "*a new heaven on earth.*")

Whatever changes the War has brought, we have not broken with our traditions. Our King is on his throne, the head of his people and their servant, and in a few weeks the whole nation will join in thanking God for his Majesty's reign and for the noble example of duty that he has set to all of us.

We are not a people who will take with any pride or enthusiasm to wearing gas-masks. We regard them for what they are—a monstrous and tragic necessity born of the prostitution of science to the service of barbarism.

The nations are not walking in the way of peace, but in most dangerous roads, which may lead to war. The Great Powers have a long history, and what happened in 1914 has not essentially changed their character.

We fought the War to make the world safe for democracy. At the end of the War many in this country and in others embraced those doctrines of nationality, self-determination and democracy as the keys of a new heaven on earth. It has been found that the keys do not fit, that Heaven is as far away as ever.

If one went through the pronouncements of the great leaders of the Totalitarian states of Germany, Russia and

---

[1] The two quotations taken from Basil Stewart, *The Witness of the Pyramid*.

[2] Report in daily paper, 9th April 1935, at Annual Assembly of the Evangelical Free Churches.

Italy he would not find that one of them conformed to the ideals which were cherished in this country of freedom of thought.

The danger of the honest-minded intellectual doctrinaire arises because his god was his intellect, and he would sacrifice every man, woman and child in the country rather than acknowledge that his theories were wrong. This was a form of terrorism that could not exist in a free country.

It could not exist and never would among our people, but it was one of the most terrible fruits of materialism and the worship of material things, and the abnegation of the rights of their God and their Creator.

Unless they would change the ideal and the purpose of a nation they would never change its practice. Where the war mind is, where fear is, where suspicion is, there will armaments be gathered together.

Now hear the preaching of Professor Ernst Bergmann, who is shaping German culture (reported July 1934):

> Everything great and supreme in modern humanity has come into existence without God and without the Church.... The national condition of international relations in the Twentieth Century will be a latent or open state of war between nations.

And Sir Austen Chamberlain, as reported 3rd May 1935, following German reintroduction of conscription in March, and her throwing off the mask from her rearmament:

> If Germany will not be a member of the European family, if instead of seeking to negotiate she intends to extort her will, she will find this country in her path again; and with this country, those great, free Commonwealths that centre around it. And she will have met a force that will once again be her master.

And can we add also "with this country" the United States?—or is Manasseh so immersed in false values—in the accumulation of gold—as to think she can stay outside the Kingdom?

Then a united Israel will stand in the path of Cain-Bel if it wills war and not peace, "in the battle of that great day," as she was intended to do 3420 years ago.[1]

[1] See p. 128.

The delay, having caused all nations to drink of *"the golden cup of Babylon,"* now drunken with her idea of the Totalitarian State and the accumulation of armaments for destruction, will necessitate—

"I will yet for this be enquired of by the House of Israel to do it for them" (Ezek. xxxvi, 37).

¶ 7. What say the prophets of "that day," "the battle of that great day of God Almighty"? And therein to search the immediate future, as prophecy becomes fulfilled in history [1]—

"Go, and proclaim these words toward the North, and say, Return, thou backsliding Israel; only acknowledge thine iniquities. [2]

In those days the House of Judah shall walk to the House of Israel,[3] *and they shall come together out of the land of the North to the land that I have given for an inheritance unto your fathers"* (Jer. iii, 12, 13, 18).

"Behold, we come unto Thee, for Thou art the Lord our God; for we have sinned against the Lord our God, we and our fathers, from our youth even unto this day, and have not obeyed the voice of the Lord our God" [4] (iii, 22, 25).

"Lo, the days come, that I will bring again the captivity of My people Israel and Judah, saith the Lord; and I will cause them to return to the land that I gave to their fathers, and they shall possess it" [5] (xxx, 3).

"Alas, for that day is great; it is even the time of Jacob's trouble, but he shall be saved out of it.

"For it shall come to pass in that day, saith the Lord,

---

[1] Morgan, *St Paul in Britain*, p. 67: "The Druidic idea of prophecy . . . is nothing but the theological term for science." And, quoting *Principles of Prediction of Gildas the Prophet*, Iolo MSS., p. 609: "He that would be a prophet of God must never rest till he has traced everything to its cause and mode of operation. He will then know what God does, for God does nothing but what should be, in the manner it should be, at the time and in the order it should be. By understanding these laws of God he will be able to see and foretell the future."

[2] To Britain, "in the north" from Palestine.

[3] From 1918–1932 four million Jews have "walked to" Israel's countries, especially into Palestine; and since then Hitler is forcing across all German Jews.

[4] *Cf.* our English Prayer Book.

[5] They now possess Palestine.

"THAT GREAT DAY OF GOD ALMIGHTY" 395

that I will break his yoke from off thy neck,[1] *and strangers shall no more serve themselves of him.*

"But they shall serve the Lord their God, and David their King, whom I will raise up unto them.[2]

"For I am with thee to serve thee: though I make a full end of all nations whither I have scattered thee, yet will I not make a full end of thee, but I will correct thee in measure, and will not leave thee altogether unpunished" (xxx, 7-11).

"And ye shall be My people, and I will be your God. In the latter days,[3] *ye shall consider it"* (xxx, 22, 25).

"Behold, the days come, that I will make a new Covenant with the House of Israel and the House of Judah.

"After those days,[4] *I will put My Law in their inward parts and write it in their hearts; and will be their God, and they shall be My people. For they shall know Me"* (xxxi, 31-34).

"In those days, and in that time, the children of Israel shall come, and the children of Judah together, and seek the Lord their God, saying, Come and let us join ourselves to the Lord in a perpetual covenant that shall not be forgotten.

"My people have been lost sheep: their shepherds have caused them to go astray, they have turned them away on the mountains, they have gone from mountain to hill, they have forgotten their resting place" (Jer. l, 4 ff.).

"Thus saith the Lord, David shall never want a man to sit upon the throne of the House of Israel"[2] (xxxiii, 17).

"In that day, shall there be an altar to the Lord in the midst of the Land of Egypt, and a pillar at the border thereof to the Lord.[5]

"And it shall be for a sign and for a witness unto the Lord of Hosts in the Land of Egypt; for they shall cry unto the Lord because of the oppressors, and He shall send them

---

[1] *Cf.* Deut. xxviii, 48: *"He shall put a yoke of iron upon thy neck"* (the Babylonian system, including Roman history and the classical way of thought).

[2] And at this Jubilee he sits more secure than ever in the affections of his people in a far-flung Empire, the bond of its unity.

[3] *I.e.* in these days, "the consummation of the age."

[4] *I.e.* during the King's Chamber period.

[5] The Great Pyramid, which has stood for over forty-five centuries, is becoming recognised as such, for only during the last century has man come again to the scientific knowledge of the laws there embodied.

a Saviour, and a great one, and he shall deliver them" (Isa. xix, 19, 20).

"In that day shall there be a highway out of Egypt to Assyria.

"In that day shall Israel be the third with Egypt and Assyria, even a blessing in the midst of the land" (xix, 23, 24).

"In that day, there shall be a root of Jesse, which shall stand for an ensign of the people: to it shall the Gentiles seek.

"And it shall come to pass in that day, that the Lord shall set His hand <u>again the second time to recover the remnant of His people, which shall be left from Assyria, and from the islands of the sea</u>.[1]

"He shall assemble the outcasts of Israel, and gather together the dispersed of Judah <u>from the four corners of the earth</u>.

"The envy of Ephraim shall depart and the adversaries of Judah shall be cut off.

"They shall lay their hand upon Edom.

"And the Lord shall utterly destroy the tongue of the Egyptian sea; and with His mighty wind shall He shake His hand over the river, and shall smite it in the seven streams,[2] *and make men go over dryshod.*

"And there shall be an highway for the remnant of His people which shall be left from Assyria" (xi, 10-16).

Thus on the one side, Israel, and on the other, the Babylonian succession of kingdoms which to-day prepare war *"to the battle of that great day of God Almighty"* (Rev. xvi, 14)—as prophecy becomes fulfilled, identity becomes revealed.

"The burden of Babylon. Lift ye up a banner upon the high mountain[3]*: I have commanded my sanctified ones.*

"The noise of a multitude in the mountains, like as of a great people, a tumultuous noise of the kingdoms of nations gathered together.

---

[1] Thus a remnant of Israel did escape from the Assyrian Captivity, and are found in the British Isles.

[2] The Nile delta's seven streams have dried to two to-day (the Rosetta and the Damietta), and these will dry up also *"in that day."*

[3] Mountain, symbolic of nation.

"THAT GREAT DAY OF GOD ALMIGHTY"

"*Howl ye, for the day of the Lord is at hand, it shall come as a destruction from the Almighty. Therefore shall all hands be faint, and every man's heart shall melt.*

"*Behold, the day of the Lord cometh, and He shall destroy the sinners thereof out of it.*

"*For the stars of heaven and the constellations thereof shall not give their light: the sun shall be darkened in his going forth, and the moon shall not cause her light to shine.*

"*And I will punish the world for their evil, and the wicked for their iniquity: and I will cause the arrogancy of the proud to cease, and will lay low the haughtiness of the terrible.*

"*Therefore will I shake the heavens, and the earth shall remove out of her place, in the wrath of the Lord of Hosts, and in the day of His fierce anger.*

"*Then shall every man turn to his own people, and flee every one into his own land.*

"*And Babylon, the glory of kingdoms, shall be as . . . Sodom and Gomorrah.*

"*Her time is near to come, and her days shall no longer be prolonged*" (Isa. xiii).

"*In that day, this proverb against the King of Babylon.*

"*How hath the oppressor ceased! the golden city ceased! The whole earth is at rest and is quiet.*

"*How art thou fallen from heaven, O Lucifer,*[1] *which didst weaken the nations!*

"*For thou hast said in thine heart, I will ascend into heaven, I will exalt my throne above the stars of God; I will sit in the sides of the North, I will ascend above the heights of the clouds, I will be like the Most High.*

"*Is this the man that made the earth to tremble, that didst shake kingdoms? That made the world as a wilderness, and opened not the houses of his prisoners?*

"*I will rise against them and cut off from Babylon the name and the remnant.*

"*I will break the Assyrian in my land and upon my holy*

---

[1] As originally deviated freewill—*i.e.* evil, Bel. And the words which follow are to-day the exact words repeated in the Antigod propaganda of the Soviet.

mountain: *then shall his yoke depart from off them, and his burden depart from off their shoulders.*[1]

"*Howl, thou whole Palestine, for there shall come from the North a smoke.*[2]

"*This is the purpose that is purposed upon the whole earth: and this is the hand that is stretched out upon all nations.*

"*For the Lord of Hosts hath purposed, and who shall disannul it?*

"*His hand is stretched out, and who shall turn it back?*" (Isa. xiv).

¶ 8. Are these the words of the local Jahwe of a wandering bedouin tribe, of some lesser category of that polytheistic galaxy derived from Anu, Ea and Bel?

Or do the words not fit the God of the highbrow rationalist who derives all wisdom from Greek philosophy?

Then watch present-day history bring forth the dog-fight which is impending.

ITALY

(Ephraim—England—is warning her, from no covetous spirit but from friendship with her late ally, that if she begins despite provocation this minor affair of Abyssinia, she but lays the bait for the real hunters.)

Sept. 1926. Treaty of Friendship signed with Yemen.

1932. Similar treaty with Ibn Saud, King of the Hedjaz. Thus their colonial policy switched attention from Abyssinia to Arabia.

May 1934. Saudi-Yemen affair, resolved by a conclusive peace. Italian ideas switched again over the Red Sea. Was Britain then not silent, that she is now suspect?

Oct. 1934. The King of Italy visits his African possessions.

Nov. Mussolini blankets his Press as to service movements: and takes over the foreign credits and bonds in the hands of his nationals.

Dec. The Ual-Ual incident (? some sixty miles inside Abyssinian territory), still debated as to the aggressor.

---

[1] Cain-Bel influence and Babylon system of economics and religion. *Cf.* Deut. xxviii, 48.

[2] The threat of invasion to Palestine.

ITALIAN ACTION

Dec. By chance it was watched by a colonel of Ephraim on the Anglo-Ethiopian Boundary Commission, an impartial referee, who reported (daily paper, 19th January 1935):

"The British members of the Commission did all they could to find an equitable solution, but they were repeatedly checkmated by the intransigent and unaccommodating attitude of the Italian officer . . . remarks as 'Accept or refuse,' 'as you like,' and by the menace that in the event of refusal he would summon 'several hundred soldiers.'"

Jan. 1935. Notes and notes.

Feb. Two Italian divisions sent as reinforcements to East Africa (marvellous how the red-brown uniforms appeared so pat!).

Mar. 29 General Baistrocchi to the Senate (as reported):

"Neither we nor anyone else can tell when war will break out, but it can safely be declared that it will break out suddenly—that is, after a few days of political tension."

And Mussolini: "No event will catch us unprepared to face it," sounds good—but silly, if also the author of the event.

Apr. 800,000 Italians under arms, and 400,000 Fascist Militia.

10,000 skilled workmen besides troops and natives building roads in Eritrea.

May 8 (A Minister reported in the Italian Chamber) "it was not right that such a slave-owning country should rule over other tribes."[1]

May 24 Mussolini says: "We do not turn back, and now that we feel that decisions are being taken we burn our boats and take our destiny firmly in our hands. Better one day as a lion than a hundred years as a sheep—that is our gospel, to which we adhere against everyone everywhere."

And within a few days he has accepted the League's counsel,[2] and delayed action for the Conciliation Commission to arbitrate.

---

[1] The moral note creeping in for "right" of action: and whence this sudden consideration? Nor have the "barbarous" acts of Abyssinia on her other neighbours been mentioned by them.

[2] The quandary which has been set the League of Nations by Italy was prophetically foreseen in September 1934.

Better a thousand years as "sheep," but, like a goat that has tried and found sand too gritty for a quick swallow, the Dictator and Minister of ten portfolios has to wait on the weather.

And on 3rd October—the anniversary eve of the Messiah's Birth (see p. 97)—hostilities began.[1]

Russia and Germany watch the ground-bait drawing, before they start fishing towards the Dead Sea.

So the three Totalitarian States are seen, each watching the other and waiting for the other to unmask force, and to be involved; then themselves to enter the dog-fight when their own chance of spoil becomes easier, with or without ultimate alliance.

When the Italo-Abyssinian War is settled up, one thing is certain in this civilisation's economics. Whichever side gains, the peace treaty (as the financial clauses of Versailles Treaty) will be so drawn up that the international financiers at Basle (or is it Baal?) may enforce debt-claims and levy interest-tribute for years on both countries.

An echo is heard over the centuries,—

"*To you, it is commanded, O peoples and nations . . . that ye worship the golden image*" (Dan. iii, 4, 5).

That story had a sequel, and so has this.

For out of the *fiery furnace* thus generated, the financial chaos of Gold Standard worship, which produces poverty out of plenty, and booms out of munitions of war, there emerges unscathed the system of money-thinking, Divinely ordained at Sinai; out of the fires of war there comes forth mutual service and goodwill among men; out of the flames of ambition and power, that was Nebuchadnezzar's, the succeeding kingdoms, "the Totalitarian States," which limit the freewill of man, there stands forth intact the ideal of a commonwealth of free nations.

Thus out of the *fiery furnace*, 666-inspired, there comes forth unharmed the three theocentric ideals.[2]

Such then, throughout the centuries of man's attempted civilisations, is the work of the Logos "*by Whom all things consist*"; for "*in the midst of the fire . . . the form of the fourth is like the Son of God*" (Dan. iii, 25).

[1] And in Christmas week the heaviest casualties. [2] See pp. 363–369.

Germany

As Clausewitz and Bernhardi, so now Ludendorff writes preparatory to war—*The Total War* (October 1935):

> The Christian nations are no longer in the fortunate position —as are the Japanese—of having a peculiar faith founded upon the unity of government and people, of people and army, and of all national life. The teaching of Christianity is a foreign doctrine in utter opposition to our racial inheritance. It kills the people's peculiar spiritual unity and leaves them defenceless.

After enumerating the economic and industrial measures necessary for the "total war," many of which Dr Schacht has already undertaken, he continues:

> The preparation of the fighting forces, industry and the people for service in the total war, begins the moment a country decides to go to war. This preparation—this mobilisation—will be conducted in accordance with carefully thought-out and exactly calculated prescriptions made in peace and renewed yearly.

These bombastic generals do signal their punches!

Hitler has indicated no aggression on, but defence of, her western frontier, by land and sea.

With von Ribbentropp's "hunting" to ally Poland as an eastern buffer, and Hungary as a stepping-stone, his eyes are directed towards south and south-east Europe, there to baptize his new swastika war flag.

And now he looks around for the appropriate starting-handle, questions whether Memel, Austria, demand for colonies, or other.

France is nervous for her ally, the Little Entente, and seeks to pull Italy out of the African venture to her side. If such should fail, is the promise of Mediterranean aid the price she will give for British help here?

Russia

Her voice was heard at Geneva in August demanding sanctions on Italy, and then was strangely silent, though having twice the Italian war strength. The echo was heard, as might be expected, at the Labour Party Congress at Brighton—such sanctions to be applied, even to war.

Then Russia in October advised at Geneva sanctions on countries which applied not sanctions—Austria, Hungary, Albania: anything to get Britain involved to waste prematurely her strength.

With most of the 80,000 French troops removed from Syria to meet the European embroilment, Russia seizes her hour—her road to the Near East lies open, a feint through Sinkiang or Afghanistan towards India, then an aeroplane attack, followed by a mass advance from the Caspian, allied with Persia and Turkey (out to regain lost spoil)—the aeroplanes are manned, the railways prepared, the horses accumulated.[1]

For in the invasion impending, the sudden surprise air attack on a limited objective, obliterating defended positions and creating disorder and panic, is followed at once by the swift armoured car and tank to destroy remaining opposition, to be followed in turn by the commissariat to occupy; and from such base similar tactics are applied on the next selected objective.

(The defence lies in imitating the Russians themselves—at Moscow before Napoleon.)

The Russian path lies parallel to the German, so in their beginnings mutually unopposed. But there lies the essence of a dog-fight in the division of the "spoil"—that Near East, the junction of the air routes and railway systems of three continents, the outlet of the oil lines, the potential granaries of the future.

The standards of the vultures will be seen in Jerusalem, scrapping on the carcase of Palestine.[2]

It will need more than the strength of British diplomacy,

---

[1] The words of a Russian officer—"We are going down into that land (Palestine) to destroy those things which are held sacred by all those great religions (Mohammedanism, Judahism, Christianity), and to demonstrate to the world that there is no God": Japan in Manchukuo having delayed them.

[2] "*For wheresoever the carcase is, there will the eagles be gathered together*" (Matt. xxiv, 28), read with Jer. xvi, 18, and Psalm lxxix, 1, 2.

seeking righteousness in international dealing, to avert conflict.

What of the League of Nations, that husk of peace with no kernel of Divine harmony?

It hath been said:

Collect together, you nations, but you shall be broken ;
Listen, lands at a distance, arm yourselves, but you shall be broken ;
Decide a united scheme, discuss a plan, it shall not succeed.
For God is with us (Isaiah viii, 9-12, Farrar Fenton's translation).

Do you deserve that, Israel-Britain?—or realise your part in—

"*I will shake all nations, and the desire of all nations shall come*" (Haggai ii, 7)

Across the centuries are heard snatches from the refrain of the Design—

"*Until Shiloh come . . . ye are My battle-axe and weapons of war . . .* [displaced to become] *the multitude of peace . . . to be a blessing to all the families of the earth.*"

Come, Ephraim (England), why so muddled and hesitant? The way has been shown by Manasseh, your brother across the Atlantic, to "come out"; by Benjamin, with the "light" of eschatology; by Zarah-Pharez of Judah, as symbol of unity; by Seth, your ancestor, in scientific symbology, mirrored in stone, for your post-1844 understanding; by your KING, Whom ye left deserted and crucified in a Cain-Bel world.

Your faith will not and can not rest content in Geneva, but only in His Sanctuary to be built in Palestine, on the site of the emergence of your race, 4000 B.C., ready as the "sixth day" ends.[1]

To you, Ephraim, having the birthright, do all the tribes look to lead.

Oppose to the totalitarian the theocentric ideal.

Use the red tape of your diplomacy, if you wish, but found the Theocentric State—the British Commonwealth of Nations, the U.S.A., the Jews; then Holland, Denmark, Norway, Sweden, Switzerland, Saxon Germany, Celtic

[1] The first Temple was founded by Solomon at the half time-distance 1000 B.C.

France and Belgium, the Samurai of Japan, with Ishmael (the Arabs) to dwell with you—to become His Kingdom on earth when He comes,[1] *"the Lion of the tribe of Judah, the Root of David"* (Rev. v, 5), *" the Lamb slain from the foundation of the world"* (v, 6, 12; xiii, 8; xiv, 1), *"Faithful and True, the Logos, the Word of God"* (xix, 11, 13), *"Lord of lords and King of kings"* (xix, 16).

Have you looked upon Nature and seen the rhythm of its action—the changing cloud, the breaking wave, the flight of bird, the leap of deer?

Then at last you will see beauty in the handling of human affairs.

¶ 9. To continue prophecy about to be fulfilled, and visualising coming history therein:

"Thus saith the Lord, Behold, a people cometh from the north country,[2] *and a great nation shall be raised from the sides of the earth. They are cruel, and have no mercy; their voice roareth like the sea, they ride upon horses, set in array as men for war against thee, O daughter of Zion"* (Jer. vi, 22, 23).

"How is the hammer broken! How is Babylon become a desolation among the nations!" (Jer. l, 23).

Dan. xi (after the fall of Greece, there appears a "horn" as the eastern half of "Babylon," then)—

"And the King shall do according to his will, and he shall exalt himself, and magnify himself above every god [the dictator will exalt the State idea of freewill, Bolshevism], *and shall speak marvellous things against the God of gods* [Antigod propaganda, *vide* cartoons in the Leningrad paper *The Atheist*, as the one of the Last Supper] *and shall prosper till the indignation be accomplished* [1939], *for that that is determined shall be done."*

"Neither shall he regard the god of his fathers [the Greek Orthodox Church], *nor regard any god* [Atheism], *nor the desire of women* [family and children] : *for he shall magnify himself above all* [the Communist party].

---

[1] *"The Sceptre shall not depart from Judah . . . until Shiloh come"* (Gen. xlix, 10). See its context, given here and p. 196.

[2] Watch the ensuing identity with Soviet Russia.

"*But in his estate shall he honour the god of forces and munitions* [the Red Army, and factories for manufacture]: *and a god whom his fathers knew not shall he honour with gold and silver, and with precious stones and pleasant things* [Communism and its commissaries].

"*Thus shall he do in the strongholds with a strange god, whom he shall acknowledge and increase with glory* [the Ogpu]; *and he shall cause them to rule over many, and shall divide the land for gain* [collective farms and slave-labour camps].

"*And at the time of the end shall the king of the south*[1] *push at him: and the king of the north shall come against him like a whirlwind, with chariots and with horsemen, and with many ships; and he shall enter into the countries, and shall overflow and pass over.*

"*He shall enter also into the glorious land,*[2] *and many countries shall be overthrown, but these shall escape out of his hand, even Edom.*

"*He shall stretch forth his hand also upon the countries, and the land of Egypt shall not escape.*

"*But he shall have power over the treasures of gold and silver and all the precious things of Egypt, and the Libyans and the Ethiopians shall be at his steps.*

"*But tidings out of the east and out of the north*[3] *shall trouble him, therefore shall he go forth with great fury to destroy, and utterly to make away with many.*

"*And he shall plant the tabernacles of his palace between the seas in the glorious holy mountain*[4]; *yet he shall come to his end, and none shall help him*" (Dan. xi, 36–45).

"*And in that time shall Michael stand up, the great prince which standeth for the children of Thy people; and there shall be a time of trouble, such as never was since there was a nation even to that same time*[5]: *and at that time Thy people shall be delivered, everyone that shall be found written in the Book.*

[1] Revealed during the last year as Italy.
[2] Palestine, and as far as Egypt.
[3] No word for N.E. in Hebrew; thus, from Japan.
[4] Mount Zion, between the Mediterranean and Dead Seas.
[5] The tribulation, 1914–1936; never such so widespread since the Sumerians appeared as the first civilisation: *vide* Matt. xxiv, 21, 22.

"And many of them that sleep in the dust of the earth shall awake, some to everlasting life, some to shame and everlasting contempt.[1]

"But thou, O Daniel, shut up the words and seal the book, even to the time of the end: many shall run to and fro,[2] *and knowledge shall be increased.*[3]

"Then said I, O my Lord, what shall be the end of these things? . . . The words are closed up and sealed till the time of the end, but the wise shall understand.[4]

"From the time that the daily sacrifice shall be taken away, and the abomination that maketh desolate set up, there shall be 1290 days. Blessed is he that waiteth and cometh to the 1335 days" (Dan. xii).

Davidson [5] gives those two dates as A.D. $1891\frac{3}{4}$ and $1936\frac{3}{4}$.

The latter time signifies the entrance date of the King's Chamber period: Biblical and Pyramid datings have coincided throughout history.

And as 1936 breaks, let prophecy continue to reveal history:

---

[1] With His Second Coming: *cf.* 2 Esdras ix, xii; Rev. xiii, 8; xx, 12, 13; xxi, 27.

[2] Cars and trains, cruises and aeroplanes.

[3] Since 1844, and especially since 1918.

[4] As prophecy becomes fulfilled, identity becomes revealed.

"For prophecy came not in old time by the will of man, but holy men of God spake as they were moved by the Holy Spirit" (2 Peter i, 21). With Christ's words, *"When ye shall see all these things, know that it is near, even at the doors"* (Matt. xxiv, 33): *"then look up and lift up your heads, for your redemption draweth nigh"* (Luke xxi, 28).

[5] Davidson, *The Great Pyramid, its Divine Message*, p. 391 and Plate LXVI.

A.D. 70–639 = a period of "abominations," when there was no "daily sacrifice" in Jerusalem: until the climax in A.D. $639\frac{1}{4}$, when the Mosque of Omar set up, and still standing, on the site of the Temple.

2520 solar years = 2520 lunar (or Mohammedan) years + 75 years (this difference called the "epact"): thus $1222\frac{1}{2}$ solar = 1260 lunar years.

A.D. $639\frac{1}{4}$ + 1290 (prophetic days = $1222\frac{1}{2}$ solar + 30 "epact") = A.D. $1891\frac{3}{4}$.

A.D. $639\frac{1}{4}$ + 1335 (prophetic days = $1222\frac{1}{2}$ solar + 75 "epact") = A.D. $1936\frac{3}{4}$.

Also $584\frac{1}{4}$ B.C. (destruction of the Temple by Babylon) + $1222\frac{1}{2}$ = A.D. $639\frac{1}{4}$: and "the lunar year 1335 of Moslem reckoning coincided with 1917 A.D., when Jerusalem was delivered" (Davidson).

And with this, read Matt. xxiv, 15: *"When ye therefore shall see the abomination of desolation, spoken of by Daniel the prophet, stand in the holy place (whoso readeth, let him understand).*"

"THAT GREAT DAY OF GOD ALMIGHTY" 407

"*Blow ye the trumpet in Zion, for the day of the Lord cometh, for it is nigh at hand. A day of darkness and gloominess, a day of clouds and of thick darkness; a great people and a strong, there have not been ever the like, neither shall be any more after it, a fire devoureth before them, and behind them a flame burneth: the land is as the garden of Eden before them, and behind them a desolate wilderness, yea, and nothing shall escape them.*

"*The appearance of them is as the appearance of horses,*[1] *like the noise of chariots, like the noise of a flame of fire, as a strong people set in battle array. Before their face, the people shall be much pained, all faces shall gather blackness.*

"*The earth shall quake before them, the heavens shall tremble, the sun and the moon shall be dark, and the stars shall withdraw their shining.*

"*For the day of the Lord is great and very terrible and who can abide it?*

"*Therefore, saith the Lord, turn ye even to Me with all your heart: and rend your heart, and not your garments, and turn unto the Lord your God.*

"*Blow the trumpet in Zion, gather the people. Let the priests, the ministers of the Lord, weep.*

"*Spare Thy people, O Lord, and give not Thine heritage to reproach, that the heathen should rule over them. Then will the Lord be jealous of His land, and pity His people*" (Joel ii, 1–18).

"*Enter into the rock, and hide thee in the dust: and they shall go into the holes of the rocks, and into the caves of the earth,*[2] *for fear of the Lord, and for the glory of His majesty when He ariseth to shake terribly the earth*" (Is. ii, 10, 19).

"*Come, My people, enter thou into thy chambers, and shut thy doors about thee: hide thyself as it were for a little moment, until the indignation be overpast. For, behold, the Lord cometh out of His place to punish the inhabitants of the earth for their iniquity*" (Is. xxvi, 20).

"*I will remove far from you the northern army, and will*

---

[1] *Cf.* the Cossacks. It has been stated that Russia has accumulated in her southern steppes multitudes of horses of wiry, hill-climbing breeds.

[2] Suggestive of dug-outs, trenches and shelters underground.

drive him into a land barren and desolate, with his face toward the east sea, and his hinder part toward the utmost sea.[1]

"*And I will shew wonders in the heavens and in the earth, blood, fire, and pillars of smoke.*

"*The sun shall be turned into darkness, and the moon into blood, before the great and terrible day of the Lord come.*

"*And it shall come to pass, that whosoever shall call on the name of the Lord shall be delivered; for in Mount Zion and in Jerusalem shall be deliverance, as the Lord hath said, and in the remnant whom the Lord shall call*" (Joel ii, 20, 30–32).

"*In those days, and in that time, when I shall bring again the captivity of Judah and Jerusalem, I will also gather the nations, and will bring them down into the valley of Jehoshaphat, and will plead with them there for My people and for My heritage Israel.*

"*Proclaim ye this among the Gentiles*[2]; *Prepare war, wake up the mighty men, let all the men of war draw near, let them come up: beat your ploughshares into swords, and your pruning hooks into spears. Assemble yourselves, and come, all ye heathen, and gather yourselves together.*

"*Let the heathen be wakened, and come up to the valley of Jehoshaphat; for there will I sit to judge all heathen round about. Multitudes, multitudes in the valley of decision, for the day of the Lord is near in the valley of decision.*

"*The sun and the moon shall be darkened, and the stars withdraw their shining.*

"*The Lord also shall roar out of Zion, and utter His voice from Jerusalem*[3]; *and the heavens and the earth shall shake: but the Lord will be the hope of His people, and the strength of the children of Israel.*

"*So shall ye know that I am the Lord your God dwelling in Zion, My holy mountain; then shall Jerusalem be holy.*

"*And all the rivers of Judah shall flow with water, and a fountain shall come forth of the house of the Lord, and shall water the valley of Shittim*" (Joel iii).

[1] The Pacific and Arctic Oceans.
[2] The Continental nations.
[3] *Cf.* 2 Esdras xiii and Rev. xiv. 1.

"THAT GREAT DAY OF GOD ALMIGHTY"

"*In that day will I make Jerusalem a burdenstone for all people: all that burden themselves with it shall be cut in pieces, though all the people of the earth be gathered together against it.*

"*In that day, saith the Lord, I will smite every horse with astonishment, and his rider with madness.*

"*The Lord also shall save the tents of Judah first. In that day shall the Lord defend the inhabitants of Jerusalem: and he that is feeble among them at that day shall be as David.*

"*It shall come to pass in that day that I will seek to destroy all nations that come against Jerusalem*" (Zech. xii).

"*Behold, the day of the Lord cometh. For I will gather all nations against Jerusalem to battle, and the city shall be taken.*

"*Then shall the Lord go forth and fight against those nations.*

"*And His feet shall stand in that day on the Mount of Olives*[1] . . . *and it shall cleave in the midst thereof toward the east and toward the west, and there shall be a very great valley, and half of the mountain shall move toward the north and half of it to the south* . . . *as the earthquake in the days of Uzziah, King of Judah, and the Lord my God shall come, and all the saints with Him.*

"*And it shall come to pass in that day, that the light shall not be clear, not dark.*

"*And it shall be in that day that living waters shall go out from Jerusalem, half of them toward the former sea, and half of them toward the hinder*[2] *sea, in summer and in winter shall it be.*

"*All the land shall be turned as a plain from Geba to Rimmon south of Jerusalem, and it shall be lifted up, and shall abide. And men shall dwell in it and there shall be no more utter destruction, but Jerusalem shall be safely inhabited*" (Zech. xiv).

Gayer[3] states that Kitchener as a sapper lieutenant

---

[1] Rev. xiv, 1.
[2] Former = toward the sun — *i.e.* the Dead Sea. Hinder = toward sunset — *i.e.* the Mediterranean.
[3] Gayer, *Heritage of the Anglo-Saxon Race*, p. 46.

found an underground hot-water lake of immense size under the Mount of Olives, the only visible exit of its waters was to the Pool of Bethesda (John v, 1–16), and that this region was one of the most volcanic, a great earthquake would let the water run towards the Mediterranean and Dead Sea.

Also the Moslems on Mount Moriah hear under their Mosque the roaring below as of a mighty subterranean river.

It is known that a geological "fault" exists, commencing south of the Caspian Sea, passing down through the Sea of Galilee (680 feet below sea-level), the Jordan valley, the Dead Sea (1250 feet below sea-level), the Wadi-el-Arabah to the Gulf of Akaba: and that twice in prehistoric times the Mediterranean has been through the Plain of Esdraelon into this gap of rock strata and thus into the Red Sea, and receded.

Now only a shallow thirty feet rise in the plain, where the Haifa-Nazareth road crosses it, and a block some few miles north of Akaba remain.

A great earthquake, as Zechariah here and Rev. xvi, 18, prophesy, which splits the Mount of Olives, will demolish such barriers, and let the Mediterranean through for the third and last time, to fulfil Zech. xiv, 10; Is. xi, 15; xxxiii, 21, and with waters from under the Mount of Olives (Ezek. xlvii, 9–10) thus fertilise Palestine.

Premonitory of such was the earthquake of July 1927, which shook the Mount of Olives and damaged thereon the late Kaiser's palace in use as Government House.

The site for the new Government House was chosen well out of the line of this cleft, and it has now been erected on the hill between Jerusalem and Bethlehem, commanding a wonderful view towards the Jordan valley.

¶ 10. Now this leads up to the one prophecy which by universal consent has never so far been fulfilled—Ezek. xxxviii, xxxix:

"*The Word of the Lord came, saying, Set thy face against Gog, the land of Magog, the chief prince of Rosh, Meschech*

"THAT GREAT DAY OF GOD ALMIGHTY"

and Tubal[1]; behold, I am against thee, and I will turn thee back, all thine army, horses and horsemen, all of them clothed with all sorts of armour, even a great company.

"Persia, Ethiopia and Libya[2] with them; Gomer and all his bands[3]; the house of Togarmah of the north quarters,[4] and all his bands, and many people with thee.

"In the latter days, thou shalt come into the land, against the mountains of Israel, that is brought forth out of the nations, and dwell safely. Thou shalt ascend and come like a storm, thou shalt be like a cloud to cover the land, thou and all thy bands, and many people with thee.

"At the same time shall things come into thy mind, an evil thought: and thou shalt say, I shall go up to the land of unwalled villages,[5] I will go to them that are at rest and dwell safely; to take a spoil, to turn thine hand upon the desolate places that are now inhabited, and upon the people that are gathered out of the nations,[6] which have gotten goods and cattle.

"And the merchants of Tarshish, with all the young lions thereof shall say unto thee, Art thou come to take a spoil, to carry away silver and gold, to take away cattle and goods?[7]

"Thus saith the Lord God, In that day when My people of Israel dwell safely, shalt thou not know it?

"Thou shalt come from thy place out of the north parts, a great company and a mighty army, against My people of Israel, as a cloud to cover the land, in the latter days.[8]

---

[1] These have been interpreted as Russia, Moscow, Tobolosk (thus the whole of Soviet territory).

[2] The ancient names Ethiopia and Libya recently revived for Abyssinia and Tripoli.

[3] The Continental Kimmerian residue—? Italy? Bulgaria? Germany.

[4] Togarmah as Asia Minor and Armenia — ? Prussians from the Hittites (same vulture emblem): ? the Turks to-day.

[5] As Palestine.   [6] Israel-Britain.

[7] On such invasion, one can imagine the doubt of enemy intentions and delay of action advised by some elements of uncertain foresight in Parliament, as in the last war.

"Merchants" = those who think in terms of money, who calculate trade by the gold standard and "wailed" at Babylon's economic fall (Rev. xviii, 11–19); who preach war as impossible without gold in the banks' vaults (as before the last war).

"Spoil" = the junction of three continents for future air and rail routes; the oilfields; the salt deposits of the Dead Sea, etc.—that which paid for success in arms in early wars, and now entices a bankrupt Russia or Germany whose culture has gone primitive.

[8] The era defined (1936–1939).

"*Art thou he*[1] *of whom I have spoken in old time by My servants the prophets of Israel, which prophesied in those days many years that I would bring thee against them?*

"*Then saith the Lord God, My fury shall come up in My face.*

"*Surely in that day there shall be a great shaking in the land of Israel: men that are upon the face of the earth shall shake at My presence, and the mountains shall be thrown down. And I will call for a sword*[2] *against him throughout all My mountains, saith the Lord God: I will plead against him with pestilence and with blood, and I will rain upon him an overflowing rain, and great hailstones, fire and brimstone*" (Ezek. xxxviii).

"*Behold I am against thee, O Gog, and I will turn thee back, and leave but the sixth part of thee; and I will smite thy bow out of thy left hand, and will cause thine arrows*[3] *to fall out of thy right hand. Thou shalt fall upon the open field, for I have spoken, saith the Lord God. And I will send a fire on Magog, and among them that dwell carelessly in the Isles.*[4]

"*So will I make My holy name known in the midst of My people, Israel; and the heathen*[5] *shall know that I am the Lord, the Holy One in Israel.*

"*Behold, it is come, and it is done, saith the Lord God, this is the day whereof I have spoken*" (xxxix, 1-8).

Read with Revelation xvi, 17, "*It is done*"—thus the time is fixed at the seventh seal.

And Revelation xvi, 16, "*And He gathered them together into a place called in the Hebrew tongue, Armageddon*[6]"—

---

[1] The connection between Cain-Bel and the invaders will be questioned by some.

[2] The sword of the Spirit, for our disarmed condition will be of little use against such a horde, with our material sword broken at the hilt at the first encounter. "*Not by might, nor by power, but by My Spirit, saith the Lord of Hosts*" (Zech. iv, 6).

[3] The bombs and the shells.

[4] Either climatic or by aeroplane attack, on Britain, "dwelling carelessly"—without thought on His purpose.

[5] The inhabitants round about—Egypt, Arabia, Iraq, India.

[6] The word is interpreted by Pascoe Goard, according to the Hebrew, "the armed fortress or stronghold of Megiddo." Now Megiddo to-day is not a fortress, entrenched by the British, but a "tell," the site of excavations which find one of Solomon's numerous summer-houses and a ranch for his horses.

"THAT GREAT DAY OF GOD ALMIGHTY"

a verse interposed as if with no context between the "halt" of xvi, 15 and the breaking of the seventh seal (xvi, 17)—is to occur at the time of Ezekiel xxxix, 1-7.

When our army is making its last stand, as it has so often done, its precarious position becomes "the armed fortress," a place of safety,[1] while destruction overtakes Gog's armies—a Divine intervention so marvellous that our army will yet teach our Church where God is:

"*So I will make My holy name known in the midst of My people, Israel*" (Ezek. xxxix, 7)—with this admonition to Britain in the coming crisis: "*I will yet for this be enquired of by the House of Israel to do it for them*" (xxxvi, 37).

"*And Israel shall go forth and burn the weapons, the bows*[2] *and the arrows . . . they shall burn them with fire seven years.*

"*In that day I will give unto Gog a place there of graves in Israel, and there shall they bury Gog and all his multitude; and seven months shall the House of Israel be burying of them, that they may cleanse the land: and they shall sever out men of continual employment to cleanse it, after the end of seven months shall they search.*

"*And the heathen shall know that the House of Israel went into captivity for their iniquity, because they trespassed against Me, therefore hid I My face from them, and gave them into the hand of their enemies, so fell they all by the sword.*[3]

"*Thus saith the Lord God, Now will I bring again the captivity of Jacob, and have mercy upon the whole House of Israel. When I have brought them again from the people and gathered them out of the enemies' lands, and am sanctified in them in the sight of many nations; then shall they know*

---

[1] In attempting to work out the strategy of this battle my photographs were lost, in circumstances that could never again happen—a personal sign to talk guardedly to the merely curious of a prophecy, which is true and coincides with the intelligence and information of to-day, when current events (the trend in the side column rather than the popular imagination of the front page) tumble into place like a jig-saw puzzle.

[2] The accumulation of armaments by the "Abmachungen" since 1918; of the recent rapid rearmament by Germany, and the concealed armament of Russia.

[3] The identity of Israel-Britain in all their migrations, revealed to all the world.

that I am the Lord their God, which caused them to be led into captivity among the heathen.

"But I have gathered them unto their own land, and have left none of them any more there. Neither will I hide My face any more from them, for I have poured out My spirit upon the House of Israel, saith the Lord God" (Ezek. xxxix, 9-29).

"And I will cleanse them, so shall they be My people, and I will be their God.

"And David My servant shall be king over them, and they shall have one shepherd; they shall also walk in My judgments and observe My statutes to do them.

"And they shall dwell in the land that I have given unto Jacob My servant, wherein your fathers have dwelt, even they and their children's children for ever: and My servant David shall be their Prince for ever.

"Moreover I will make a covenant of peace with them: it shall be an everlasting covenant with them, and will set My sanctuary in the midst of them for evermore.

"My tabernacle shall be with them, yea, I will be their God, and they shall be My people.

"And the heathen shall know that I the Lord do sanctify Israel, when My sanctuary shall be in the midst of them for evermore" (Ezek. xxxvii, 23-28).[1]

"I will make a new covenant with the House of Israel and with the House of Judah. . . . After those days, saith the Lord, I will put My Law in their inward parts, and write it in their hearts, and will be their God, and they shall be My people.

"And they shall teach no more every man his neighbour, and every man his brother, saying, Know the Lord; for they shall know Me, from the least unto the greatest of them" (Jer. xxxi, 31-34).

Followed by that emphatic emphasis in vv. 35-38, which parallels that of Christ's after speaking of His Second Coming, "Heaven and earth shall pass away, but My words shall not pass away" (Matt. xxiv, 35).

"Therefore the days come, saith the Lord, that they shall

[1] The quotation of verses postponed from p. 163.

THE THEOCENTRIC KINGDOM

no more say, The Lord liveth which brought up the children of Israel out of the land of Egypt: but, <u>the Lord liveth which brought up and led the seed of the House of Israel out of the north country, and from all the countries whither I had driven them</u>; and they shall dwell in their own land.

"*Behold, the days come, that I will raise unto David a righteous Branch, and a King shall reign and prosper, and shall execute judgment and justice on the earth.*

"*In His days Judah shall be saved, and Israel shall dwell safely: and this is His name whereby He shall be called, The Lord our Righteousness*" (Jer. xxiii, 5–8).

"*And it shall come to pass in that day, that the Lord shall beat off from the channel of the River unto the stream of Egypt, and ye shall be gathered one by one, O ye children of Israel.*

"*<u>And they shall come which were ready to perish in the land of Assyria, and shall worship the Lord in the holy Mount at Jerusalem</u>*" (Isa. xxvii, 12, 13).

"*Therefore, thus saith the Lord God, Behold, I lay in Zion for a foundation, a stone, a tried stone, a precious cornerstone a sure foundation (he that believeth shall not make haste)*" (Isa. xxviii, 16).

¶ 11. To follow such majestic flow of language with mere words to clarify the thread of an argument is but vain quibbling.

Let the late Professor Piazzi Smyth,[1] Astronomer Royal of Scotland, interpret the prophecy in words written fifty-five years ago:

Judah will make advances to Ephraim, and both together return to Palestine.[2]

And now the time is near approaching when Jerusalem is to be visited by Divine favour once again ... the Anglo-Israelites meanwhile guarding the frontiers of the country for the Jews.

Calamities following calamities such as the world has never

---

[1] Professor Piazzi Smyth, *Our Inheritance in the Great Pyramid*, pp. 632–634, 1880.
[2] *I.e.* the Jews to England: *vide* the Balfour Declaration during the War.

known will befall them,¹ at what time the seventh vial of the wrath of God having been poured out, the seventh trumpet shall begin to sound and bring the overflowing hosts of the north (verified on p. 591 as the Russians) upon them even in the sacred city itself.

Ruin and destruction will then stare them in the face, nothing human will be able to save either them or the Anglo-Israelite defenders of their borders, from utter earthly perdition.

But precisely then it will be that the Saviour, whom Jews have rejected so long, will descend both to their and Israel's succour, with His armies of glorified Saints and Angels . . . and they, the Jews, shall look on "*Him whom they pierced,*" in a moment be convicted, convinced and turned into the deepest grief for their traditional sins.

That instant . . . will be the occasion of the . . . recognition of Judah and the rest of Israel . . . the beginning of our Lord's visible reign amongst men and over the world's varied nations, during which . . . "*the earth shall be filled with the knowledge of the Lord, as the waters cover the sea.*" ²

To each, the signs have been given—"*Behold I have told you before*" (Matt. xxiv, 25); and the injunction to "watch" the course of events move, in quiet certainty of attuned freewill, "*lest coming suddenly he find you sleeping*" (Mark xiii, 36).

The Biblical prophecies and the Pyramid symbology coincide.³

At the climax of the impending crisis, both religious and political, by Christ's own act to intervene, so that His Presence is felt and universally recognised by all, even by the heathen round about, Israel-Britain is delivered.

Britain emerges after the second "*stooping*" (since 1928)

---

[1] In recent years, to the Jews in Russia, in Eastern Europe, and now expelled from Germany: and to Israel-Britain, the 1914–1936 "tribulation."

[2] Piazzi Smyth has here conceived Matt. xxiv, 30, as one event, what may be a series of events over an unknown period: (1) "*the sign,*" (2) "*mourn,*" (3) "*see.*" In this context, "*as a thief*" (xxiv, 43) coincides with *the sign*, and meaning: Personal action in earth environment perceived but Presence concealed; and "*I come quickly*" meaning such Action (and Visibility as "*see*") as of sudden onset.

[3] To interpret prophecy as it becomes fulfilled in these times, attention is directed to the weekly *National Message* (Covenant Publishing Co.), especially the articles by A.R.H., not a religious crank, but one who has known action against "Richthofen's circus" and Goering himself in air combat.

THE THEOCENTRIC KINGDOM

to stand upright—to "*look up and lift up your heads, for your redemption draweth nigh*" (Luke xxi, 28)—in the King's Chamber period, 1936–1953: there to stand at last in the central plane of the Pyramid, under the rejected Apex Stone, and to look into an empty coffer, at last to know, without any philosophic jargon, for Creation's displacement is now restituted, that His Resurrection is actual.

Britain will know that it is her duty to become the Theocentric State, and with her Commonwealth of Nations as the nucleus, to be prepared under a new Covenant (Jer. xxxi, 31–34) to widen her freedom under His rule of Righteousness to become His Kingdom on earth.

"*And the Lord shall be King over all the earth; in that day shall there be one Lord and His name one.*

"*And every one that is left of all the nations which came against Jerusalem shall even go up from year to year to worship the King, the Lord of Hosts, and to keep the Feast of Tabernacles.*[1]

"*In that day shall there be upon the bells of the horses, Holiness unto the Lord.*

"*And in that day there shall be no more the Canaanite in the House of the Lord of Hosts*" (Zech. xiv, 9, 16, 20, 21).

"*And I will pour upon the House of David and upon the inhabitants of Jerusalem the spirit of grace and of supplications: and they shall look upon Me whom they have pierced and they shall mourn for Him as one mourneth for his only son*" (Rev. i, 7).

"*Yea, many people and strong nations shall come to seek the Lord of Hosts in Jerusalem, and to pray before the Lord.*

"*In those days it shall come to pass, that ten men out of all the languages of the nations shall take hold of the skirt of him that is a Jew,*[2] *saying, We will go with you, for we have heard that God is with you*" (Zech. viii, 22, 23).

---

[1] That is, Christmas. For His birth was on the Feast of Tabernacles (15th Tizri) in 4 B.C. (see p. 97).

[2] The Jews will recognise and accept the risen Christ. Marvellous to contemplate—united with Israel again in a new and combined Covenant, the change of heart of the really religious Jew ("*a heart of flesh and not of stone*"), with his centuries of experience of finance, will inspire to that

"*At that time they shall call Jerusalem the throne of the Lord, and all the Nations shall be gathered unto it, to the name of the Lord, to Jerusalem: neither shall they walk any more after the imagination of their evil heart*"[1] (Jer. iii, 17).

"*And it shall come to pass in the last days that the mountain of the Lord's House shall be established in the top of the mountains,*[2] *and shall be exalted above the hills, and all nations shall flow unto it.*

"*And many people shall go and say, Come ye, let us go to the mountain of the Lord, to the House of the God of Jacob: and He will teach us of His ways and we will walk in His paths: for out of Zion shall go forth the Law, and the Word of the Lord from Jerusalem.*

"*And He shall govern the Nations, and shall rebuke many people; and they shall beat their swords into ploughshares, and their spears into pruning hooks: nation shall not lift up sword against nation, neither shall they learn war*[3] *any more.*

"*O House of Israel, come ye, and let us walk in the Light of the Lord*" (Isa. ii, 2–4).

"*And in that day, the Lord of Hosts shall reign in Mount Zion and in Jerusalem*" (Isa. xxiv, 23).

"*And He shall speak peace unto the heathen: and His*

Divine economic system in which gold and money are no longer accumulated for wealth but merely an exchange for goods and service given.

Even now is "*the budding of the fig-tree*" (Matt. xxiv, 32–34; fig-tree in prophecy is Judah):

July 1928, at Hamburg, 153 delegates from eighteen nations at the meeting of the International Hebrew Christian Association; the Bible being translated into Yiddish; and in 1930 the *National Message* gave the number of Jews on the Continent who joined the Christian Church as 224,000.

[1] *Cf.* Gen. vi, 5 (p. 34).

[2] See Micah iv, 1–3; Esdras xiii; Rev. xiv, i: "*Lo, a Lamb stood on the mount Zion and with Him an 144,000, having His Father's name written in their foreheads.*"

[3] For with his reign, Spiritual power over matter is restored to, and the use of cosmic energy placed in the hands of, only attuned Freewill, that it may not be misdirected to cause man's destruction, but to be "*a blessing to all the families of the earth.*"

Thus every bomb and shell shall be made a "dud," every war-plane to fall, every torpedo and submarine to sink, and every poison gas neutralised.

Thus man shall have peace on earth, and learn that "*the earth is the Lord's* [not Bel's] *and the fulness thereof.*"

THE THEOCENTRIC KINGDOM

dominion shall be from sea even to sea, and from the river even to the ends of the earth" (Zech. ix, 10).

"Of the increase of His government and peace there shall be no end, upon the throne of David, and upon His Kingdom, to order it and to establish it with judgment and with justice from henceforth even for ever.

"The zeal of the Lord of Hosts will perform this" (Isa. ix, 7).

Thus is fulfilled Luke i, 32, 33: *"He shall be great, and shall be called the Son of the Highest; and the Lord God shall give unto Him the throne of His father, David.*

"And He shall reign over the House of Jacob for ever, and of His Kingdom there shall be no end."

Exekiel (xl to xlviii) describes the future Palestine, how the land is to be ruled; the area reserved for the Holy Oblation (xlviii, 8–20), within the midst of it on the topmost mountain His sanctuary to be erected; the land on each side of this area to be reserved for "the Prince" (our Royal House) (xlv, 7, 8).

And from the Mount of His Sanctuary *"waters run out. These waters* [1] *issue out to the east country and go down into the desert, and go into the sea, which being brought forth into the sea, the waters shall be healed, and everything shall live whither the river cometh.*

"And it shall come to pass that the fishers stand upon it, there fish shall be as the fish of the great sea, exceeding many.

"And by the river upon the bank thereof, on this side and on that, shall grow all trees, whose leaf shall not fade; it shall bring forth new fruit according to his months; because their waters issued out of the sanctuary, the fruit thereof shall be for meat, and the leaf thereof for medicine" (xlvii, 1–4).

And the new city of Jerusalem—*"the name of the city from that day shall be, The Lord is there"* (xlviii, 35).

---

[1] Identical with the river that *"went out of Eden to water the garden"* (p. 21), in these times underground below the Mount of Olives and Mount Moriah (p. 410), revived to restore the land which in 4000 B.C. was Eden, and now Palestine, so that *"they shall say, This land that was desolate is become like the Garden of Eden"* (Ezek. xxxvi, 35).

"And He will cause to come down for you the rain, the former rain, and the latter rain in the first month" (Joel ii, 23).

"There in Jerusalem, the glorious Lord shall be unto us a place of broad rivers and streams, not wanting in oar-driven ships and proud vessels to sail past" [1] (Isa. xxxiii, 21 : Ferrar Fenton's translation).

"I will open rivers in high places, and fountains in the midst of the valleys ; I will make the wilderness a pool of water, and the dry land springs of water" (Isa. xli, 18).

With the restoration of Palestine to fertility and colonisation, by the changes in its configuration and with the coming of rain to its parched soil, His Law of Righteousness will go forth from Zion.

It will be no use for financiers to seek their wealth, nor for profiteers in land values, within that country revived by Divine hand—

"They shall not hurt nor destroy in all My holy mountain . . . righteousness shall be the girdle of His loins."

It may be that here our unemployed will find again work for their hands, and health to their spirits.

It will be Britain's first duty as a Theocentric State by finance and organisation to recognise such claim.

It may be that Christ—knowing this result of Chaos, to which the Babylon economic system would lead—has so prepared a restored Palestine (the land of our forefathers, promised to Abraham) for His Israel-Britain.

And as the river which went out of Eden is identical with, and revived as, the river which ran out of the Sanctuary—is there here some indication that its first head as *"Pison* [the Jordan rift], *which compasseth the land of Havilah where there is gold, and the gold of that land is good"* (Gen. ii, 11, 12 ; so that it has never been used as wealth) will reveal this gold [2] to end this money chaos,

---

[1] As the result of the great earthquake the Dead Sea connects with the Mediterranean as it did in prehistoric times.

[2] The national issue of these physical counters as money by a scientific department of the Theocentric State would be regulated to keep the price-level constant, according to the production and consumption of the community. Then at last would the banks be solvent, would the risk of false accountancy be avoided, and would end the frenzied shipments of gold back and forth to raise money-value here and depress it there.

neutralising War Debts and the hoarding of nations, so that restored in circulation is man's medium of exchange for "services given and goods produced," never again to be utilised as wealth?

"*And there shall go forth a rod out of the stem of Jesse, and a Branch shall grow out of his roots.*

"*With Righteousness shall He govern the poor, and reprove with equity the meek of the earth; and He shall smite the earth with the rod of His mouth, and with the breath of His lips shall He slay the wicked.*"[1]

"*And Righteousness shall be the girdle of His loins, and Faithfulness the girdle of His reins.*

"*They shall not hurt nor destroy in all My holy mountain: for the earth shall be full of the knowledge of the Lord, as the waters cover the sea*" (Isa. xi, 1–9).[2]

"*Behold, God is my salvation; I will trust and not be afraid, for the Lord Jehovah is my strength and my song.*

"*In that day shall ye say, Praise the Lord, call upon His name, declare His doings among the people: for He hath done excellent things, for great is the Holy One of Israel in the midst of thee*" (Isa. xii).

"*In that day shall this song be sung in the land of Judah: Thou wilt keep him in perfect peace, whose mind is stayed on Thee. Trust ye in the Lord for ever, for in the Lord Jehovah is everlasting strength*" (Isa. xxvi, 1–4).

"*The wilderness and the solitary place shall be glad for them, and the desert shall rejoice and blossom as the rose.*

"*Then the eyes of the blind shall be opened, and the ears of the deaf unstopped.*

"*For in the wilderness shall waters break forth, and streams in the desert; and the parched ground shall become a pool, and the thirsty land springs of water.*

"*And an highway shall be there, and it shall be called, The Way of Holiness.*

"*And the ransomed of the Lord shall return, and come to Zion with songs and everlasting joy upon their heads; they*

---

[1] Also 2 Esdras xiii.
[2] Also Jer. xxxiii, 14–16; xxiii, 5, 6.

shall obtain joy and gladness, and sorrow and sighing shall flee away" [1] (Isa. xxxv).

"For, behold, I create new heavens and a new earth ; and the former shall not be remembered.

"But be ye glad and rejoice for ever in that which I create ; for, behold, I create Jerusalem a rejoicing, and her people a joy.

"And Mine elect shall long enjoy the work of their hands.

"And it shall come to pass that before they call, I will answer them, and while they are yet speaking, I will hear" [2] (Isa. lxv, 17–25).

¶ 12. The King's Chamber (1936–1953) denotes the period of Divine Assessment of the white races (descent from Adam, and distributed as in Gen. x).

With Armageddon and Gog's destruction, and the cleansing of world thought from Cain-Bel influence, the nucleus of His Kingdom, Britain and her Empire, has been extended to include America, the Jews, and associated nations—as the Scandinavian countries, that part of Germany whose Church opposes the neo-paganism of the Nazis, probably France when she gives up the worship of gold, and Roman Catholicism that can recognise her true Head.

In the period of sixty-five years, 1936–2001, falls the restitution of all things (Acts iii, 21).

1953–2001 [3] is the *"cleansing of the Sanctuary"* (Dan. viii, 13, 14). The sacred sites in Eden, in and around Jerusalem, having been cleansed of Mohammedanism, Judaism and paganised Christianity, the true Sanctuary of God on earth is built in the height of the mountains.[4]

And into the kingdom come the other races of evolved

---

[1] This is meant literally in this world—for a restored land of Palestine, and a happy Britain without the spectre of unemployment; for in the "second world" *"there are no more seas."*

[2] Again verifying as this world—*cf.* Deut. xxviii, 20, 67.

[3] The Pyramid chronology coinciding with prophecy: 300 B.C.—fall of Alexander's Empire (the he-goat, Dan. viii, 8, which smote the ram with two horns, Medo-Persia, viii, 7) — +2300 "evenings-mornings" (viii, 26) as complete days, or solar years = A.D. 2001.

[4] See pp. 384, 403.

THE THEOCENTRIC KINGDOM

man, "*all the families of the earth*"—"*the heathen,*" the Mongol and the Negro—to bring in Rev. vii, 9, the Millennial Era, when the King of Righteousness leads His people into their sabbath of rest on the eve of the Feast of Trumpets (17th September) in A.D. 2001, to reign from pole to pole.

"*Six days shalt thou labour and do all thy work, but the seventh day is the Sabbath of the Lord thy God.*"

In six days He created the Universe, and "*He rested on the seventh and sanctified it.*"

"*One day is with the Lord a thousand years*" (2 Pet. iii, 8). 4000 B.C.–A.D. 2001 is thus six days, in which the earth travaileth; on the seventh day, A.D. 2001–3001, the Millennial Era (Rev. xx, 4–6), the earth rests in peace, with evil driven underground (*vide* Subterranean Chamber and passage symbology).

Then evil as a world force is again let loose (xx, 7), to test mankind as to Freewill in operation whether in Divine Harmony, and finds plenty that will deviate (xx, 8).

"*And Satan shall go out to deceive the nations which are in the four quarters of the earth . . . Gog and Magog, to gather them together to battle, the number of whom is as the sand of the sea.*"

Finally comes the annihilation ("*eternal destruction from the presence of the Lord,*" 2 Thess. i, 9), of all that which deviates in the earth below and the heavens above; and the end of this world.

We are dependent on the light and heat of the sun, and these cannot remain for ever as they are:

(1) the sun may run into another star, any asteroid may hit another and be deflected . . . none likely.

(2) the sun is perilously near to the left-hand edge of the main-sequence; it may cross this edge and proceed to contract precipitately to the white dwarf state . . . as the faint companion of Sirius.

(3) "novæ" occasionally appear in the sky, temporarily emitting anything up to 25,000 times the radiation of the sun . . . on the average each star becomes a nova once in 400 million years . . . our sun does not seem to have done so

for the last 1000 million years¹ (Jeans, *The Universe Around Us*, pp. 365, 366).

With the end of this planetary system, Matter will not matter, Spirit alone remains.

Thus the work of Christ, the Logos, "*Who verily was foreordained before the foundation of the world, . . . all things were created by Him and for Him, and by Him all things consist . . . according to His eternal purpose which He purposed in Christ Jesus our Lord,*" to restore Freewill, from the lowest spiritual realms to the highest, to Love and Unity in all God's Creation, is accomplished.

"*Then cometh the end, when He shall have delivered up the Kingdom to God, even the Father; when He shall have put down all rule and all authority and power. For He must reign, till He hath put all enemies under His feet. The last enemy that shall be destroyed is death*"² (1 Cor. xv, 23–26).

And the next "world," where the Freewill of all Spirit is attuned—"*that God may be All in All.*"

"*And I saw a new heaven and a new earth, for the first heaven and the first earth were passed away, and there was no more sea.*

"*And I heard a great voice saying, Behold, the tabernacle of God is with men, and He will dwell with them, and they shall be His people, and God Himself shall be with them, and be their God.*

"*And the city had no need of the sun, neither of the moon, to shine in it, for the glory of God did lighten it, and the Lamb is the Light of it.*

"*Behold, I make all things new. It is done. I am Alpha and Omega, the beginning and the end.*

"*He that overcometh shall inherit all things, and I will be his God, and he shall be My son*" (Rev. xxi, xxii).

. . . .
. . . .
. . . .

---

[1] For in Him (the Logos) the continuance of all things consist, until His work is accomplished.

[2] See p. 28.

THE LOGOS SUMS UP HIS-STORY

Let Christ Himself sum up A.D. history, from the view of Spirit, not of Matter (for these words are no man's invention, be he disciple or saint)—and therein to visualise the signs given coming to pass ("*whoso readeth, let him understand*"), and therein an injunction to our age more than to any other, to His Israel-Britain as a God-given Empire, to prepare herself as the nucleus of His Kingdom on earth (verified also by the Divine Message from Seth, unread before so direct to our race to-day, come to the last Chamber, that of the Open Coffer), to WATCH:—

"*As He sat on the Mount of Olives,*" on which one day "*His feet shall stand,*"[1] overlooking Jerusalem, "*O Jerusalem, how often would I have gathered thy children together as a hen gathereth her chickens under her wings and ye would not.*[2] . . . *Ye shall not see me henceforth, till ye shall say, Blessed is He that cometh in the name of the Lord*"[3] (Matt. xxiii, 37–39); and after foretelling the fall of Jerusalem, He is asked by His disciples:

"*When shall these things be? and what shall be the signs of Thy coming, and of the consummation of the age?*"[4]

"*Take heed, lest any man deceive you, for many shall come in my name, saying, I am Christ.*

"*When ye shall hear of wars and rumours of wars, be ye not troubled, for such things must needs be,*[5] *but the end is not yet.*

"*For nation shall rise against nation, and kingdom against kingdom: there shall be famines and earthquakes in divers places; but all these things are the beginning of travail.*

"*But before all these things,*[6] *they shall lay their hands on you, and kill you; and ye shall be hated of all nations for My name's sake.*

---

[1] Page 409, when, as the Logos, He intervenes to stop this present civilisation's Chaos.

[2] Not only in the three years, but as the Logos, 1000 B.C., when Solomon built the Temple, and "*Before Abraham, I am.*"

[3] Pages 413–416.

[4] Matt. xxiv; Mark xiii; Luke xxi: Davidson has it well in three parallel columns, with dating (*The Great Pyramid*, Table XXXI).

[5] For deviation of Freewill from Divine Harmony means discord, and such between nations means war.

[6] *I.e.* before the final travail or tribulation.

"Settle it therefore in your hearts, not to meditate before what ye shall answer, for I will give you a mouth and wisdom which all your adversaries shall not be able to gainsay.

"But there shall not an hair of your head perish [1] *; in your patience possess ye your souls.*

"And because iniquity shall abound, the love of many shall wax cold. [2]

"And this gospel of the Kingdom shall be preached in all the world for a witness, and then shall the end come. [3]

"But when ye shall see the abomination of desolation, spoken of by Daniel the prophet, standing where it ought not, (whoso readeth, let him understand [4]*) : then let them which be in Judea flee into the mountains.*

"And they shall fall by the edge of the sword, and shall be led away captive into all nations ; and Jerusalem shall be trodden down of the Gentiles, until the Times of the Gentiles be fulfilled. [5]

"For in those days [6] *shall be great tribulation such as was not from the beginning of the creation which God created unto this time, no, nor ever shall be.*

"And except that the Lord had shortened those days, no flesh should be saved ; but for the elect's sake, whom He hath chosen, He hath shortened the days. [7]

"But take ye heed ; behold, I have foretold you all things. [8]

[1] He was about to perform that action, when for three days in the heart of the earth, on the atomic make-up, that Freewill, attuned to Divine Spirit after a martyr's death, had control of such force created to re-fashion an "incorruptible body."

[2] Because Freewill in deviation seems to bring material rewards, the attuning of Freewill to Divine Harmony, whose first law is Love, wanes.

[3] In all the world it has been so preached that there is scarce a corner unreached by the Missionary Societies: while the British and Foreign Bible Society, since founded in 1804, has published 464 million copies, in 692 languages, and where there was no language has made one for the Bible to be read.

[4] The injunction to watch Daniel's prophecy, where in xii, 11, 12, the years given are A.D. $1891\frac{3}{4}$ and $1936\frac{3}{4}$ (see p. 406), not merely the immediate future, A.D. 70.

[5] Jerusalem so delivered in A.D. 1917 (see p. 323), and the Times of the Gentiles end in their final year, A.D. 1936 (see p. 316).

[6] Thus verified as the 1917 period.

[7] Verified by Pyramid symbology in the Antechamber as Truce in Chaos (1918-1928), between the first (1914-1918) and second (1928-1936) low passages (see pp. 321, 323). The *"elect whom He hath chosen"* (see p. 98, and the number 153).

[8] The injunction again, here to emphasise the impending Event.

THE LOGOS SUMS UP HIS-STORY

"And there shall be signs in the sun, and in the moon, and in the stars [1] *; and upon the earth distress of nations, with perplexity* [2] *; men's hearts failing them for fear and for looking after those which are coming on the earth,* [3] *for the powers of heaven shall be shaken.* [4]

"For as the lightning cometh out of the east and shineth even to the west, so shall also the coming of the Son of Man be. [5]

"For wheresoever the carcase is, there will the eagles be gathered together. [6]

"Immediately after the tribulation of those days [7] *shall the sun be darkened, and the moon shall not give her light, and the stars shall fall from heaven, and the powers of the heavens shall be shaken.*

"And when these things come to pass, then look up and lift up your heads, [8] *for your redemption draweth nigh.*

"And then [9] *shall appear the Son of Man in heaven: and then shall all the tribes of the earth mourn: and they shall*

---

[1] Ask the astronomer, as to certain unexplained happenings—*e.g.* the moon some miles off its calculated course; June 1918, a new star, Aquila, blazed out; and in April 1928, "Is the star, Nova Pictoris, blowing itself to pieces?"

Abnormal weather conditions attributed to Sun changes every 60 and 262 years—the two cycles reached their maximum intensity at the same time, 1927–1930, never known to have occurred before (R. E. Delary, Observatory at Ottawa), coinciding with Jupiter at perihelion in 1928.

The new conception of astrophysics and the philosophy of the Space-Time continuum are of to-day's date; and probably there are other signs in the heavens that the intellect of our scientists has not yet reached.

[2] In the fewest words, how better could the League of Nations be described?

[3] Typical of present times—disorders and strikes, political confusion and revolutions, treaties and conferences, problems adjusted but never settled; broadcast talks on subjects innumerable, *The Way to God, Whither Britain?* (see p. 373).

[4] Suggesting that deviation in Spiritual realms, seeing the end of Chaos leading to its inevitable destruction, is "panicking," so forces the issue in Spirit affairs within Matter (see pp. 352, 357, 386, 408).

[5] The warning of His actual Presence in the earth atmosphere (see p. 416)—a challenge to the scientists to find the "sign."

[6] The standard of the eagles *"to take a prey"* will be seen in Jerusalem (see p. 402). Also suggestive of the touting element after the political, financial and industrial plums of a restored Palestine (see p. 420).

[7] That is marked by Pyramid symbology at September 1936 (see p. 321) (Isa. xiii, 10).

[8] Now standing upright in the King's Chamber period, 1936–1953, in the central plane below the displaced Apex Stone and before an Open Coffer (see pp. 379, 383).

[9] The era has thus been defined, but not the year.

see the Son of Man coming in the clouds of heaven with power and great glory.[1]

"And He shall send His angels with a great sound of a trumpet,[2] *and they shall gather together His elect from the four winds, from the uttermost part of earth to the uttermost part of heaven.*[3]

"Now learn a parable of the fig-tree[4]*: When his branch is yet tender and putteth forth leaves, ye know that summer is nigh.*

"So likewise ye, when ye see these things come to pass,[5] *know ye that the Kingdom of God is nigh at hand.*

"Verily, I say, this generation shall not pass away,[6] *till all things be fulfilled.*

"Heaven and earth shall pass away: but My words shall not pass away.

"But of that day and hour knoweth no man, no, not the angels in heaven, neither the Son, but My Father only.

"And take ye heed to yourselves, lest at any time your hearts be overcharged with surfeiting and drunkenness and cares of this life, and so that day come upon you unawares.[7]

"For so in the days that were before the Flood, they were eating and drinking, marrying and giving in marriage, until the day that Noah entered into the ark, and knew not until the Flood came and took them all away[8]*; so also shall the coming of the Son of man be.*

---

[1] Dan. vii, 13, 14; 2 Esdras xiii; Rev. i, 7.

[2] 1 Cor. xv, 51–56, and breaking of the seventh seal (pp. 250, 387).

[3] *Vide* the prophetical "gather" (pp. 141–150), Isa. xi, 12; Ezekiel xxxvii, with the vision of the valley of "dry bones" (pp. 163, 164)—the dead whose freewill attuned (p. 250) and the living Israel-Britain (pp. 372–374).

[4] The fig-tree is symbolic of Judah: the "budding" of the Jews (footnote, p. 418). Christ gives a word as symbol, and expects man by spiritual intuition to interpret.

[5] Again the injunction to "watch" these signs. And if the time misinterpreted, He has given the Pyramid as a symbol in stone wherewith to verify (Isa. xix, 19, 20). Does any man dare prophesy events 1900 years ahead to such accuracy?

[6] The Jews will still be found as Jews, even after the siege of A.D. 70.

[7] As the era is defined, especially to these times: spiritually intuitive of the post-War effect on the social code (see pp. 357, 358).

[8] The illustration given of the destruction of the deviating civilisation of the selected Adamic race, so that it is now unknown. Yet it must have been higher than the derived Sumerian civilisation of 3500 B.C. (see p. 40 ff); must have contained the scientific knowledge of Seth embodied in the Pyramid (see p. 85 ff)—such is given to His "chosen" Israel race

THE LOGOS SUMS UP HIS-STORY

"Watch ye therefore, that ye may be accounted worthy to escape all these things that shall come to pass, and to stand before the Son of Man.[1]

"Who then is a faithful and wise servant, whom his lord made ruler over his household?

"When he cometh, he shall make him ruler over all his goods.

"But if that servant shall say in his heart, My lord delayeth his coming, and shall smite his fellow-servants, and shall eat and drink with the drunken[2] *; the Lord shall come in an hour that he is not aware of, and cut him asunder."*

Then follows the parable of the ten virgins (Matt. xxv, 1–13), ten being here symbolic of the Nation and Church of Britain without the Bridegroom (pp. 372–374).

Then, the parable of the talents (xxv, 14–30) (p. 375).

Then, the parable of the sheep and the goats (xxv, 31–46) (pp. 375, 376).

And after His Resurrection, the disciples, still puzzled, ask:

"Lord, wilt Thou at this time restore again the Kingdom to Israel? And He said unto them, It is not for you to know the times or the seasons which the Father has put into His own power. . . . Ye shall receive power . . . and ye shall be My witnesses[3] *. . . unto the uttermost part of the earth.*

"And when He had spoken these things, while they beheld He was taken up, and a cloud received Him out of their sight.

"And two men stood by them in white apparel, which also

as a warning to its own high civilisation, deviating under a Babylon economic-*cum*-political system (pp. 334–339, 354–359).

[1] The impending crisis and its "shaking" out of Israel-Britain of all trace of the "golden cup of Babylon," brewed by Cain-Bel, of which all nations have drunken; so that she may become the Theocentric State (p. 383), the nucleus of His Kingdom, and stand in His Presence.

[2] Isa. v, 19–22; 2 Peter iii, 3, 4 (pp. 135, 358, 377).

[3] The Most High, who created the Space-Time continuum (pp. 1, 3, 260), foreknowing the beginnings and its endings, and the result of Freewill deviated to chaos and destruction, is all-powerful to order all such effects and laws, in His Perfect Oneness of Harmony. Such course is seen in the evolution of the earth, and the history of mankind (from Cain-Bel to the Totalitarian State, from Abraham to the British Empire).

But He yet requires Freewill attuned, to receive and emanate Divine Love, "to be His witness," so had foreordained, and has acted to redeem such Freewill from inevitable annihilation (pp. 23–28, 247–259, 263, 270).

said, Ye men of Galilee, why stand ye gazing up into Heaven? This same Jesus, which is taken up from you into Heaven, shall so come in like manner as ye have seen Him go into Heaven" (Acts, i, 6–11).

. . . .
. . . .
. . . .

Britain, awake!—to a greater Armada, your Armageddon.

Ring out the old cry with a new spirit, "Trâ Môr, Trâ Brython." [1]

"*I will yet for this be enquired of by the House of Israel to do it for them.*"

"*Watch ye that ye may stand before the Son of Man.*"

"*Thy Kingdom come. Thy Will be done in earth, as it is in heaven.*"

THE KINGDOM OF GOD IS AT HAND.

---

[1] In the impending crisis, spiritual and physical—translated "As long as there are seas [for in the second world "*there is no more sea,*" Rev. xxi, 1] so long shall there be Brythons (men of the New Covenant) as Israel (ruling with God),"—(from Rudyard Kipling's *Recessional*; apologies, R.K., for the intrusion of the last two lines)

God of our fathers, known of old,
  Lord of our far-flung battle-line,
Beneath whose awful Hand we hold,
  Dominion over palm and pine,
Lord God of Hosts, be with us yet,
  Lest we forget—lest we forget!

The tumult and the shouting dies,
  The Captains and the Kings depart.
Still stands thine ancient sacrifice,
  An humble and a contrite heart,

Lord God of Hosts, our aid hath met,
  Ne'er at Thy Hill can we forget.

INDEX

PRINTED IN GREAT BRITAIN BY NEILL AND CO., LTD., EDINBURGH.

Printed in the United States
978500001B